AQA

Make the Grade!

Understanding GCSE Geography

for AQA Specification A

Ann Bowen
John Pallister

With contributions from
John Hopkin
Emma Rawlings Smith

www.heinemann.co.uk

✓ Free online support
✓ Useful weblinks
✓ 24 hour online ordering

0845 630 22 22

Heinemann is an imprint of Pearson Education Limited, a company incorporated in England and Wales, having its registered office at Edinburgh Gate, Harlow, Essex, CM20 2JE. Registered company number: 872828

www.heinemann.co.uk

Heinemann is a registered trademark of Pearson Education Limited

First published 2009

12 11
10 9 8 7 6 5

British Library Cataloguing in Publication Data
A catalogue record for this book is available from the British Library

ISBN 978 0 435353 30 8

Designed by Kamae Design
Typeset by HL Studios, Long Hanborough
Original illustrations © Pearson Education Ltd
Illustrated by HL Studios, Long Hanborough
Cover design by Pearson Education Ltd
Picture research by Helen Hope and Helen Reilly
Cover photo © G.M.B Akash/Panos Pictures
Printed in Malaysia, CTP-KHL

Acknowledgements
The authors and publisher would like to thank the following individuals and organisations for permission to reproduce photographs:

p.5 Andrew Woodley/Alamy; **p.7** Dewitt Jones/Corbis UK Ltd; **p.11** Tony Wharton/Frank Lane Picture Agency; **p.13** National Geographic/Getty Images; **p.16** Ullstein/LS-PRESS/Still Pictures; **p.20** J Lightfoot/Robert Harding World Imagery/Corbis; **p.26** (**Figure 2**) Dick Makin/Alamy; **p.29** Chris Howes/Wild Places Photography/Alamy; **p.30** Robert Harding World Imagery/Robert Harding Picture Library Ltd/Alamy; **p.34** (**Figures 1** and **2**) Geoscience Features Picture Library; **p.38** Vincent Lowe/Leslie Garland Picture Library/Alamy; **p.49** University of Dundee, www.sat.dundee.ac.uk; **p.51** (**Figure 2**) University of Dundee, www.sat.dundee.ac.uk; **p.52** Getty Images/Jeremy Walker; **p.53** Stephen Hird/Reuters/Corbis; **p.56** NASA/Corbis; **p.61** Mark Edwards/Still Pictures; **p.62** Woodfall Wild Images/David Woodfall; **p.68** (**Figure 3**) Woodfall Wild Images; **p.74** Earth Satellite Corporation/Science Photo Library; **p.84** Layne Kennedy/Corbis UK Ltd; **p.90** Shehzad Noorani/Still Pictures; **p.102** David Robinson/Snap2000 Images/Alamy; **p.111** (**Figure 2**) Prof. B. Booth/Geoscience Features Picture Library; **p.111** (**Figure 3**) iStockphoto/Anna Pustovaya; **p.115** Mike Page; **p.116** (**Figure 3**) Matt Cardy/Getty Images; **p.123** The lighthouse for Education; **p.126** John Giles/PA/Empics; **p.127** APS UK; **p.128** (**Figure 2**) John butcher/Alamy; **p.145** Karen Kasmauski/Corbis UK Ltd; **p.159** Sefton Samuels/Rex Features; **p.163** (**Figure 2**) Alistair Berg /Alamy; **p.165** Cia de Foto; **p.166** (**Figure 3**) Rainer Kzonsek/Das Fotoarchiv./Still Pictures; **p.168** (**Figure 1**) Sipa Press/Rex Features; **p.168** (**Figure 2**) Raf Makda/View Pictures/Rex Features; **p.173** www.firsthouse.co.uk; **p.178** Robert Harding Picture Library Ltd/Alamy; **p.180** Nigel Hicks/Alamy; **p.183** David J. Green/Alamy; **p.184** Cultura/Alamy; **p.185** Cultura/Alamy; **p.186** Eye ubiquitous/Robert Harding; **p.189** Sipa Press/Rex Features; **p.193** Angelo Cavalli/Robert Harding; **p.199** Mary Saunders/Oxfam; **p.201** Digital Vision; **p.203** Chris Fredriksson/Alamy; **p.204** Buddy Mays/Alamy; **p.205** Philip Wolmuth/Alamy; **p.207** (**Figure 3**) Seaman Ash Severe U.S. Navy/Handout/Navy Visual News Service (NVNS)/Corbis; **p.207** (**Figure 4**) Danita Delimont/Alamy; **p.213** Peter Macdiarmid/Getty Images; **p.243** Corbis; **p.245** Images of Africa Photobank/Alamy; **p.247** Gallo Images/Alamy; **p.248** Danita Delimont/Alamy; **p.249** DLILLC/Corbis.

Pages 175 and **177**, Pearson Education Ltd/Emma Rawlings Smith. **All remaining photographs**, Pearson Education Ltd/John Pallister.

Extracts, diagrams and maps:
pp.38, **86**, **89**, **96**, **97**, **103**, **112**, **123**, **130**, **131**, **175** (**Figure 2**), **177**, **179** (**Figure 3**) Ordnance survey on behalf of HMSO (Her Majesty's Stationery Office); **p.142** (**Figure 2**) *Financial Times*, 13 February 2006; **p.150** (**Figure 2**) *Daily Telegraph*, 23 April 2005 (reproduced by permission of the Government Actuary's Department and *Daily Telegraph*); **p.179** (**Figure 4**) Controller of HMSO (Her Majesty's Stationery Office); **p.181** Controller of HMSO (Her Majesty's Stationery Office); **p.189** (**Figure 4**) *New Internationalist*, July 2001 (Issue 336); **p.194** (**Figure 1**) 80:20 Educating and Acting for a Better World; **p.200** (**Figure 1**) World Resources Institute; **p.201** (**Figure 4**) World Bank Publications; **p.206** (**Figure 1**) World Bank Publications; **p.216** (Figure 2) *Daily Telegraph*, 24 January 2005 (reproduced by permission of the Telegraph Media Group Ltd); **p.217** *Sunday Telegraph*, 21 November 2004 (reproduced by permission of the Telegraph Media Group Ltd); **p.222 (Figure 2)** BP Statistical Review of World Energy 2008; **p.224** (**Figure 1**) *Guardian*, 19 September 2005 (reproduced by permission of Guardian News and Media Ltd); **p.225** (**Figure 4**) British Wind Energy Association; **p.229** (**Figure 4**) World Wildlife International (WWF Data Support Sheet for Education 24).

Every effort has been made to contact copyright holders of material reproduced in this book. Any omissions will be rectified in subsequent printings if notice is given to the publishers.

Heinemann Understanding GCSE Geography for AQA A website
You can find weblinks for further research and more on our bespoke website www.contentextra.com/aqagcsegeog. Your user name is: aqagcsegeog. Your password is: GeographyAQA111004.

Contents

Introduction

This book has been written specifically to support you during your GCSE Geography AQA A course. Resources for this course include this student book with ActiveBook, a Teacher Guide and an ActiveTeach CD-ROM.

How to use this book

This new edition of Understanding GCSE Geography has been completely updated to offer well-structured and thorough coverage of the new specification. It is practical and easy to use, containing all you need to know for your course. Case studies are integrated within the text as separate pages with a yellow background to show real-life, up-to-date examples.

Understanding GCSE Geography website

Our bespoke AQA Geography website provides opportunities for further research as well as additional exam questions for foundation level. The website will also include additional case studies in the future. See p.ii for further details.

How to use the ActiveBook CD-ROM

In the back of your copy of the student book you will find your FREE ActiveBook CD-ROM. This is for individual use and includes a digital copy of the book on screen, plus the unique Grade Studio and Exam Café.

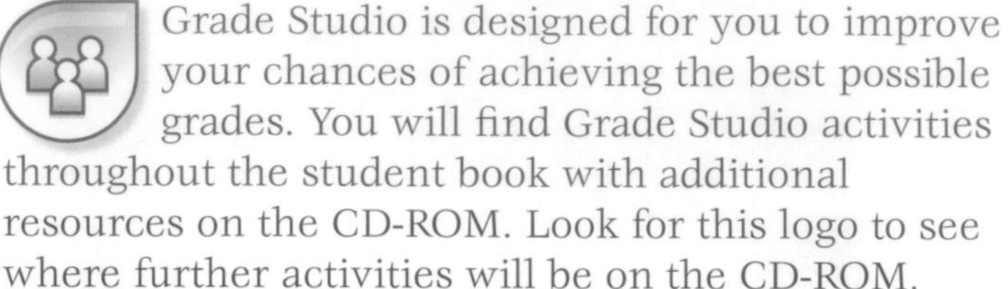

Grade Studio is designed for you to improve your chances of achieving the best possible grades. You will find Grade Studio activities throughout the student book with additional resources on the CD-ROM. Look for this logo to see where further activities will be on the CD-ROM.

Exam Café is to be used when revising and preparing for exams. Like Grade Studio, Exam Café is in the student book with additional resources on the CD-ROM. Look for this logo to see where further activities will be on the CD-ROM.

Thought-provoking starter questions to encourage you to really think about the content right from the beginning of each lesson.

Important points to remember for the exam.

UNDERSTANDING GCSE GEOGRAPHY

Granite

What features distinguish granite scenery from that of other rocks? What are the opportunities and problems for people and economic activities?

All the granite rocks in the UK are found to the north and west of the Tees–Exe line (see **Figure 4** on page 27). Granite formed underground hundreds of millions of years ago in dome-shaped masses of magma, known as batholiths. The rocks on top of it have been eroded to expose the granite as surface rock.

Landscape features

In south-west England granite gives relatively flat-topped moorland plateaus with frequent rock outcrops, which from time to time form rock blocks called **tors** (**Figure 1**). Tors are some 5–10 metres high and are surrounded by weathered materials of all sizes from boulders to sand. On the higher parts of the moorlands there are many areas of standing surface water, which form marshes and bogs. The many surface streams have cut deeply into the upland block of Dartmoor to form deep and steep V-shaped valleys, especially where rivers such as the Dart go over the edge of the plateau. Dartmoor has a radial pattern of drainage, with rivers flowing outwards in all directions from its high centre (**Figure 3**).

Dramatic coastal scenery occurs where granite and Atlantic breakers meet, as in Cornwall at Land's End (see **Figure 2** on page 26). In Scotland the granite peaks in the Grampians and on Goat Fell in Arran are rocky and frost-shattered, although where the land is relatively flat, such as on Rannoch Moor, extensive bogs occur.

Figure 1 Bowerman's Nose, a tor on Dartmoor.

Granite is a hard rock, resistant to erosion, which is why it forms areas of high relief inland and cliffs along the coast. It is an impermeable rock, which explains why there is so much surface water. Another reason for the presence of so many bogs is the high precipitation in western upland areas.

Formation of tors

Tors occur where joints in the granite are wider apart (**Figure 2**). Freeze–thaw weathering (see **Figure 1** on page 28) can operate more effectively where the joints are close together, because there are more cracks in the rock for water to fill. Each time the water freezes and expands within a joint, more pressure is put on the surrounding rock. Where there are fewer joints, it takes longer for the blocks of rock to be broken off and the blocks are left upstanding as tors.

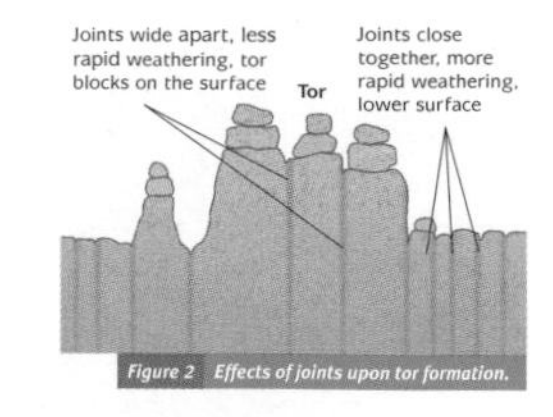

Figure 2 Effects of joints upon tor formation.

Land use and economic uses

On the higher areas, bog, marsh and moorland produce some of the least useful land in the UK. In some places there may be opportunities for water storage. At lower levels there may still be nothing better than poor grazing land suitable only for sheep and cattle (and, on Dartmoor, also for ponies). Soils are acidic and infertile; it is only around the edges of the uplands that the pastures improve sufficiently to allow grazing by dairy cattle.

Granite is a fine building stone. Aberdeen is known as 'the granite city' since so much use was made by builders of locally available supplies of stone. It is also often used for headstones in graveyards.

30

Granite rock is susceptible to attack by chemical weathering and in some places it has decomposed. This has resulted in the feldspar in the granite being converted into clay minerals such as china clay (kaolin). China clay is best known as the raw material for the pottery and porcelain industries, and much is sent to the Potteries region around Stoke-on-Trent. It is also used in the manufacture of paper and is an ingredient in paint, toothpaste, skin creams and many other products.

EXAM PREPARATION

Dartmoor
- Dartmoor has a high rainfall and is known for its mists and fogs.
- Much of the land is covered by heather.
- The many boggy areas contain a rich variety of plant life.
- The central upland block was enclosed within a National Park in 1951.
- The Park covers almost 100 000 hectares and over 30 000 people live inside it.
- Up to 8 million people visit or pass through the Park each year.
- Most of the towns, such as Tavistock, Okehampton and Ashburton, are located around the edges of the central block.
- Places popular with visitors include Buckfast Abbey, Haytor, Becky Falls and Lydford Gorge.
- Some of the remains of old woodlands have been preserved as nature reserves.

Key: Moorland; Agricultural land; Woodland; National Park; Tourist information centre; Popular tourist places; Roads; Reservoirs; Peaks and tors

Figure 3 Map of Dartmoor.

ACTIVITIES

1 (a) Describe the pattern of land uses in Dartmoor shown in **Figure 3**.
(b) Explain why opportunities for farming improve away from the centre of Dartmoor.

2 (a) Name two different economic uses of granite.
(b) Explain their importance.

GradeStudio

1 Study **Figure 1**. Make a frame and draw a sketch from the photograph. Label the landscape features and land uses shown. (4 marks)

2 a Name and locate an example of a tor. (2 marks)
b Explain how a tor is formed. (4 marks)

Exam tip
- Look at the number of marks as a guide e. for number of labels needed in **1** and what might be ne ded in **(2a)**.
- When explaining landforms, always explain w the processes forming them operate. Which pro s needs to be explained in **(2b)**?

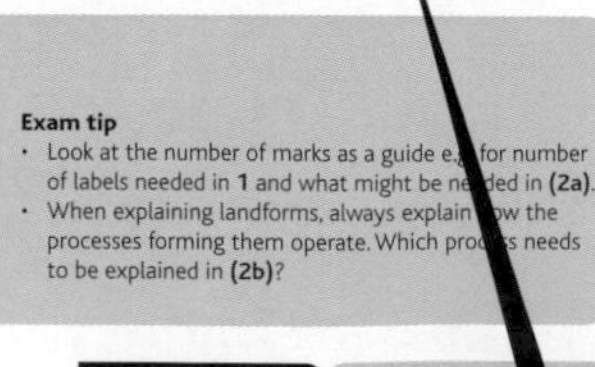

FURTHER RESEARCH

Find out more about the centre of Dartmoor visiting the weblink www.contentextra.com/aqagcsegeog.

ROCKS, RESOURCES AND SCENERY

Questions that will help you improve your performance in the exam. You can find further Grade Studio advice at the end of each chapter and on page 256.

Questions that will help you develop a thorough understanding and a broad range of skills.

Chapter 1

The Restless Earth

Plymouth, the capital city of the island of Montserrat in the Caribbean, was destroyed by the massive eruption of the Soufrière Hills volcano in 1997. More than ten years later, and the volcano still erupts from time to time. Why could this not happen in Plymouth, UK?

QUESTIONS

- **Why are plate margins so important – for landforms and people?**
- **Why do high fold mountain ranges, active volcanoes and main earthquake zones hug plate margins?**
- **How can the risks to life and property be reduced in countries located on plate margins? Does it matter whether the countries are rich or poor?**
- **Why are people in countries away from plate margins not completely safe? Remember the Asian tsunami of 2006? What could happen if a supervolcano were to blow its top?**

The Earth's crust is unstable

What happens at different plate margins?

We live on a thin skin of cool rock – the **crust** (**Figure 1**). At the centre of the Earth is the **core** surrounded by a large mass of molten rock called the **mantle**. There are two main types of crust: **oceanic crust**, which is denser and about 5 kilometres thick; and **continental crust**, which is lighter but 30 or more kilometres thick.

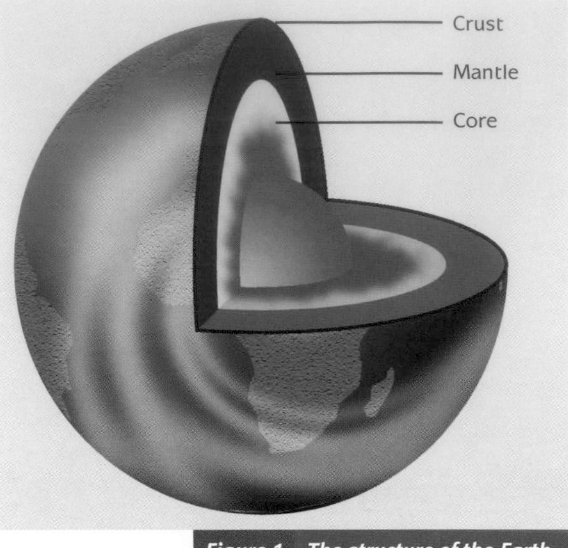

Figure 1 The structure of the Earth.

The Earth's crust is not one continuous layer but is made up of seven large tectonic **plates** and many smaller ones. **Figure 2** shows the distribution of the main tectonic plates. The Earth's crust is unstable because the plates are moving in response to rising hot currents called **convection currents** within the mantle. The movement of the plates has greatest impact at the plate margins, where two tectonic plates meet. The centres of the plates, away from the margins, tend to be stable and distant from major tectonic activity. How far is the UK from the nearest plate margin?

Plates may move apart, or closer together, or slide past each other. These movements lead to earthquakes and the formation of fold mountains and volcanoes.

Destructive plate margins

Plates move together. **Figure 3** shows what happens. The Nazca plate is made of oceanic crust, which is denser than the continental crust of the South American plate. The Nazca plate is forced to sink below the South American plate. The oceanic crust sinks into the mantle where it melts in the **subduction zone**, an **oceanic trench**. Energy builds up in the subduction zone – at certain times this may be released as an earthquake. The molten rock, called magma, may rise upwards, causing volcanic eruptions and leading to the creation of **composite volcanoes**.

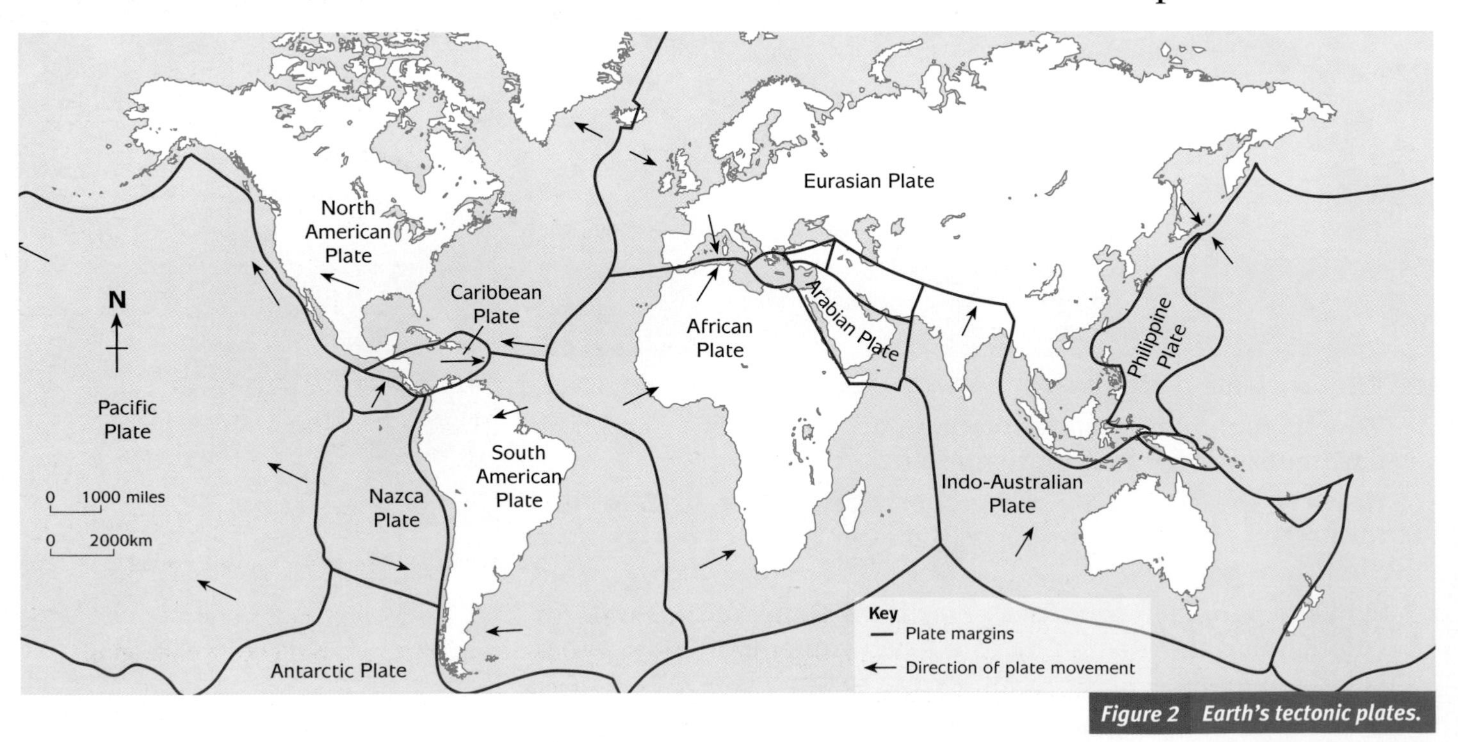

Figure 2 Earth's tectonic plates.

The lighter continental crust stays at the surface but sediment becomes crumpled into fold mountains. The Andes are the fold mountains that have formed along the west coast of South America.

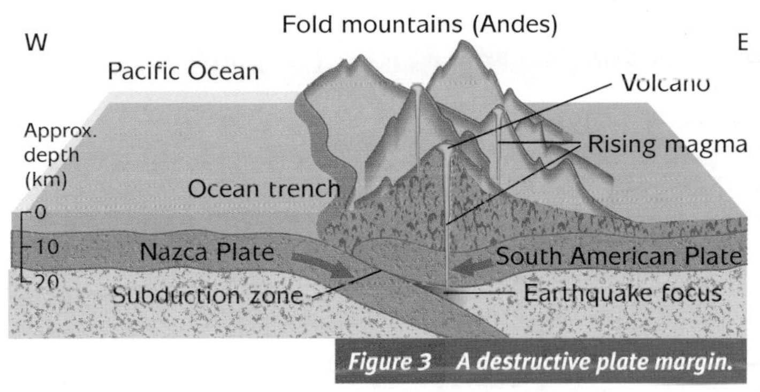

Figure 3 A destructive plate margin.

Constructive plate margins

Some plates, like the North American and Eurasian plates, are moving in opposite directions, away from each other. This type of movement mostly happens under the oceans. As the plates move apart, the gap is filled by magma rising up from the mantle below. The rising magma creates **shield volcanoes** which, if they become high enough, form volcanic islands, such as Iceland. So much magma is poured out that ridges are built up from the sea bed, like the Mid-Atlantic ridge shown in **Figure 4**, upon which Iceland is located.

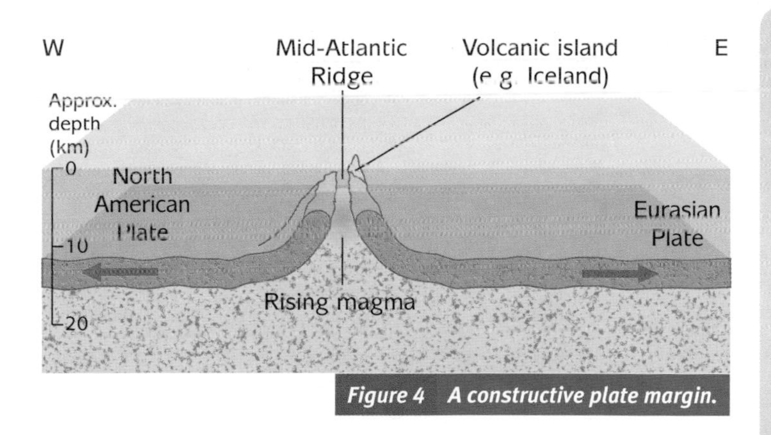

Figure 4 A constructive plate margin.

Conservative plate margins

At the San Andreas fault in California, the North American plate and the Pacific plate are sliding past each other. They are moving in the same direction but the North American plate is moving slightly faster. Pressure builds up along the fault until one plate jerks past the other, causing an earthquake. The movement has also caused the land to become ridged and crumpled, as shown in **Figure 5**.

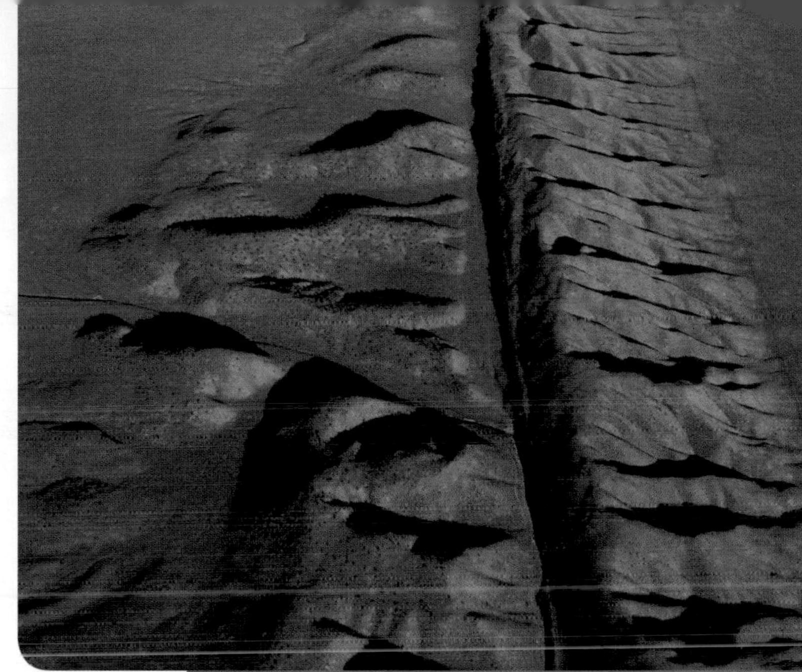

Figure 5 The San Andreas fault, which passes through San Francisco Bay where 7 million people live.

EXAM PREPARATION

Trace the main plate margins in **Figure 2.** Keep the tracing at hand for use when studying:

- Fold mountains
- Volcanoes
- Earthquakes.

ACTIVITIES

1 (a) What are the differences between (i) plates made of oceanic and continental crust (ii) destructive and constructive margins?
   (b) How is a conservative margin different?

2 Copy and complete a table like the one below for the three types of plate margins:

Type of plate margin	Examples of plates	Features produced	Example country/area

3 Study **Figure 2.**
   (a) What does it show about the likelihood of tectonic activity in the UK?
   (b) Tectonic activity in Europe is concentrated in Iceland and southern Italy. State one tectonic similarity and one difference between Iceland and Italy.

4 Scientists warn that San Francisco is more than 60 per cent likely to experience a damaging earthquake by 2038. Why?

Landforms at plate margins – fold mountains and ocean trenches

Highest mountain 8848m. Deepest ocean 11 022m. How can tectonic activity explain both?

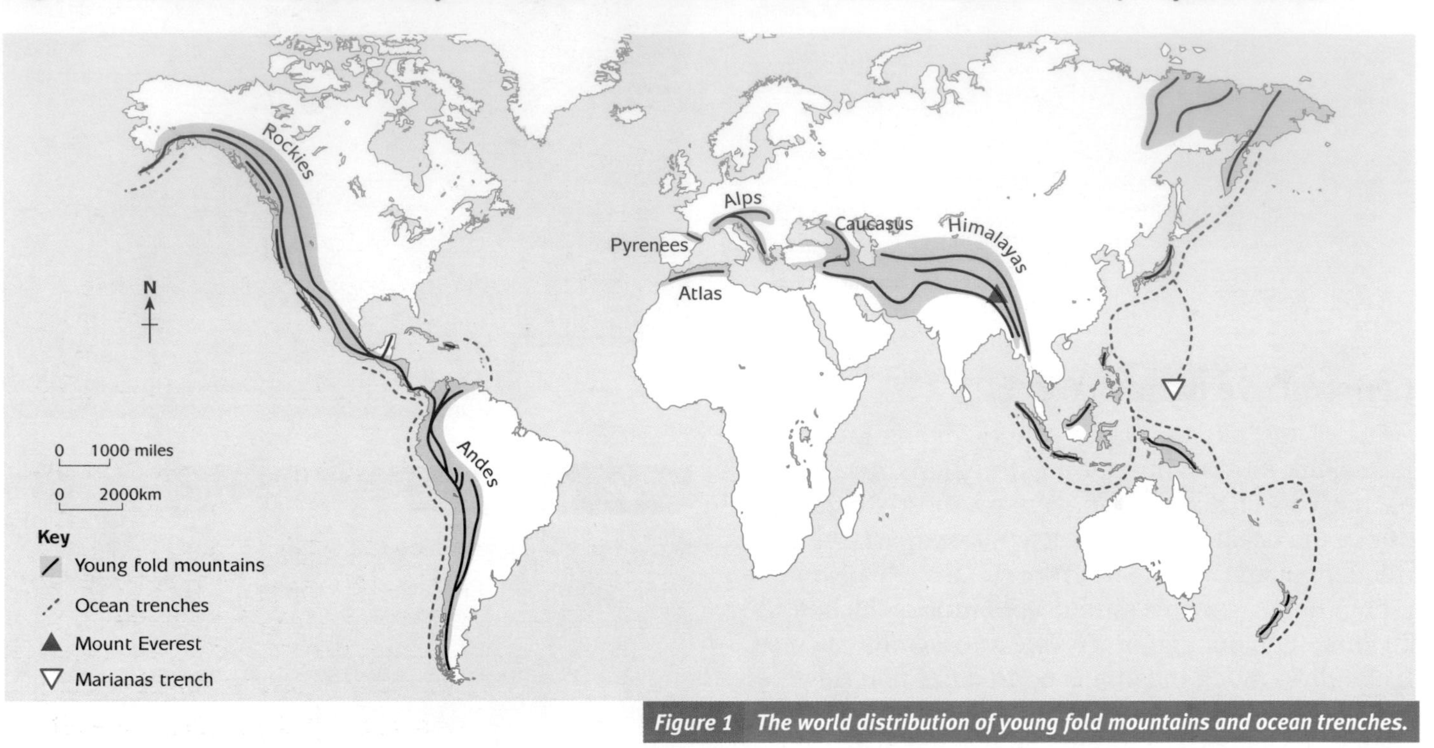

Figure 1 *The world distribution of young fold mountains and ocean trenches.*

Fold mountains

Young fold mountains are found in many parts of the world (**Figure 1**) and a glance back at **Figure 2** on page 6 shows that they form along the plate margins where great Earth movements have taken place.

Figure 2 (opposite) shows the formation of fold mountains. There were long periods of quiet between Earth movements during which sedimentary rocks, thousands of metres thick, formed in huge depressions called **geosynclines**. Rivers carried sediments and deposited them into the depressions. Over millions of years the sediments were compressed into sedimentary rocks such as sandstone and limestone. These **sedimentary rocks** were then forced upwards into a series of folds by the movement of the tectonic plates. Sometimes the folds were simple upfolds (**anticlines**) and downfolds (**synclines**), as shown in **Figure 3**. In some places the folds were pushed over on one side, giving overfolds.

Fold mountains have been formed at times in the Earth's geological history called mountain-building periods. Recent mountain-building movements have created the Alps, the Himalayas, the Rockies and the Andes, some of which are still rising. For this reason many of these ranges are called young fold mountains.

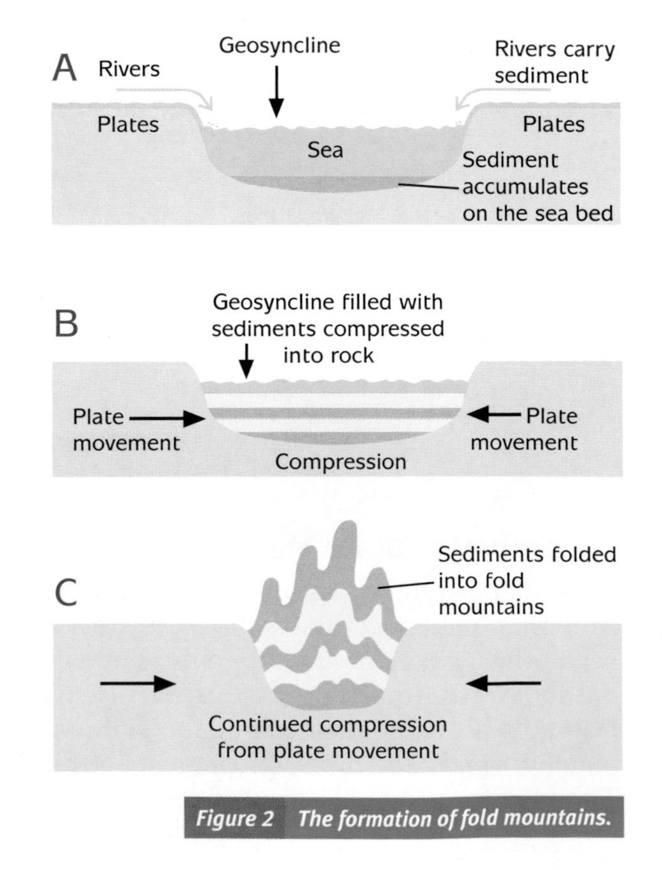

Figure 2 *The formation of fold mountains.*

UNDERSTANDING GCSE GEOGRAPHY 1

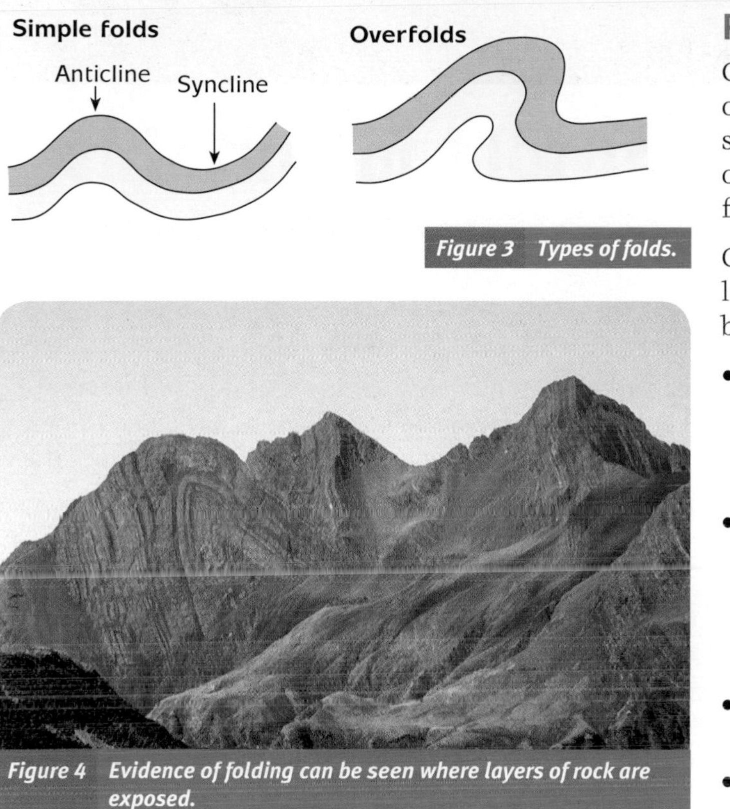

Figure 3 Types of folds.

Figure 4 Evidence of folding can be seen where layers of rock are exposed.

Ocean trenches

Notice from **Figure 1** that the majority of ocean trenches are located around the sides of the Pacific Ocean. Take another look at **Figure 2** on page 6 – with which type of plate margin are the ocean trenches associated?

The subduction zone is an ocean trench (**Figure 3** on page 7). One wall is formed by subducted ocean plate (the Nazca plate in this **Figure**), the other by the overriding continental plate (the South American plate). These ocean trenches are very deep, typically 5000–10 000 metres.

Figure 5 The Earth's surface from highest to lowest.

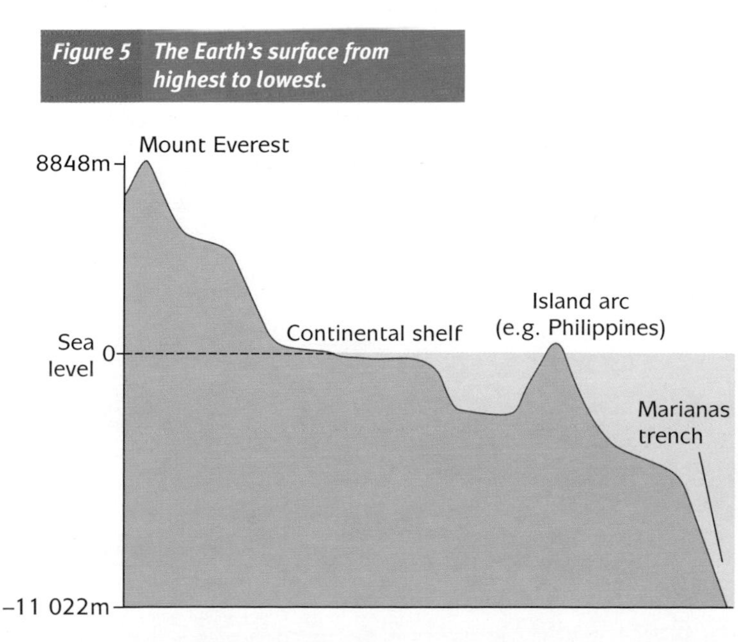

Possibilities for human use

Ocean trenches are inaccessible to humans. The ocean area of greatest importance is the continental shelf, the shallow zone less than 200 metres deep off the coast. The main opportunities here are for fishing and drilling for oil and gas.

On land, most high fold mountains are places with low densities of population. Physical problems can be formidable:

- Relief – mainly high and steep. Rock outcrops are frequent and many mountain valleys are narrow and gorge-like. There is little flat land for farming and building settlements.
- Climate – with increasing height the climate becomes colder, windier and wetter, and more of the precipitation falls as snow. The growing season is short and it is often impossible to grow crops at high levels.
- Soils – mountain soils are typically stony, thin and infertile.
- Accessibility – roads and railways are expensive and difficult to build; travel on them is frequently disrupted by rock falls, avalanches and bad weather. High mountains in inland locations, such as the Himalayas, are the least accessible of all.

There are exceptions. Parts of the Andes are well settled. Countries such as Chile, Bolivia and Peru are rich in mineral resources including silver and gold. Volcanic soils can be very fertile (pages 14–15).

ACTIVITIES

1 Study **Figure 1**.
   (a) Describe the two main directions in which ranges of young fold mountains are aligned.
   (b) For one range of fold mountains, name the two tectonic plates responsible for its formation. **Figure 2** on page 6 will help.
   (c) Explain why fold mountains are formed along destructive plate margins.

2 Draw a simple labelled sketch to show the evidence that the mountains in **Figure 4** are fold mountains.

3 Describe how tectonic activity explains the big difference between highest and lowest surfaces on the Earth.

FURTHER RESEARCH

Calculate the height difference between the highest mountain and lowest ocean on Earth. Use the weblink www.contentextra.com/aqagcsegeog to help you.

Case study of a range of fold mountains – the Alps

The Alps were formed about 35 million years ago by the African plate pushing north against the Eurasian plate. As it moved, sediments which had accumulated in the sea of Tethys (the geosyncline between the plates) were squeezed upwards to form a fold mountain range. Today the Alps form the border between Italy and the neighbouring countries of France, Switzerland, Austria and Slovenia. The highest peak is Mont Blanc near the Franco-Italian border at 4810 metres, but there are many other peaks above 3800 metres (**Figure 1** on page 102). From above, the Alps look like a wasteland of rock, ice and snow (**Figure 1**): only the valleys give a hint that human settlement may be possible.

Human activities in the Alps

These vary according to height, which creates vertical zones of land use. **Figure 2** summarises typical land uses in an Alpine valley. Many Alpine valleys are aligned west to east so that better opportunities for settlement, farming and other economic activities are created on sunny south-facing slopes. Land uses are less varied on more shaded north-facing slopes.

1. Farming and forestry

Most farms are located on the sunnier and warmer south-facing slopes. The traditional pattern of farming is dairy farming using a system called transhumance, the seasonal movement of animals. In summer the cattle are taken up to the high alp to graze, which allows hay and other fodder crops to be grown on the small fields on the flat land in the valley floor. Here summers are warmest and soils are deepest and most fertile. In winter the animals return to the farm on the valley floor, where they are kept in cattle sheds and stall-fed on the fodder crops grown in summer.

Over the years there have been many changes to this traditional system of farming.

- Cable cars (built for skiers and tourists) are now used to bring milk to the co-operative dairies down on the valley floor. In the past the farmers stayed with the cattle all summer and turned the milk into butter and cheese (which keep for longer) on the high alp.
- Farmers buy in additional feedstuffs, so that they and their cattle can stay on the valley floor farm all year.

Figure 1 Aerial view over the Alpine fold mountain range.

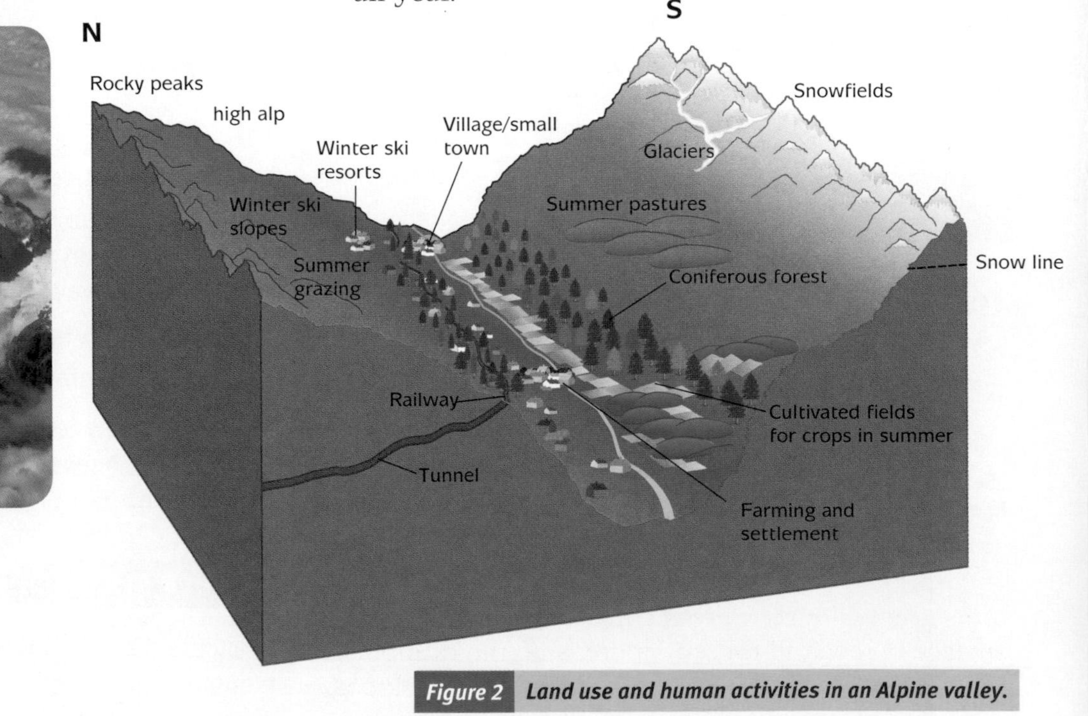

Figure 2 Land use and human activities in an Alpine valley.

Figure 3 *Alpine view – the Jungfrau region of the Lauterbrunnen valley in Switzerland.*

Coniferous trees cover many of the slopes, especially north-facing ones. Wood, as a plentiful local resource, has always been the main building material and winter fuel in Alpine lands. Most sawmills are located on the valley floors near to rivers; timber that cannot be used for construction is made into pulp and paper.

2. Tourism

The Alps have physical advantages for tourism all year round, attracting skiers, climbers and walkers as well as those who simply want to admire the spectacular scenery. These activities are a major industry.

A For winter tourism (examples of resorts are St Moritz, Chamonix):

- snow for skiing and other winter sports; in between the days with heavy snowfall are many sunny, crisp and clear days
- flatter land on the high-level benches (high alp) for easy building of hotels, restaurants, ski-lifts and other facilities
- steep slopes above the resorts for ski runs amid great mountain views.

B For summer tourism (examples of resorts are Interlaken and Garda):

- large glacial lakes on valley floors
- beautiful mountain scenery with snow-capped peaks.

The main worry is that Alpine winters are warming up and becoming less snowy than they used to be. More people are skiing on worn slopes, damaging the vegetation and the surface below and thereby increasing the number of bare surfaces and the risk of soil erosion on steep slopes.

3. Hydro-electric power (HEP) and industry

The steep slopes, high precipitation and summer melting of the glaciers produce fast-flowing rivers that are ideal for generating HEP. The narrow valleys are easy to dam and there are lakes in which to store water. Some of the cheap HEP is used by industries which require a high input of electricity, such as sawmills, electrochemicals and fertiliser manufacture and aluminium smelting. Some of the electricity is also exported to other regions to supply towns and cities.

How do human activities in the Alps compare with other fold mountains?

The Alps are not mineral-rich like the Andes. However, compared with most other fold mountain ranges, Alpine areas are well populated. Because the countries are rich, they have the money and technology to overcome high mountain transport problems. Modern road tunnels, such as the St Bernard and Mont Blanc, have replaced old routes over the high passes. Many electrified railways link the Alpine valleys to the cities; rail tunnels under the Alps include the Brenner and St Bernard. Mountain cog railways, cable cars and chair lifts link the valley floors to high-level benches and ski slopes above them.

ACTIVITIES

1 Study **Figures 1, 2** and **3.**
   (a) List the problems for (i) settlement (ii) farming (iii) transport in the Alps.
   (b) Describe the attractions of the Alps for tourism.
   (c) Explain why most human activities in the Alps are concentrated on valley floors.

2 How have people tried to reduce the problems
   (a) in farming (b) for transport?

3 Using **Figure 2** on page 8 and information about the Alps, draw labelled diagrams to show how the Alps were formed.

FURTHER RESEARCH

Find out about the Andes on the weblink www.contentextra.com/aqagcsegeog.

Landforms at plate margins – volcanoes and supervolcanoes

Where are active volcanoes found? Why are not all volcanoes the same? What is different about supervolcanoes?

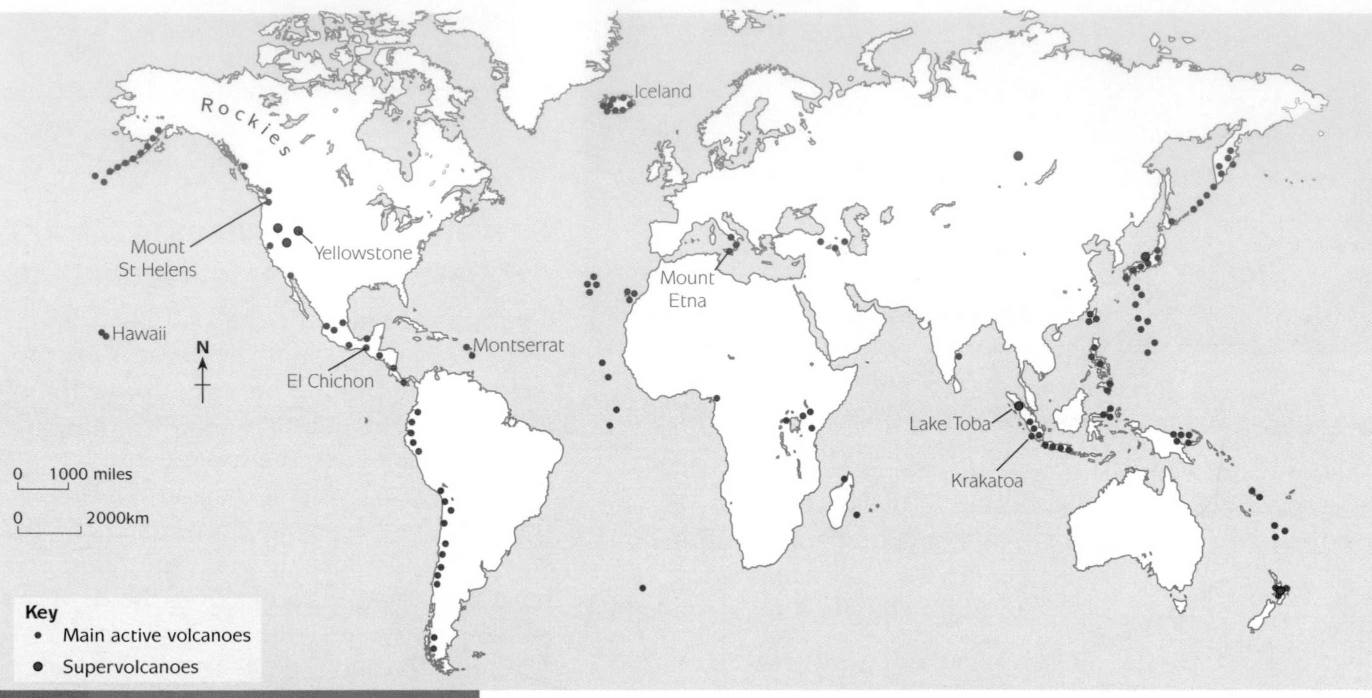

Figure 1 *World distribution of active volcanoes.*

A **volcano** is a cone-shaped mountain formed by surface eruptions from a **magma chamber** inside the Earth. The magma that reaches the surface in an eruption is called **lava**, and is one of the many different products that can be thrown out, including ash, cinders, pumice, dust, gases and steam. The world distribution of active volcanoes (**Figure 1**) shows an almost perfect fit with the locations of the tectonic plate margins (see **Figure 2** on page 6).

How are volcanoes formed?

Volcanoes form where magma escapes through a **vent**, which is a fracture or crack in the Earth's crust. This happens most often at plate margins. Lava and other products are thrown out from the circular hole at the top called the **crater**. Each time an eruption takes place, a new layer of lava is added to the surface of the volcano; since more lava accumulates closer to the crater during every eruption, a cone-shaped mountain is formed (**Figure 2**).

Different types of volcanoes

Volcanoes are divided into two main types, depending upon the material thrown out in an eruption and the form (height and shape) of the volcanic cone produced. These differences are shown in **Figure 3**. Basically the division is between volcanoes formed along constructive plate margins and along destructive margins, because of the different types of lava emitted. Along constructive margins the **basic lava** that has come from within the mantle has a low silica content: it pours out easily, is runny and flows long distances, building up shield volcanoes. However, along destructive margins the **acid lava** has a high silica content, which makes it more viscous so that it travels shorter distances before cooling; these are more explosive volcanoes. After an eruption the vent becomes blocked, which results in great pressure building up before the next eruption. During explosive eruptions lava is shattered into pieces so that bombs, ash and dust are showered over a wide area.

Figure 2 *The Osorno volcano in Chile, an almost perfect cone shape. Last eruption, 1869.*

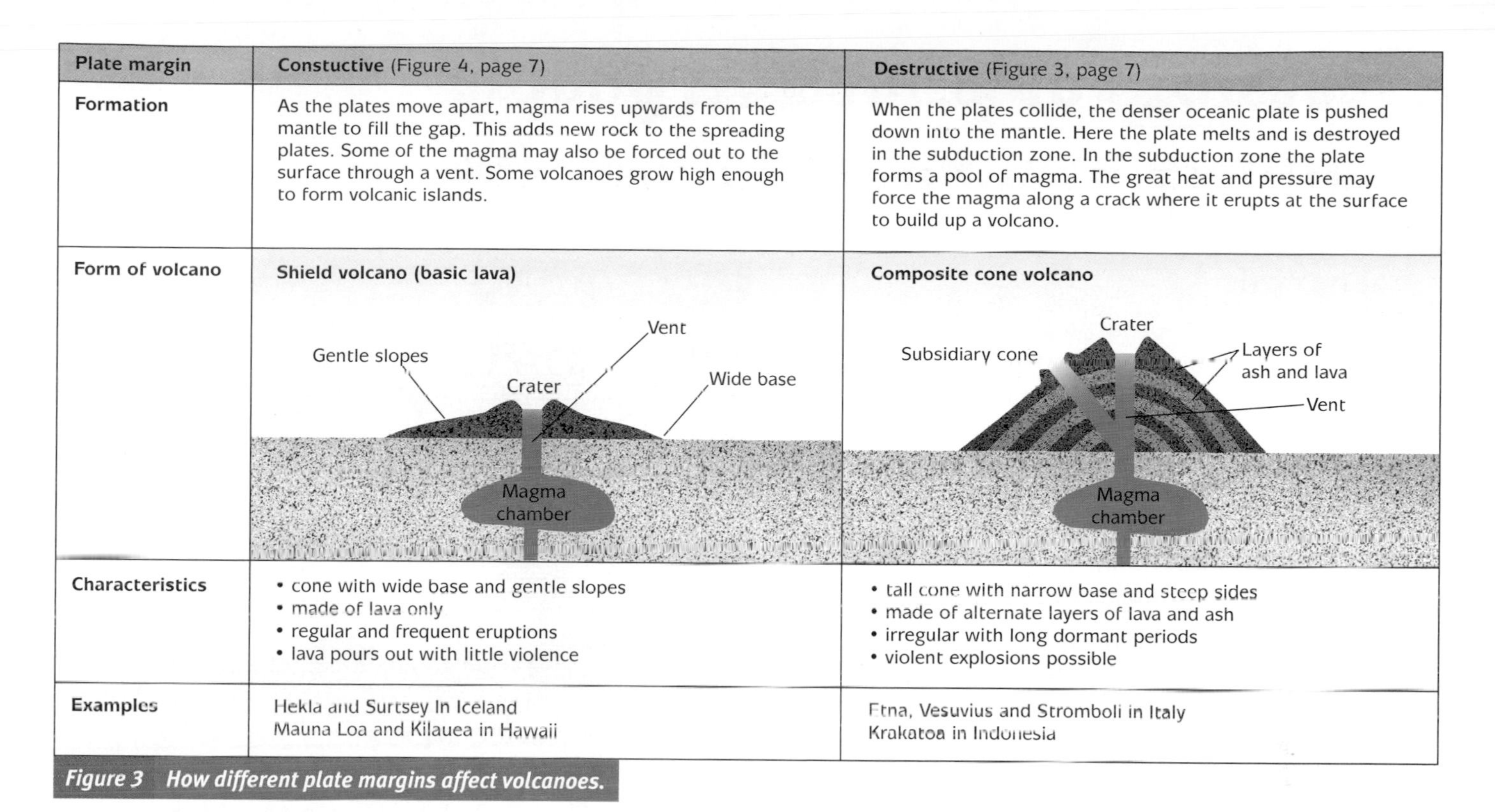

Plate margin	Constuctive (Figure 4, page 7)	Destructive (Figure 3, page 7)
Formation	As the plates move apart, magma rises upwards from the mantle to fill the gap. This adds new rock to the spreading plates. Some of the magma may also be forced out to the surface through a vent. Some volcanoes grow high enough to form volcanic islands.	When the plates collide, the denser oceanic plate is pushed down into the mantle. Here the plate melts and is destroyed in the subduction zone. In the subduction zone the plate forms a pool of magma. The great heat and pressure may force the magma along a crack where it erupts at the surface to build up a volcano.
Form of volcano	**Shield volcano (basic lava)**	**Composite cone volcano**
Characteristics	• cone with wide base and gentle slopes • made of lava only • regular and frequent eruptions • lava pours out with little violence	• tall cone with narrow base and steep sides • made of alternate layers of lava and ash • irregular with long dormant periods • violent explosions possible
Examples	Hekla and Surtsey in Iceland Mauna Loa and Kilauea in Hawaii	Etna, Vesuvius and Stromboli in Italy Krakatoa in Indonesia

Figure 3 *How different plate margins affect volcanoes.*

Supervolcanoes

A **supervolcano** is a volcano that erupts with a massive volume of material, much more than from a normal volcano – at least 1000km^3 of magma. To give you some idea of the great volume, the big eruption of Mount St Helens in the USA in 1980 produced 1km^3. A supervolcanic eruption alters the landscape over hundreds, if not thousands, of kilometres. So much dust is circulating in the atmosphere that it can lead to a 'volcanic winter' – lower temperatures on Earth (i.e. global cooling) because less sunlight reaches the surface. All the world would be affected. Think of the likely effects on nature and people. What about farming and food supply? The last known supervolcano eruption was Toba in Indonesia, about 75 000 years ago.

EXAMPLE: Yellowstone supervolcano

Figure 4 *Millions visit Yellowstone National Park for its surface hot pools and geysers, but how many of them know what lies under the surface and gives the heat?*

Last eruption – about two million years ago. Volcanic ash from this eruption covered more than half of North America.

Next eruption? Probability during the next few thousand years is thought to be low.

GradeStudio

1 a State two differences between each of the following pairs:
  (i) Composite and shield volcanoes (2 marks)
  (ii) Acid and basic lava (2 marks)
  (iii) Volcano and supervolcano (2 marks)

2 Give reasons why some volcanic eruptions are more violent than others. (3 marks)

3 Is the Osorno volcano in **Figure 2** a shield or composite volcano? Explain your answer. (3 marks)

Exam tip

When stating differences, make sure that you mention both – not just one of them.

FURTHER RESEARCH

Find out more about the Yellowstone supervolcano at the weblink www.contentextra.com/aqagcsegeog.

Effects of volcanoes on people

How great a hazard to people are volcanoes? Why do an estimated 500 million people live next to active volcanoes? How safe are these people?

Natural hazards

Natural hazards have both primary and secondary effects for people living in the surrounding area (see the Exam Preparation Box). Volcanoes are no exception. Loss of life can be high if the eruption is large, explosive and happens with little warning; people may be hit by falling debris, suffocated by poisonous gases or buried under mud flows after the heat of the eruption melts snow at the top of the volcano. Whilst the human and physical effects of most volcanoes are limited to the local area, a few massive eruptions have had global effects. Dust from the great eruption of Krakatoa in 1883 circled the world for several years, reducing world temperatures. Nearly a century later, the eruption of El Chichon in Mexico in 1980 reduced the average temperature in the northern hemisphere by 0.25 °C for a year, during a time of global warming.

No matter how gentle, however, a volcanic eruption is always destructive (**Figure 1**). Buildings in the path of the lava are destroyed and farmland is covered. It will be many years before the lava visible in **Figure 1** weathers into fertile soil that can be used again for farming. Homes, farms, animals and crops are lost; people lose their livelihoods and are forced to migrate, often with nothing. Roads are blocked and electricity and telegraph poles are brought down, further disrupting normal life and economic activities.

Loss of life is always more likely in volcanic eruptions along destructive plate margins. Living close to shield volcanoes is safer. Lives are rarely threatened when only lava is being erupted, even from a composite cone, because people have plenty of time to move out of the way of the lava flows.

Figure 1 Results of a lava flow from the 2001 eruption of Mount Etna, which covered vineyards and orchards without killing anyone.

EXAM PREPARATION

Primary and secondary effects of natural hazards

Primary effects – the immediate impact
- people injured and killed
- buildings, property and farmland destroyed
- communications and public services (transport, electricity, telephones etc.) disrupted.

Secondary effects – the medium- and long-term impact
- shortages of drinking water, food and shelter
- spread of disease from contaminated water
- economic problems from the cost of rebuilding and the loss of farmland, factories, tourism and other economic activities
- social problems from family losses and stress.

Volcanoes: hazard or blessing?

Why do people choose to live near active volcanoes when an eruption could happen at any time? The answer is that living in volcanic areas has its attractions. After the lava weathers, volcanic soils are some of the world's most fertile soils. Often they are much better quality than the soils in surrounding mountainous areas. Soils from the many eruptions of Vesuvius on the Plain of Campania are the best in southern Italy (**Figure 2**).

Volcanic areas also offer a variety of attractions for tourists, including bathing in the hot springs and mud pools, watching geysers, and volcano walking up to and around the crater. The supplies of hot water have economic uses, either as domestic heating or for generating electricity (geothermal power) as in Iceland. In addition, there are valuable minerals, notably sulphur, borax and pumice.

Campania is a region in southern Italy. It includes the major city of Naples and the famous volcano, Mount Vesuvius (1198 metres). Vesuvius is a composite cone volcano, which is dormant at present but has been very destructive in the past. Its most notable eruption was in AD 79 when the towns of Pompeii and Herculaneum were destroyed. Thousands of people were killed by the poisonous gases and the area was buried under metres of ash. So why do so many people choose to live close by?

Excavations at Pompeii and Herculaneum, trips to the crater on Vesuvius and visits to hot springs have brought a thriving tourist industry employing many local people. The fine ash is very fertile. Wheat, maize, peaches, almonds, vines and especially tomatoes are all grown intensively. Yields are five times higher than the national average. The fertility of the plain contrasts with the dry, arid, thin soils on the limestone uplands. West of Naples is the Phlegraean Fields, a wasteland with hot springs, geysers and sulphur domes, which is useless for farming, although the sulphur has industrial uses.

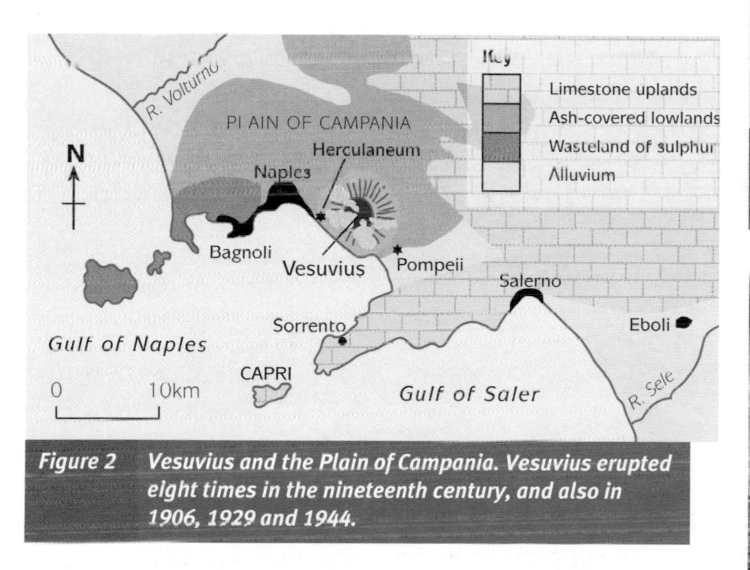

Figure 2 **Vesuvius and the Plain of Campania. Vesuvius erupted eight times in the nineteenth century, and also in 1906, 1929 and 1944.**

Can volcanic eruptions be predicted?

The unhelpful answer is 'Yes and No'. Usually there are advance warning signs that a volcano is about to come back to life. These include small earthquake shocks, increased emissions of steam and gases and visual signs of bulging around the crater. Dormant volcanoes close to large centres of population in rich developed countries, such as the Italian volcanoes, are constantly monitored (**Figure 4**). Electronic **tiltmeters** measure very small changes in the profile of the mountain; they work like a spirit level. Because of the difficulties and dangers of putting instruments in the volcano's crater, satellites are used to measure infra-red radiation and look for any sudden changes in heat activity.

Do poor developing countries have the money and technology needed for monitoring? Who would have thought of monitoring the Montserrat volcano before 1995 when it had not erupted for 350 years? Who could have predicted in 1995 that it would still be erupting 13 years later? In theory there should be no loss of life from volcanoes; people can be evacuated from the danger areas in time. But what will happen when scientists predict that Vesuvius is about to erupt again? Can half a million people be moved in a few days? Will they leave their farms, homes and possessions? Eruptions from volcanoes on destructive margins can be very violent and explosive, and are notoriously unpredictable.

Figure 3 **Italian tourists in a hot mud pool at the foot of Vulcano, a volcano on one of the Lipari Islands off the north coast of Sicily.**

Figure 4 **Monitoring equipment in the crater of Vulcano.**

ACTIVITIES

1 Draw two spider diagrams to show the advantages (positive effects) and disadvantages (negative effects) of volcanoes.

2 (a) Describe two ways of monitoring volcanoes to predict when they will erupt.
(b) Explain why they are used more in rich than in poor countries.
(c) In 2008, the volcano Chaiten in Chile erupted without warning. Why is this still possible?

3 (a) Explain the dangers for people living near Mount Vesuvius, a dormant composite volcano.
(b) Why, despite the dangers, is the area around Vesuvius one of the most densely populated parts of Italy?

Soufrière Hills volcano, Montserrat

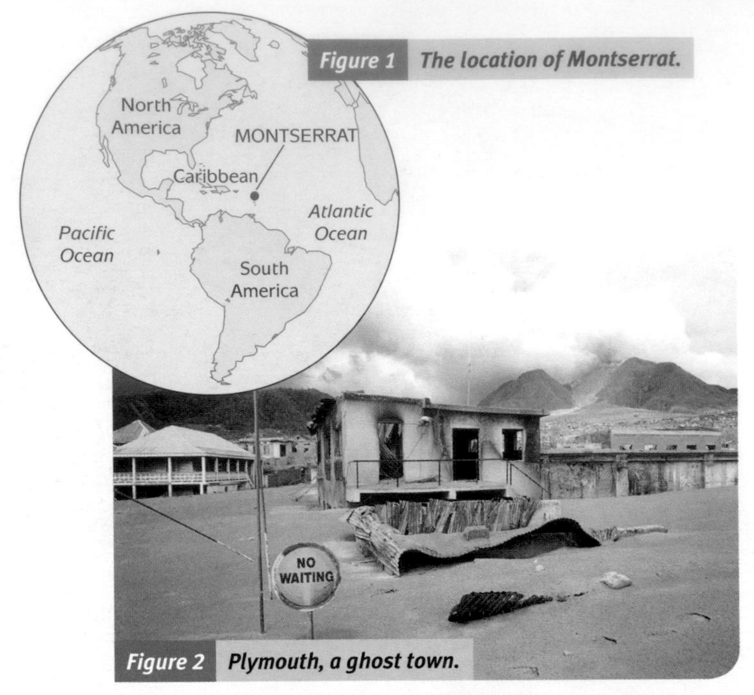

Figure 1 The location of Montserrat.

Figure 2 Plymouth, a ghost town.

Montserrat (**Figure 1**) is a small island in the Caribbean that is still a British colony. The island is mountainous and wooded, earning it the name 'Emerald Island of the Caribbean'. It has been popular with many wealthy British people including Paul McCartney and Sting – exclusive villas and hotels line the coast. However, many of the residents of Montserrat are quite poor, living in small villages and practising subsistence farming. Before the eruption the population was 12 000, 50 per cent of whom lived in the capital city, Plymouth, in the south of the island.

In July 1995 the Soufrière Hills volcano erupted for the first time in 350 years. One month later 50 per cent of the population were evacuated to the north of the island away from the danger zone. In April 1996, as the eruptions continued, Plymouth became a ghost town (**Figure 2**) as more and more people were evacuated. The eruptions became more explosive and the lava and ash caused great damage to the island. In June 1997 another eruption destroyed villages in the centre of the island, killing 23 people. Of the island's 103 square kilometres only 39 square kilometres in the north of the island were considered safe. Over 5000 people left Montserrat, most to settle on nearby islands such as Antigua or to move to the UK. Study **Figure 3**, which shows the impact of the eruptions.

Those people who stayed on the island suffered very harsh conditions. The south of the island was the most developed, with the main towns, communications and services. In the north there were few roads and settlements. Many of the evacuees were forced to live in makeshift shelters with inadequate sanitation; there were few schools and no proper hospital, and living conditions were very poor. The country's tourist industry stopped with the closure of the airport, and other industries suffered with the restricted port activities. The processing of imported rice and the assembly of electronics products declined. The country was forced to rely upon aid from London.

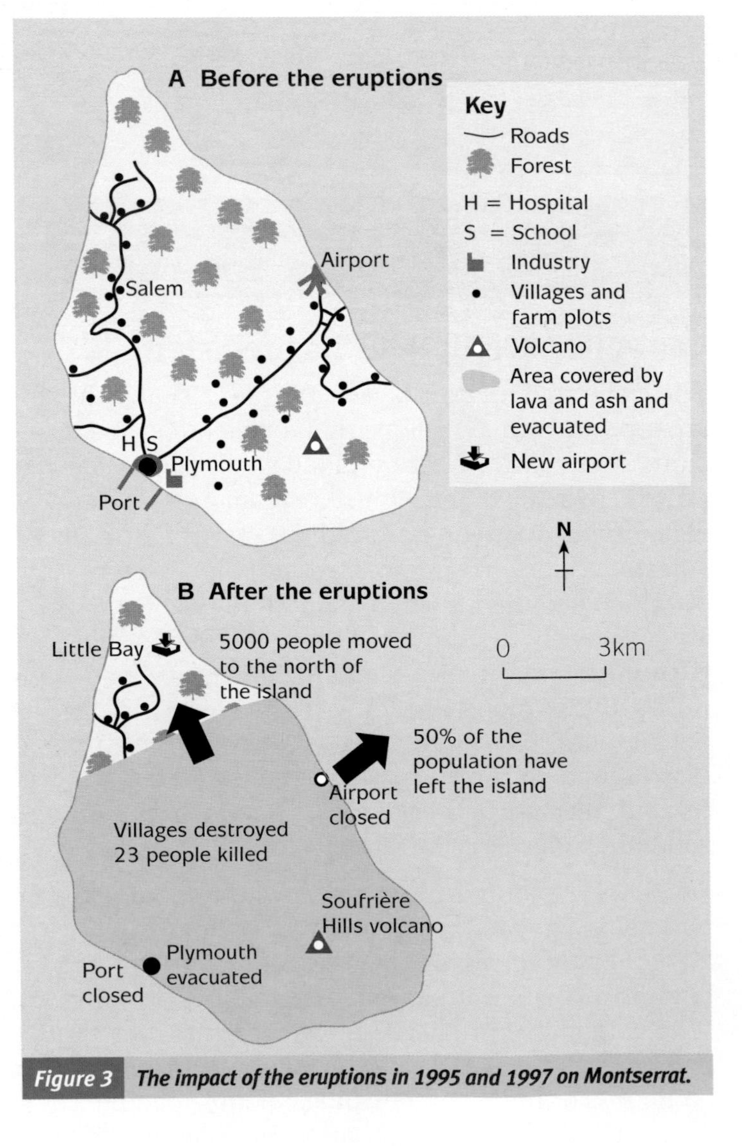

Figure 3 The impact of the eruptions in 1995 and 1997 on Montserrat.

After the eruptions, which forced the evacuation of most of the island, the people called for the British government to pay compensation and to rebuild Montserrat. Aid totalling £41 million has been offered to redevelop the north of the island and £10.5 million to relocate refugees. In 1997, £2400 was offered to each adult over 18 wanting to leave the island. There was rioting on the island because the local people felt the British government was not offering enough help. The Montserratians were demanding £20 000 per person. Can the rebuilding of the island be justified for a population of only about 4000 people? Also, scientists cannot predict if and when the volcano may erupt again. Money may be invested in rebuilding, only to be wiped out by another eruption, which destroys the whole island.

Has the volcano gone quiet?

There was another major eruption in 2003 and there has been intermittent activity since. The south still looks like a barren moon landscape (page 5). Government policy is to rebuild in the 'safe' north. A new town at Little Bay, to replace Plymouth, has new government buildings, shops, banks, tourist board offices and homes. Close by is the new airport, opened in 2005, with regular links to the international airport in Antigua. The rebuilding, backed by the determination of the Montserrat people, does not come cheap. More than £200 million has been spent by the UK government so far.

Attracting tourists back to the island and restarting profitable economic activity is difficult. The airport is limited to 20-seater planes. A few people, who had migrated, are trickling back, but socially they find a different island. Formerly there was just one old people's home; now there are four. The community is no longer there to look after them in old age.

Figure 4 *Northern limit of the exclusion zone, which covers about two-thirds of the island.*

Figure 5 *Housing hurriedly built on unused land in the north after 1997 by families forced to leave Plymouth. How good does the area look to be for the usual Caribbean activities of tourism and farming?*

ACTIVITIES

1 Put together information for the case study of a volcano – Montserrat.
   (a) Draw a sketch map to show the location of Montserrat.
   (b) Explain with the aid of diagrams why the volcanic eruption occurred on Montserrat. (**Figures 2** and **3** on pages 6–7 and **1** and **3** on pages 12–13 will help.)
   (c) Describe (i) the primary effects and (ii) the secondary effects of the eruption.
   (d) Describe the responses:
      (i) short term (until 1997)
      (ii) medium term and
      (iii) longer term (since 2003).
   (e) FAQs in Montserrat: 'When can we go back to Plymouth?' 'Where are the jobs?' 'When can we get back to normal?' Explain why the government has no answers for its people.

FURTHER RESEARCH

Find out more about the Montserrat Volcano Observatory on the weblink www.contentextra.com/aqagcsegeog.

The earthquake hazard

Why do more than 90 per cent of strong earthquakes occur at destructive plate margins? Why do earthquakes kill many more people than volcanoes do?

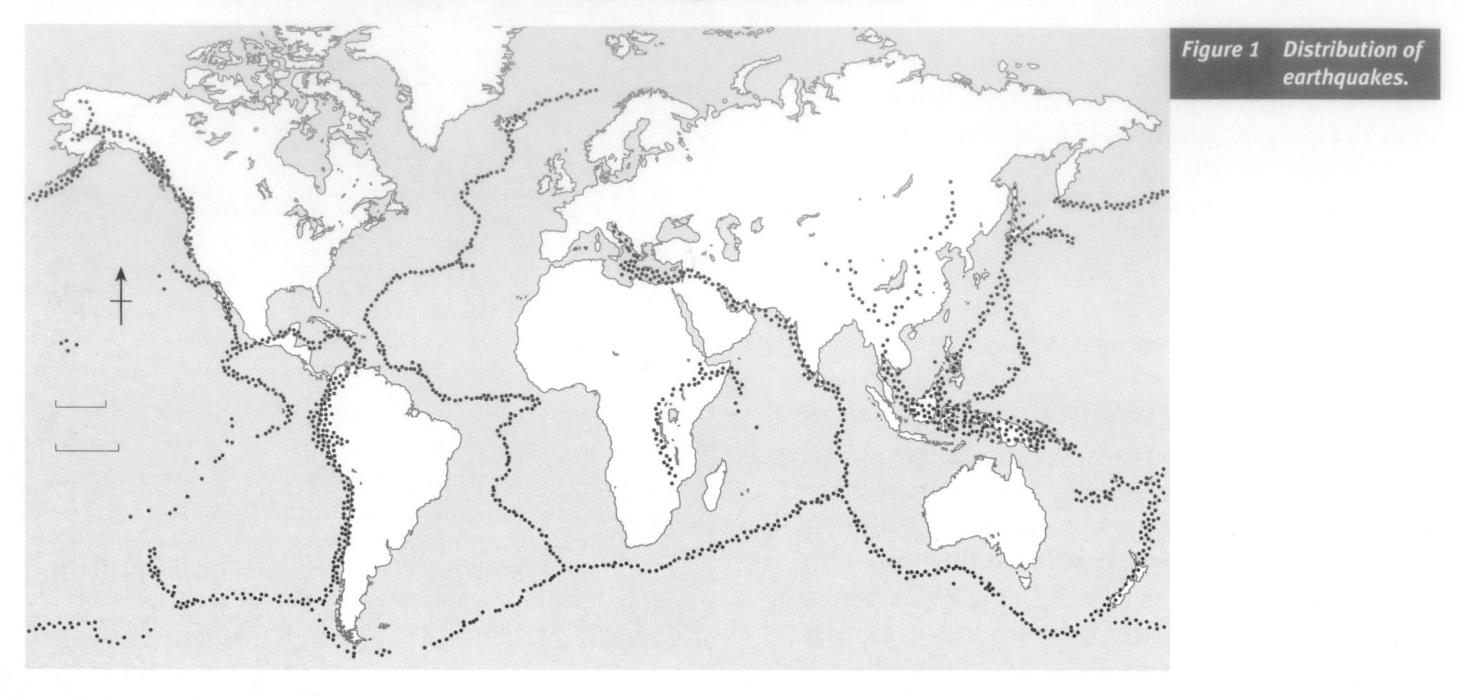

Figure 1 Distribution of earthquakes.

Earthquakes are vibrations in the Earth's crust that shake the ground surface. They are sudden, and, because they happen without warning, often lethal. About two million people have died as a result of earthquakes since 1900. Earthquakes are common events; there are around 50 000 detectable earthquakes each year, although most are too small for people to notice. They occur in well-defined zones (**Figure 1**). How closely do the earthquake zones match the tectonic plate margins shown in **Figure 2** on page 6?

Between 15 and 20 earthquakes a year are magnitude 7 or more on the **Richter scale**. These are strong enough to have devastating effects on life and property. The magnitude of an earthquake is measured by an instrument called a **seismograph** and given a value between 1 and 10. The Richter scale is logarithmic: an earthquake measured at 7 is 10 times stronger than one measured at 6 and 100 times stronger than one measured at 5.

The **Mercalli scale** is used to indicate the intensity of an earthquake. It classifies the effects of an earthquake on a scale using Roman numerals from I to XII, and takes into account the effects on the Earth's surface, people and buildings.

I	Is rarely ever felt by people
II–IV	Feeble to moderate effects; felt by people, but no damage
V–VII	Strong effects, causing panic but only slight damage
VIII	Destructive to poorly built structures but only slight damage to those well designed
IX–XII	Ruinous, disastrous, very disastrous to catastrophic with almost total destruction

The number on the scale is obtained by making a judgement; it is not measured by a machine.

Why do earthquakes happen?

Over 90 per cent of earthquakes occur where plates are colliding at destructive plate margins. Great stresses build up in the subduction zone as one plate is forced down below the other. Energy builds up, and is released in an earthquake. The point at which the earthquake happens below the ground surface is called the **focus** (see **Figure 3** on page 7). **Shock waves** radiate out in all directions, gradually becoming less strong as they get further from the epicentre.

The **epicentre** is the point on the surface directly above the focus, where the greatest force of the earthquake is felt. Earthquakes also occur along conservative plate margins like the San Andreas fault (see **Figure 5** on page 7).

The effects of earthquakes

Primary effects are the immediate damage caused by the earthquake, such as collapsing buildings, roads and bridges. People are killed by being trapped in their homes, places of work and cars. The severity of the primary effects is determined by a mixture of physical and human factors (**Figure 2**). The chance element is time of day – were there fewer people close to the epicentre when the earthquake struck than at other times?

Physical	• High magnitude on Richter scale • Shallow focus (near the surface) • Sands and clays vibrate more (e.g. Mexico City)	• Low magnitude (below 5) • Focus deep underground • Hard rock surface (e.g. Seattle)	Physical
	Great (or total) damage High number of deaths and injuries Mercalli scale VII–XII	**Superficial damage to buildings Few casualties Mercalli scale I–VI**	
Human	• High density of population • Residential area of a city • Self-built housing • Lack of emergency procedures (e.g. Gujarat in India)	• Low density of population • Urban area with open spaces • Earthquake-proof buildings • Regular earthquake drills	Human

Figure 2 ***Factors controlling the effects of an earthquake.***

Secondary effects are the after-effects, such as fires, tsunamis, landslides and disease.

- Fires are caused by earthquakes fracturing gas pipes and bringing down electricity wires. Fires spread quickly in areas of poor-quality housing.
- Tsunamis are giant sea waves caused by an earthquake on the sea floor and are really dangerous for people living along low-lying coasts.
- Landslides are most likely on steep slopes and in areas of weak rocks such as sands and clays.
- Diseases such as typhoid and cholera spread easily when burst pipes lead to shortages of fresh water and to contamination from sewage.

The impact of earthquakes is often much more severe in poorer countries, where earthquake-resistant buildings are often considered too expensive to build. Even where building regulations exist, they are frequently ignored, because builders want to make more money and people can only afford cheap houses. People and authorities are also less well prepared.

The responses to earthquakes

Irrespective of whether people are rich or poor, immediate emergency aid is desperately needed everywhere after a strong earthquake strikes. Specialist rescue teams with sniffer dogs and lifting equipment, and medical teams with field hospitals can be expected to be airlifted within hours in rich countries, thanks to advance preparations. The poorer the country, the greater its reliance upon short-term aid from overseas.

In the medium term, the need is for a quick return to normal life (or as near normal as possible) by repairing and replacing what has been lost and restarting economic activity. The focus needs to be switched from disaster aid to development aid.

Earthquakes cannot be predicted. They happen without warning. For people living in known earthquake zones, the only effective long-term response is to be prepared. The emergency services need to be trained and ready: disaster relief operations require practice. Tall buildings are more likely to remain standing if built using the methods shown in **Figure 3**.

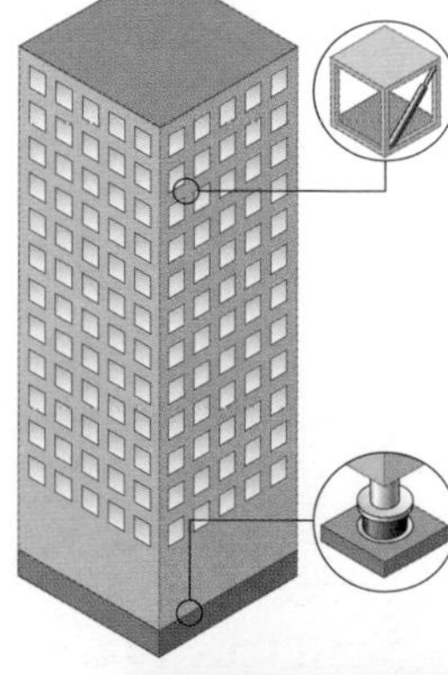
Damping and bracing systems to help absorb shocks
Foundation piles made out of alternative layers of steel and rubber to make the skyscraper flexible in an earthquake – to make it stiff vertically, but flexible horizontally

Figure 3 ***How to make a skyscraper resist earthquake shocks.***

GradeStudio

1 Describe the global pattern of earthquakes shown in **Figure 1**. (4 marks)

2 (a) Explain why many strong earthquakes occur along destructive plate margins. (4 marks)

(b) Why are tectonic stresses lower along constructive plate margins? (2 marks)

3 (a) State the difference between the primary and secondary effects of an earthquake. (2 marks)

(b) Explain why the primary effects of an earthquake are usually more severe in urban than in rural areas. (3 marks)

4 Two earthquakes in December 2003, both of magnitude 6.5 on the Richter scale, were:

California	a clock tower fell, killing three people
Iran	a large part of the city of Bam was flattened, killing an estimated 30 000 people

State and explain two factors which could have accounted for the different numbers of deaths. (4 marks)

Exam tip

Describing a pattern from a map:

- State the main feature of the pattern.
- Name places where there are many patterns.
- Also refer to areas where there are few patterns.

Gujarat, India, 2001 – Seattle, USA, 2001 – Indonesia and the Asian tsunami, 2004

Gujarat, India, 2001

The state of Gujarat in northern India is an active earthquake zone. In January 2001 a devastating earthquake occurred here, measuring 7.9 on the Richter scale. Tectonic stresses across northern India result from the pressure of the Indo-Australian plate pushing northwards against the Eurasian plate. The epicentre was close to Bhachau (**Figure 1**). As much as 48 hours elapsed before rescue efforts began; there were many complaints about the authorities' lack of preparedness for a natural disaster of this kind. In some places, rescue workers had no better equipment than shovels. The earthquake left at least 20 000 dead, 160 000 injured and 600 000 homeless. Direct economic losses were estimated at US$ 2–4 billion.

Seattle, USA, 2001

Seattle lies on a fault line near the plate margin between the Pacific and North American plates. They are moving past each other slowly, but erratically. The earthquake described in **Figure 2** was the largest to hit the areas for more than half a century.

Earthquake rocks Seattle

An earthquake measuring 6.8 on the Richter scale rocked the Pacific Coast American city of Seattle last night, sending workers onto the streets in panic. One said it was 'like a bomb going off'. Office workers used fire escapes to evacuate skyscrapers in downtown Seattle. The hotel where Microsoft founder Bill Gates was addressing a conference was evacuated. The quake shattered windows, caused skyscrapers to sway, buckled some roads and railway lines and led to electricity blackouts.

In the city, one woman died of a heart attack and more than 250 were injured. Many of the injuries were caused by falling debris. One hospital reported treating several people for anxiety attacks. Extensive damage was caused to property in the centre, which could cost millions of dollars to repair. Although transport was disrupted and people were sent home from factories as damage was checked out, within 48 hours the city was back to working normally.

Experts said that one of the reasons for the lack of deaths and serious injuries was that the earthquake's focus was much deeper than usual, 50km underground, in solid rock. Also a lot of money had been spent in the last 50 years to make sure that buildings could withstand major shocks. One of the city's best-known landmarks, the Space Needle, was built to withstand a 9.1 magnitude earthquake. It shuddered and rolled violently during the tremor, but remained undamaged. An hour or so after the quake it was open again for visitors.

Figure 2 **The Space Needle in Seattle, with observation deck 184 metres above ground.**

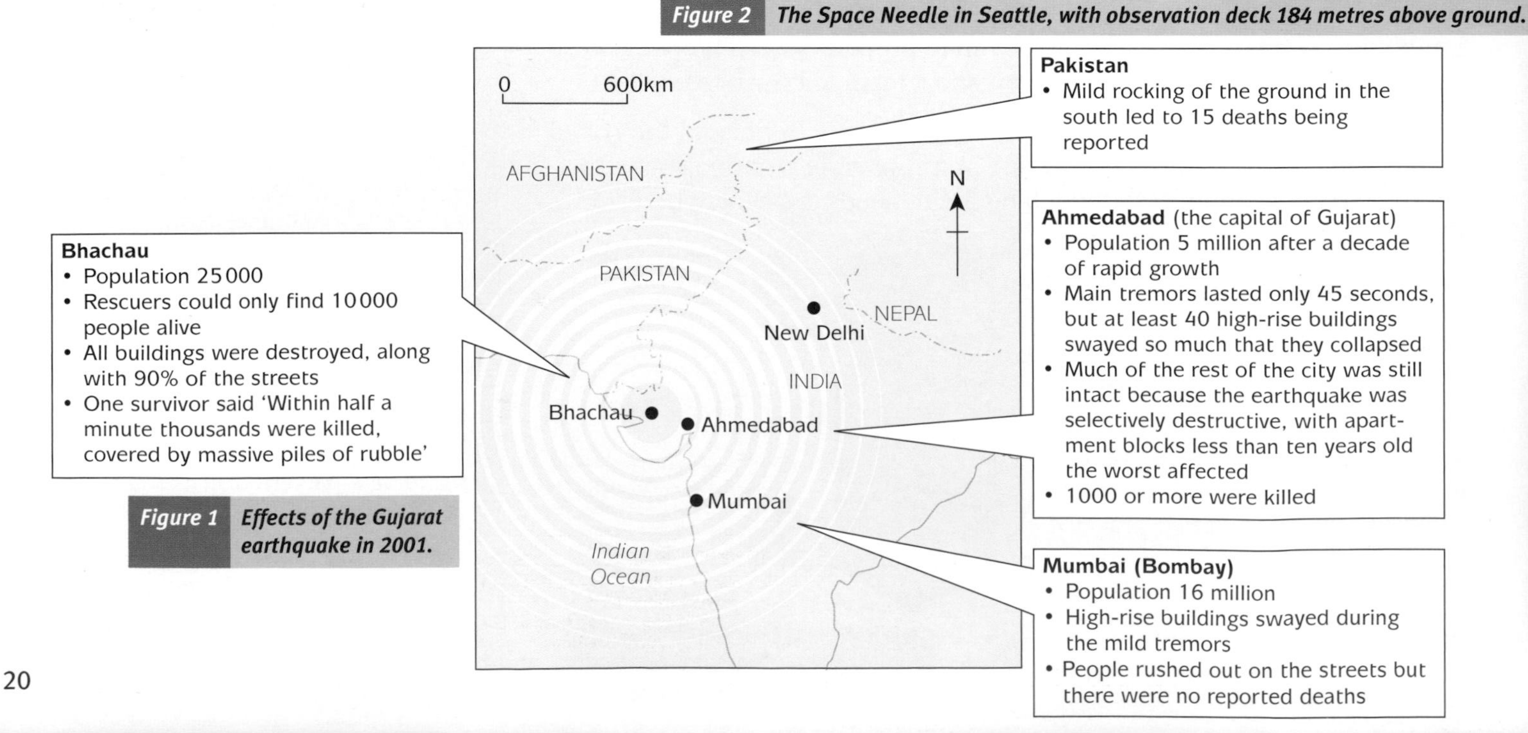
0 600km
AFGHANISTAN
PAKISTAN
N
NEPAL
New Delhi
INDIA
Bhachau
Ahmedabad
Mumbai
Indian Ocean
Pakistan
• Mild rocking of the ground in the south led to 15 deaths being reported
Ahmedabad (the capital of Gujarat)
• Population 5 million after a decade of rapid growth
• Main tremors lasted only 45 seconds, but at least 40 high-rise buildings swayed so much that they collapsed
• Much of the rest of the city was still intact because the earthquake was selectively destructive, with apartment blocks less than ten years old the worst affected
• 1000 or more were killed
Mumbai (Bombay)
• Population 16 million
• High-rise buildings swayed during the mild tremors
• People rushed out on the streets but there were no reported deaths
Bhachau
• Population 25 000
• Rescuers could only find 10 000 people alive
• All buildings were destroyed, along with 90% of the streets
• One survivor said 'Within half a minute thousands were killed, covered by massive piles of rubble'

Figure 1 **Effects of the Gujarat earthquake in 2001.**

Indonesia and the Asian tsunami, 2004

The 8.9 magnitude earthquake in the sea off the coast of Indonesian island of Sumatra, caused by pressure from the Indo-Australian plate pushing under the Eurasian plate, was the fifth-strongest earthquake ever recorded. The secondary effects from the wave of a size never previously known were devastating (see Information box). Within four hours, many densely populated coastal communities in Indonesia, Thailand, India, Sri Lanka and elsewhere were left in ruins (**Figure 3**).

INFORMATION

Asian tsunami, December 2004

Speed	500km/h
Highest recorded wave	34m
Distance travelled from epicentre	4000km
Number of people dead or missing	220 000+
Number of people displaced	about 2 million
Houses destroyed	Over 500 000

Human responses to the disaster

Within a week over £450 million had been pledged from all over the world. The world's biggest ever emergency relief operation began immediately. What happened in Aceh province in northern Sumatra was similar to elsewhere in Asia. Cargo planes from all over the world brought blankets and medicines. Trucks laden with food, medicines and body bags reached places still accessible by road. Air drops to coastal communities that were cut off from the outside world provided some relief to people stranded among the debris. Troops using bulldozers helped to clear the dead bodies into mass graves to reduce the risk of disease spreading.

The one good thing to come out of this horrendous event is the international tsunami warning system between countries. Previously none existed. Now governments and people are aware of potential dangers from strong earthquakes in the oceans. Soon after an 8.4 earthquake off Indonesia in September 2007, the authorities in far-away Kenya evacuated people from the coastal resort of Mombasa and warned local fishermen not to go out to sea.

ACTIVITIES

1 (a) State one physical and one human reason why loss of life was lower in the Seattle earthquake than in the Gujarat earthquake.
(b) Explain how the effects of the Gujarat earthquake varied with distance from the epicentre.

2 (a) Why are the primary effects of an earthquake usually more serious for people than the secondary effects?
(b) When and why can the secondary effects of an earthquake be greater?

3 Make notes, suitable for exam use as a case study for one or more of the earthquakes on these pages. Use the headings: Location, Causes, Effects and Responses.

4 View of one scientist: 'Earthquakes are inevitable, but death from earthquakes is not'.
(a) Describe more fully this scientist's view of earthquakes and their effects.
(b) Do the three examples of earthquakes mentioned above support this view? Explain your answer.

FURTHER RESEARCH

Find out more about the Asian tsunami on the weblink www.contentextra.com/aqagcsegeog.

Figure 3 *Places affected by the Asian tsunami of 2004.*

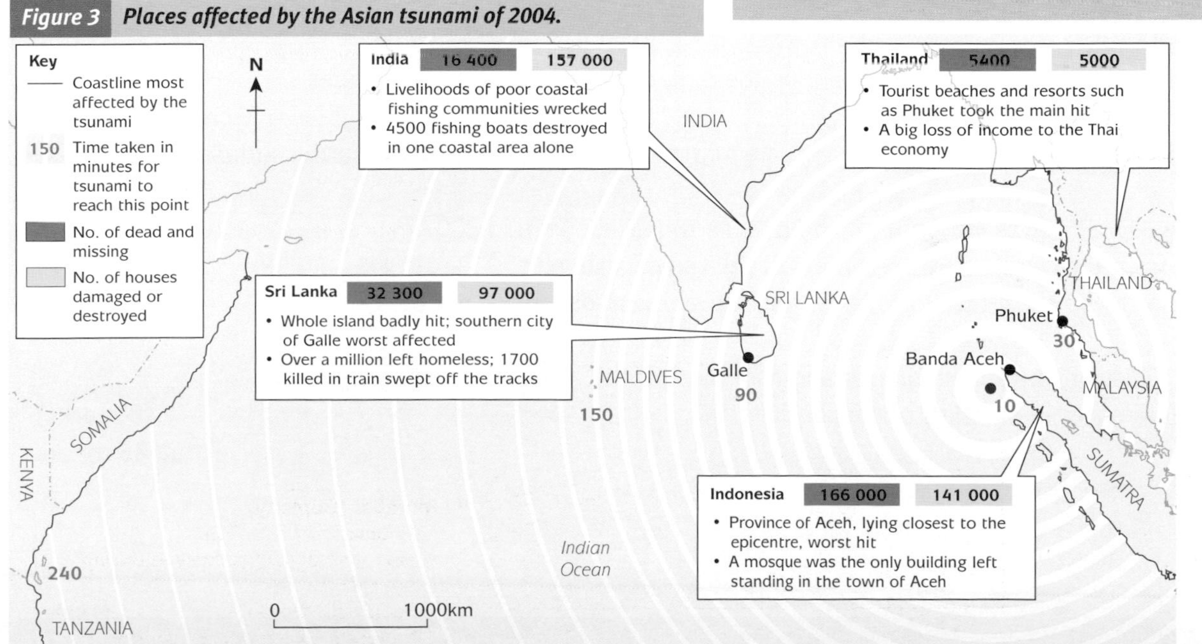

Key
Coastline most affected by the tsunami
150 Time taken in minutes for tsunami to reach this point
No. of dead and missing
No. of houses damaged or destroyed
N
India 16 400 157 000
• Livelihoods of poor coastal fishing communities wrecked
• 4500 fishing boats destroyed in one coastal area alone
Thailand 5400 5000
• Tourist beaches and resorts such as Phuket took the main hit
• A big loss of income to the Thai economy
Sri Lanka 32 300 97 000
• Whole island badly hit; southern city of Galle worst affected
• Over a million left homeless; 1700 killed in train swept off the tracks
Indonesia 166 000 141 000
• Province of Aceh, lying closest to the epicentre, worst hit
• A mosque was the only building left standing in the town of Aceh
INDIA
SRI LANKA
THAILAND
Phuket
30
Banda Aceh
10
MALAYSIA
SUMATRA
Galle
90
MALDIVES
150
SOMALIA
KENYA
240
TANZANIA
Indian Ocean
0
1000km

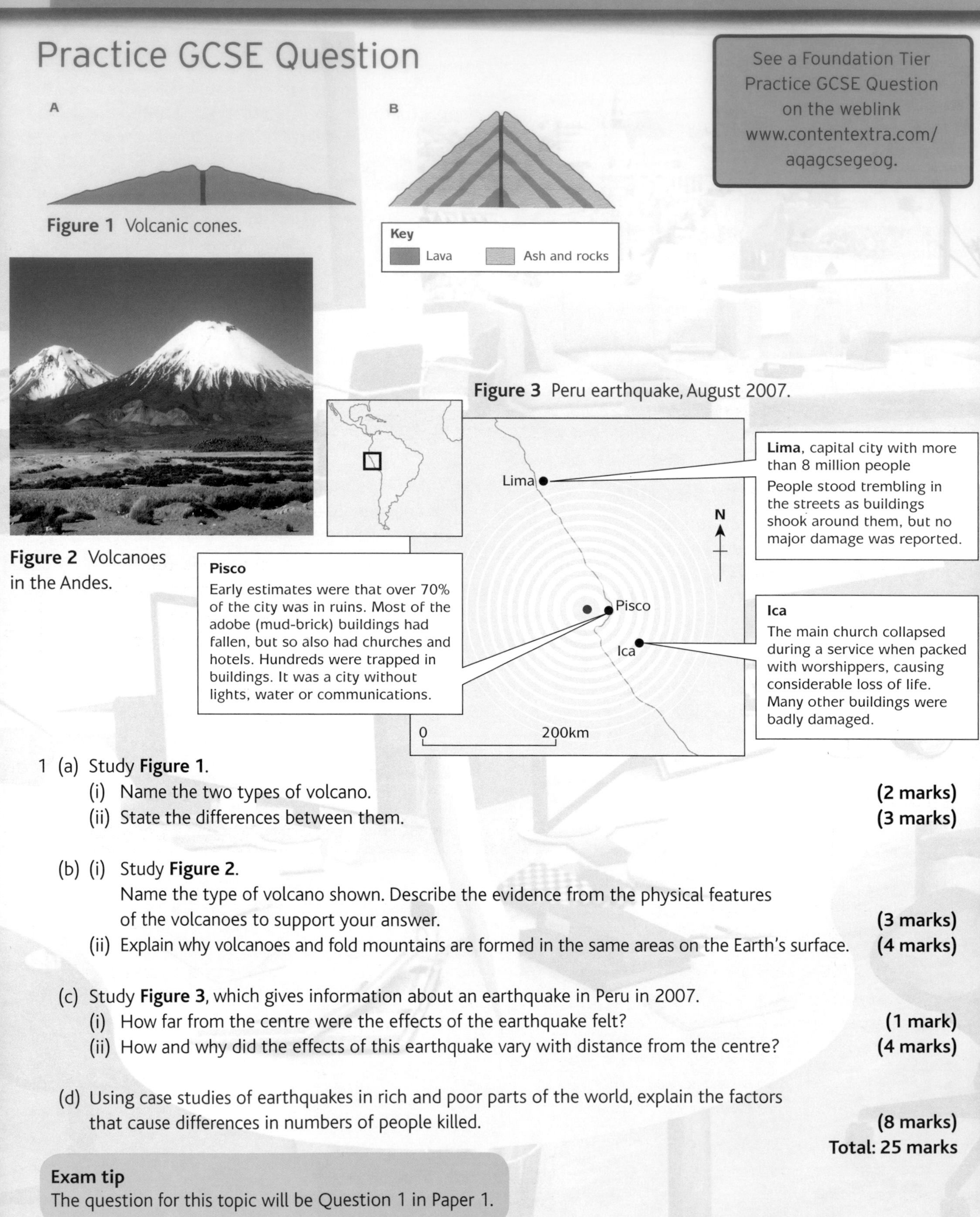

GradeStudio

Practice GCSE Question

See a Foundation Tier Practice GCSE Question on the weblink www.contentextra.com/aqagcsegeog.

A

B

Key
Lava
Ash and rocks

Figure 1 Volcanic cones.

Figure 2 Volcanoes in the Andes.

Figure 3 Peru earthquake, August 2007.

Lima
Pisco
Ica
N
0
200km

Lima, capital city with more than 8 million people
People stood trembling in the streets as buildings shook around them, but no major damage was reported.

Pisco
Early estimates were that over 70% of the city was in ruins. Most of the adobe (mud-brick) buildings had fallen, but so also had churches and hotels. Hundreds were trapped in buildings. It was a city without lights, water or communications.

Ica
The main church collapsed during a service when packed with worshippers, causing considerable loss of life. Many other buildings were badly damaged.

1 (a) Study **Figure 1**.

(i) Name the two types of volcano. **(2 marks)**

(ii) State the differences between them. **(3 marks)**

(b) (i) Study **Figure 2**.
Name the type of volcano shown. Describe the evidence from the physical features of the volcanoes to support your answer. **(3 marks)**

(ii) Explain why volcanoes and fold mountains are formed in the same areas on the Earth's surface. **(4 marks)**

(c) Study **Figure 3**, which gives information about an earthquake in Peru in 2007.

(i) How far from the centre were the effects of the earthquake felt? **(1 mark)**

(ii) How and why did the effects of this earthquake vary with distance from the centre? **(4 marks)**

(d) Using case studies of earthquakes in rich and poor parts of the world, explain the factors that cause differences in numbers of people killed. **(8 marks)**

Total: 25 marks

Exam tip
The question for this topic will be Question 1 in Paper 1.

Improve your GCSE answers

Show the differences between primary and secondary effects – immediate and long-term responses.

A Difference between **primary** and **secondary effects** of a natural hazard

1. Look at **Figure 4**. Can you separate out the primary from the secondary effects of a volcano? The final score should be Primary 5 Secondary 4.

2. Which one is different from all the others because it is a **positive** rather than **negative** effect?

- deaths of people
- loss of income from tourism
- high economic costs of rebuilding and recovery
- electricity pylons and wires brought down
- farmland destroyed
- roads blocked by lava flows
- volcanic dust and ash deposits make very fertile soils
- shortages of food and shelter
- buildings buried under thick layer of ash

Figure 4 Effects of volcanic eruptions on people.

B Difference in **immediate responses** to a natural hazard between **rich** and **poor countries**

1. Look at **Figure 5**. Separate out likely responses between rich countries with a high earthquake risk such as Japan and USA, and poor countries such as India, Iran and Pakistan.

2. What is the main aim of these immediate responses?

- Army called in but it lacks specialist knowledge and equipment
- People follow emergency drills already practised
- Rescue workers with heavy lifting equipment move in
- International aid teams with sniffer dogs fly in
- Relatives and friends try to reach people trapped using their hands and basic tools
- Supplies of tents, blankets, food and water are brought in from warehouse stores
- Emergency services (hospital, fire and rescue) are immediately alerted
- People move out and sleep out of doors in open spaces

Figure 5 Examples of immediate responses to an earthquake.

C Difference between **immediate** and **long-term responses** to a natural hazard

1. Look at the responses in **Figure 6**. Their purpose is to **monitor**, **predict**, **protect** and **prepare**. Can you identify a good example of each purpose from the list?

2. Which responses are more appropriate for (a) volcanoes (b) earthquakes?

3. Why are these responses more likely to be implemented in rich than in poor countries?

- set up scientific observatory with instruments e.g. seismographs
- use modern construction methods for earthquake-proof structures
- recruit and train emergency teams
- hold regular practice drills and educate people
- place tiltmeters and heat sensors in and around the crater
- forbid new building in high-risk areas
- tougher building regulations and stricter enforcement

Figure 6 Examples of long-term responses to volcano and earthquake threats.

ExamCafé

REVISION

Key terms from the specification

Exam Café

Earthquake – shaking of the ground
Effects – primary (first effects) and secondary (later effects), positive (good) and negative (bad)
Fold mountains – long, high mountain range formed by upfolding of sediments
Hazard – natural hazards are short-term events that threaten lives and property
Plates – large rock areas that make up the Earth's crust
Responses – immediate actions after the event or in the long-term
Tectonic activity – movement of the large rock plates of the Earth's crust
Tsunami – giant sea wave travelling at high speed
Volcano – cone-shaped mountain formed by surface eruptions of magma from inside the Earth

Checklist

	Yes	If no – refer to
Can you name several large plates forming the Earth's crust?		page 6
Do you know the differences between destructive, constructive and conservative plate margins?		pages 6–7
Can you name two landforms at destructive margins and explain their formation?		pages 6–7
Do you know the main features of the world distribution of fold mountains, active volcanoes and earthquakes?		pages 8, 12, 18
Can you state two differences between supervolcanoes and other volcanoes?		pages 12–13
Can you give the effects of volcanoes on people using these headings – primary, secondary, positive, negative?		pages 14–15
Do you know the two different ways of measuring earthquakes?		page 18
Can you explain why earthquakes cause more loss of life in poor than in rich countries?		pages 19–21

Case study summaries

Fold mountain range	Volcanic eruption	Earthquake in a rich country	Earthquake in a poor country	Tsunami
Name and location	Cause	Location and strength	Location and strength	Cause
Formation	Effects (primary and secondary)	Cause	Cause	Places affected
Physical problems for people	Impacts on people (positive and negative)	Effects	Effects	Effects
Human activities	Responses of people (immediate and long-term)	Responses	Responses	Responses

Chapter 2

Rocks, Resources and Scenery

Limestone pavement near Malham in the Yorkshire Dales National Park. This landscape feature forms only in areas of Carboniferous limestone rocks. Do any other rock types produce unique landscape features?

QUESTIONS

- **How are different types of rocks formed? What happens to them once they outcrop on the surface?**
- **Why do different rocks create such contrasting landforms and landscapes?**
- **Why are rocks a valuable natural resource for people?**
- **What conflicts and issues arise from quarrying rocks?**

Types of rock and where they are found in the UK

How are igneous rocks different from sedimentary and metamorphic rocks? Why is the Tees-Exe line an important rock divide in the British Isles?

Types of rocks

Although there are many different types of rock on the Earth's surface, there are only three groups of rocks. Rocks can be igneous, sedimentary or metamorphic.

Igneous rocks are 'formed by fire'; they begin as magma in the interior of the Earth. Some are formed by lava cooling on the Earth's surface after being thrown out by a volcanic eruption. For example, the basic lava that flows from constructive margins and forms shield volcanoes cools to form *basalt* rock. This has been eroded into hexagonal blocks at the Giant's Causeway in Northern Ireland (**Figure 1**). Others are formed by magma cooling underground after having been intruded into other rocks without reaching the surface. *Granite* is an example of this type of igneous rock. It is intruded into giant **batholiths** (dome-shaped masses of magma) during the building of fold mountains along destructive plate margins. Granite outcrops on the surface after erosion of the rocks above it over millions of years. Today granite is exposed in many places in Scotland (**Figure 4**) and forms most of the moorlands of Devon and Cornwall as well as the dramatic cliffs at Land's End (**Figure 2**).

Figure 1 The Giant's Causeway in Northern Ireland is made of basalt.

Sediments are small particles of rock transported by water, ice and wind. Most of these particles eventually reach the sea bed where successive layers of sediments accumulate over the years. The weight of materials above compresses the sediments below into **sedimentary rocks**. These rocks are laid down in layers, or beds, with lines of weakness, or bedding planes, between layers (**Figure 3**). When sand is compressed, *sandstone* rock is formed. *Clay* forms from the accumulation and compression of deposits of mud. *Limestone* and *chalk* consist of calcium carbonate, which comes from the remains of plants and animals. For example, the shells of sea creatures are made of calcium carbonate; when these animals die, masses of shells accumulate on the sea floor, building up layers of limestone rock. A lot of limestone was formed during the Carboniferous period (**Figure 5**) because at that time much of Britain was a warm shallow sea, rich in plant and animal life.

Metamorphic rocks are those that have been changed in shape or form. They begin as either igneous or sedimentary rocks but are later altered by heat or pressure. This happens, for example, along destructive plate boundaries and fault lines. Heat and pressure change limestone into *marble* and clay into *slate*. Both marble and slate are harder forms of the original rocks, and have greater economic value. Marble is widely used in building and for floors in Mediterranean countries such as Italy. Slate splits easily into sheets and, until recent times, was the main roofing material used in the UK.

Figure 2 Land's End in Cornwall is made of granite. Notice the many vertical joints.

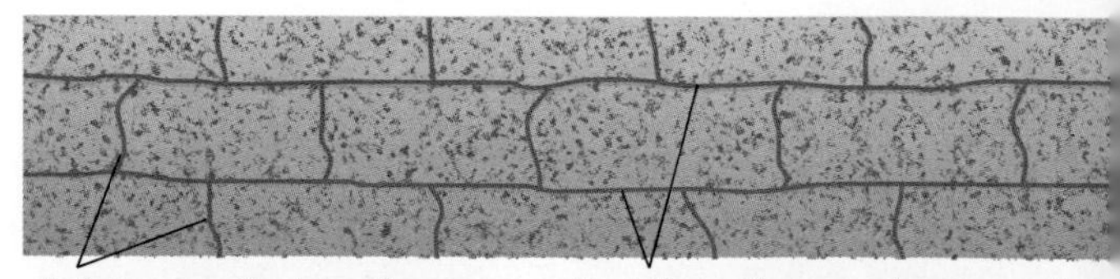

joints – vertical weaknesses within the layers of rock
bedding planes – horizontal weaknesses between the layers

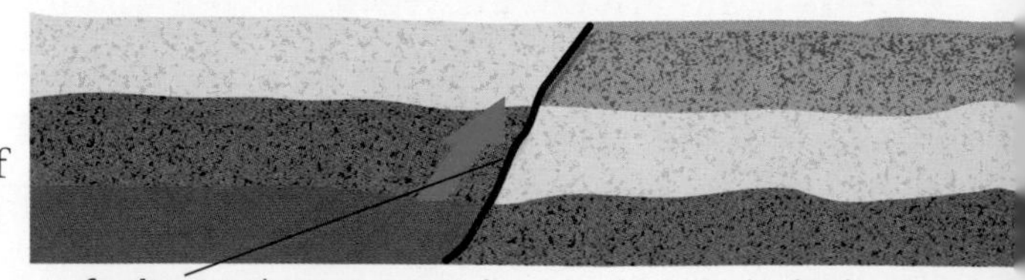

fault – earth movements have broken up the beds of rock

Figure 3 Rock weaknesses. These are important because they are the first points to be attacked by processes of weathering and erosion.

Distribution of rock types within the UK

The distribution of rocks reflects the geological history of the UK. It is customary to divide the country into two parts using a line running from the mouth of the River Tees to the mouth of the River Exe, separating Highland from Lowland Britain (**Figure 4**).

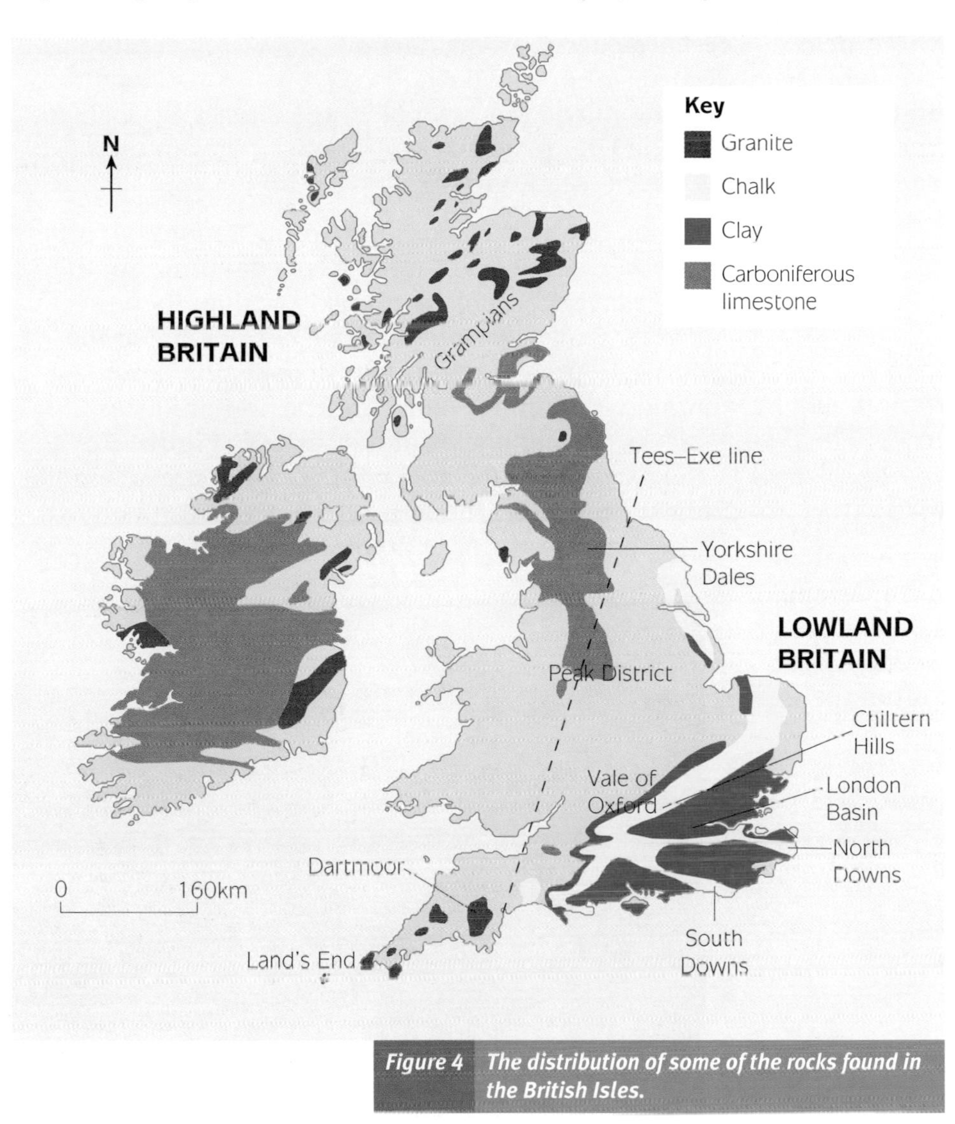

Figure 4 *The distribution of some of the rocks found in the British Isles.*

Era Period	Millions of years	Rock type
QUATERNARY	2	
TERTIARY	65	CY
CRETACEOUS	140	CH
JURASSIC	195	CY
TRIASSIC	230	
PERMIAN	280	
CARBONIFEROUS	345	CL
DEVONIAN	395	
SILURIAN	445	
ORDOVICIAN	510	G
CAMBRIAN	570	
PRE-CAMBRIAN		

Key
CY Clay
CH Chalk
CL Carboniferous limestone
G Granite

Figure 5 *Geological time scale – when the rocks in Figure 4 were formed.*

The geology of Highland Britain is dominated by old and hard rocks, mainly metamorphic and igneous. Granite formed at different times in Britain, but most is about 500 million years old. Hard rocks have resisted erosion and therefore form the upland and mountainous parts of the country.

In Lowland Britain, the geology is dominated by younger sedimentary rocks, less than 200 million years old. There is much low-lying and flat land, such as in the clay vales. Chalk, however, is more resistant to erosion than many of the other sedimentary rocks that surround it. This is why chalk ridges and scarps, such as the North and South Downs, appear to be high and steep. Yet chalk still forms lower, gentler and more rounded landscapes than the rocks in the uplands of Highland Britain.

ACTIVITIES

1 Describe the differences between Highland and Lowland Britain for (a) rocks (b) relief (height and shape of the land).

2 Make a large table to show differences between igneous, sedimentary and metamorphic rocks. Use these headings.
- Brief definition.
- Where they were formed.
- How they were formed.
- Rock types found in the UK.
- Main areas in the UK where they occur.

Weathering and the breakdown of rocks

Why do some rocks break down faster than others? What eventually happens to all the rocks on the Earth's surface?

Weathering is the breakdown of rock at or near the surface. This is mainly caused by weather conditions, such as changes in temperature. The rocks are broken down *in situ*, which means that no movement is involved (unlike erosion, which is caused by the movement of water, ice and wind). There are three types of weathering: **mechanical**, **chemical** and **biological**.

Mechanical (physical) weathering

This leads to the breakdown of rock without any change in the minerals that form the rock. The type varies with climate.

In cold climates the most widespread type is frost shattering or **freeze–thaw** (**Figure 1**). The volume of water expands when it freezes, so each time that water freezes and expands within a crack or joint in the rock, more pressure is put on the surrounding rock and the crack widens. The more often the temperature fluctuates above and below freezing point during the year, the more effective the frost shattering is at breaking off pieces of rock. The sharp-edged (or angular) pieces of broken-off rock form **scree**, which can be seen below rock outcrops in all upland areas. Some of the largest scree slopes in the UK are on the side of Wast Water in the Lake District (see **Figure 5** on page 105).

In hot, dry climates **exfoliation** is a more important weathering process. In the fierce daytime sun the outer layers of rock heat up faster than the inner layers. Expansion produces cracks parallel to the surface. At night outer layers cool down more rapidly than inner ones. Contraction produces cracks at right angles to the surface (**Figure 3**). Repeated heating and cooling creates stresses in the rock; eventually the top layer of the rock peels off like the outer layer of an onion.

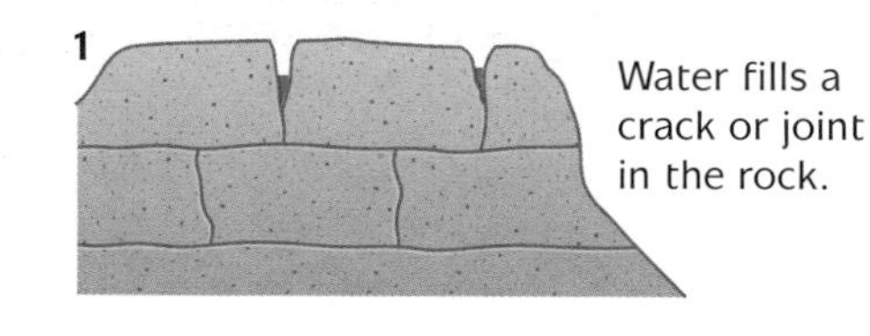

1
Water fills a crack or joint in the rock.
2
Water freezes and the crack is widened.

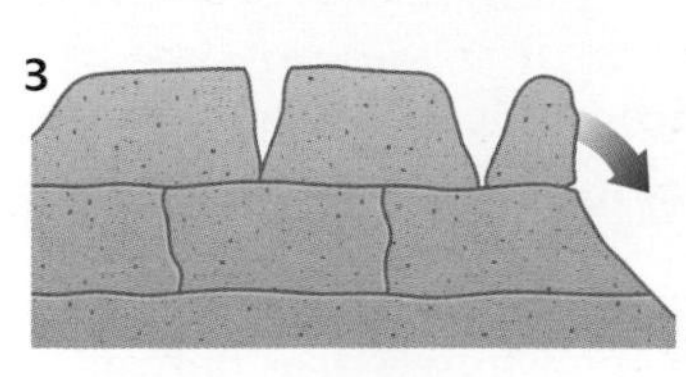

3
Repeated freeze–thaw action increases the size of the crack until the block of rock breaks off.

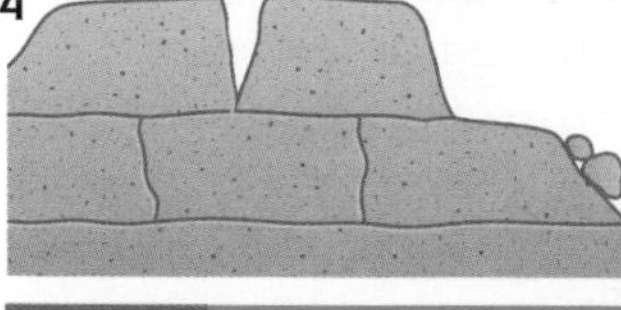

4
Loose blocks of rock are called scree.

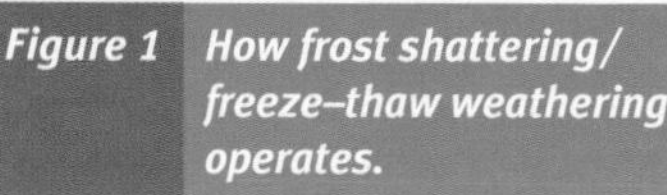

Figure 1 How frost shattering/freeze–thaw weathering operates.

Figure 2 A: Rock under attack from freeze–thaw. B: A nearby rock outcrop already broken down by freeze–thaw.

Figure 3 How exfoliation weathering operates.

Figure 4 *A different world awaits exploration underground in areas of Carboniferous limestone.*

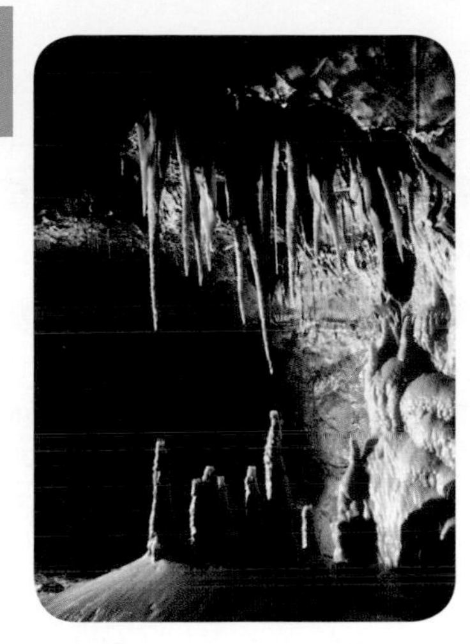

Chemical weathering

This happens when the rock's mineral composition is changed, leading to the disintegration of the rock. Granite (pages 30–31) is one type of rock that is vulnerable to chemical weathering. Feldspar, one of the minerals that make up granite, is converted into clay minerals such as kaolin (china clay).

The distinctive landforms both above and below the ground in areas of Carboniferous limestone (pages 32–3) owe their origins to **limestone solution**. This type of chemical weathering is also called **carbonation**, because the dissolving of the limestone changes calcium carbonate into calcium bicarbonate.

Unlike rock salt, which simply dissolves, limestone is hardly affected by pure water. However, rainwater is slightly acidic and contains some carbon dioxide. Rainwater and carbon dioxide in the atmosphere combine to form carbonic acid, in which calcium carbonate (of which limestone is made) slowly dissolves. Limestone is changed into calcium bicarbonate, which is removed. The limestone is very vulnerable to attack from chemical weathering because of its many lines of weakness, both horizontal (bedding planes) and vertical (joints). The joints visible between the surface blocks of the limestone pavement shown on page 25 were widened by limestone solution to form **grykes** (page 32).

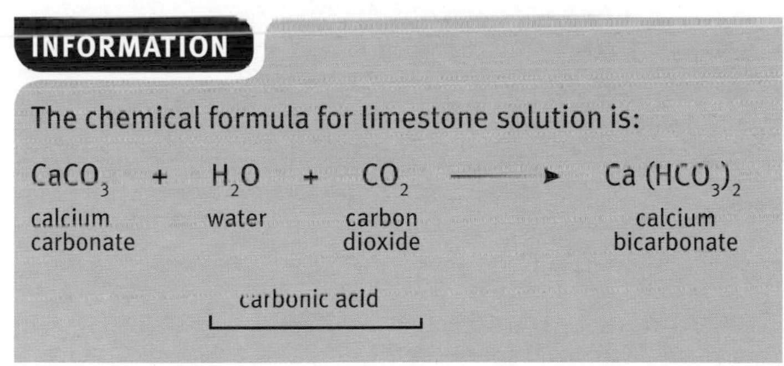

INFORMATION

The chemical formula for limestone solution is:

$$CaCO_3 + H_2O + CO_2 \longrightarrow Ca(HCO_3)_2$$

calcium carbonate + water + carbon dioxide → calcium bicarbonate

(water + carbon dioxide = carbonic acid)

Surface streams disappear underground down joints widened by limestone solution. They pass through an amazing underground world of **caves** and **caverns** dissolved out of the limestone. Water charged with lime seeps into the roofs and walls of these caves. The lime (calcium carbonate) builds up to form **stalactites**, **stalagmites**, **curtains** and **pillars** within caves and caverns.

Biological weathering

This is caused by plants and animals. They can speed up mechanical weathering; for example, roots of trees can widen joints in rock, and burrowing animals like rabbits can help break down small pieces of rock. Organic acids released by the vegetation speed up chemical weathering. Where is this happening to the limestone in the photograph on page 25?

GradeStudio

1 a Define weathering. (2 marks)
  b State two similarities between freeze–thaw and exfoliation. (2 marks)

2 a Draw labelled sketches to show rock characteristics in **Figures 2A** and **2B**. (4 marks)
  b Explain how freeze–thaw weathering produces results like these. (4 marks)

Exam tip

About **labelled sketches** from photographs:

- Labels earn most of the marks
- Focus entirely on what can be seen in the photograph.

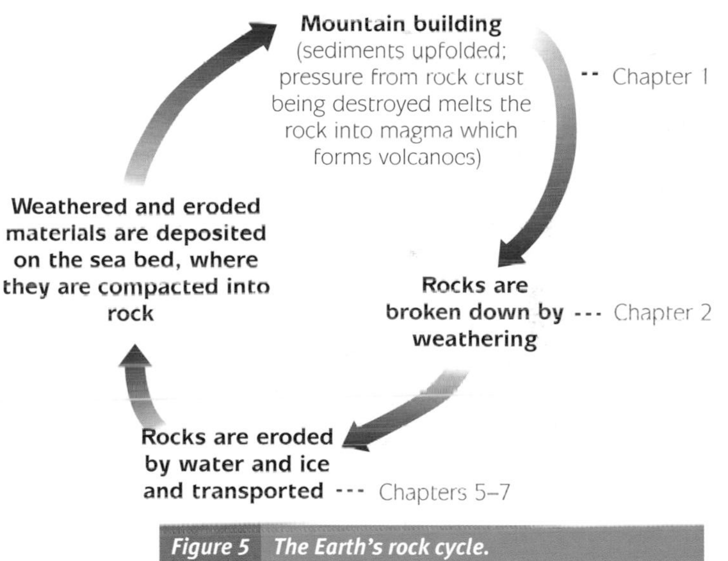
Mountain building
(sediments upfolded; pressure from rock crust being destroyed melts the rock into magma which forms volcanoes)
Chapter 1
Rocks are broken down by weathering
Chapter 2
Rocks are eroded by water and ice and transported
Chapters 5–7
Weathered and eroded materials are deposited on the sea bed, where they are compacted into rock

Figure 5 *The Earth's rock cycle.*

Why are weathering processes important?

Weathering loosens and breaks off pieces of rock. It is the first stage in wearing away the rocks of the Earth's surface after mountains have been formed. It speeds up rates of erosion because the loose pieces of rock can be picked up and carried away by rivers, waves and glaciers, which use them as tools for wearing away rock surfaces over which they pass. Therefore, weathering is an important part of the rock cycle.

Granite

What features distinguish granite scenery from that of other rocks? What are the opportunities and problems for people and economic activities?

All the granite rocks in the UK are found to the north and west of the Tees–Exe line (see **Figure 4** on page 27). Granite formed underground hundreds of millions of years ago in dome-shaped masses of magma, known as batholiths. The rocks on top of it have been eroded to expose the granite as surface rock.

Landscape features

In south-west England granite gives relatively flat-topped moorland plateaus with frequent rock outcrops, which from time to time form rock blocks called **tors** (**Figure 1**). Tors are some 5–10 metres high and are surrounded by weathered materials of all sizes from boulders to sand. On the higher parts of the moorlands there are many areas of standing surface water, which form marshes and bogs. The many surface streams have cut deeply into the upland block of Dartmoor to form deep and steep V-shaped valleys, especially where rivers such as the Dart go over the edge of the plateau. Dartmoor has a radial pattern of drainage, with rivers flowing outwards in all directions from its high centre (**Figure 3**).

Dramatic coastal scenery occurs where granite and Atlantic breakers meet, as in Cornwall at Land's End (see **Figure 2** on page 26). In Scotland the granite peaks in the Grampians and on Goat Fell in Arran are rocky and frost-shattered, although where the land is relatively flat, such as on Rannoch Moor, extensive bogs occur.

Figure 1 Bowerman's Nose, a tor on Dartmoor.

Granite is a hard rock, resistant to erosion, which is why it forms areas of high relief inland and cliffs along the coast. It is an impermeable rock, which explains why there is so much surface water. Another reason for the presence of so many bogs is the high precipitation in western upland areas.

Formation of tors

Tors occur where joints in the granite are wider apart (**Figure 2**). Freeze–thaw weathering (see **Figure 1** on page 28) can operate more effectively where the joints are close together, because there are more cracks in the rock for water to fill. Each time the water freezes and expands within a joint, more pressure is put on the surrounding rock. Where there are fewer joints, it takes longer for the blocks of rock to be broken off and the blocks are left upstanding as tors.

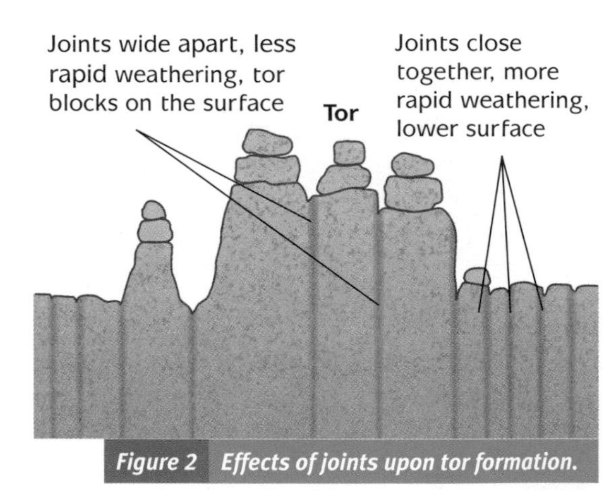

Figure 2 Effects of joints upon tor formation.

Land use and economic uses

On the higher areas, bog, marsh and moorland produce some of the least useful land in the UK. In some places there may be opportunities for water storage. At lower levels there may still be nothing better than poor grazing land suitable only for sheep and cattle (and, on Dartmoor, also for ponies). Soils are acidic and infertile; it is only around the edges of the uplands that the pastures improve sufficiently to allow grazing by dairy cattle.

Granite is a fine building stone. Aberdeen is known as 'the granite city' since so much use was made by builders of locally available supplies of stone. It is also often used for headstones in graveyards.

Granite rock is susceptible to attack by chemical weathering and in some places it has decomposed. This has resulted in the feldspar in the granite being converted into clay minerals such as china clay (kaolin). China clay is best known as the raw material for the pottery and porcelain industries, and much is sent to the Potteries region around Stoke-on-Trent. It is also used in the manufacture of paper and is an ingredient in paint, toothpaste, skin creams and many other products.

EXAM PREPARATION

Dartmoor

- Dartmoor has a high rainfall and is known for its mists and fogs.
- Much of the land is covered by heather.
- The many boggy areas contain a rich variety of plant life.
- The central upland block was enclosed within a National Park in 1951.
- The Park covers almost 100 000 hectares and over 30 000 people live inside it.
- Up to 8 million people visit or pass through the Park each year.
- Most of the towns, such as Tavistock, Okehampton and Ashburton, are located around the edges of the central block.
- Places popular with visitors include Buckfast Abbey, Haytor, Becky Falls and Lydford Gorge.
- Some of the remains of old woodlands have been preserved as nature reserves.

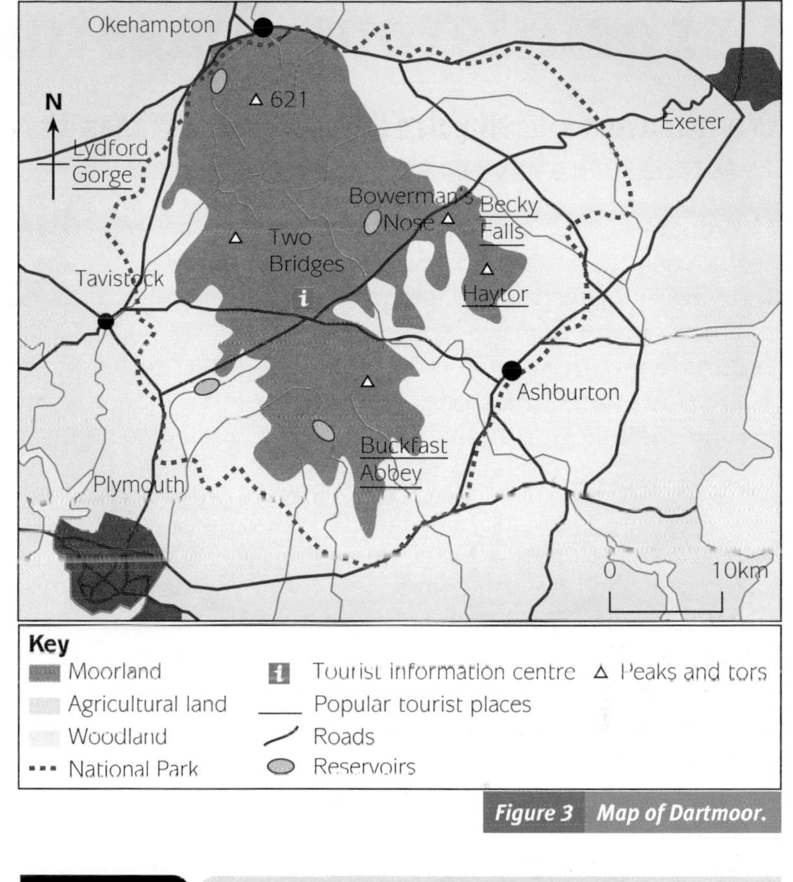

Figure 3 **Map of Dartmoor.**

ACTIVITIES

1 (a) Describe the pattern of land uses in Dartmoor shown in **Figure 3**.
  (b) Explain why opportunities for farming improve away from the centre of Dartmoor.

2 (a) Name two different economic uses of granite.
  (b) Explain their importance.

GradeStudio

1 Study **Figure 1**. Make a frame and draw a sketch from the photograph. Label the landscape features and land uses shown. (4 marks)

2 a Name and locate an example of a tor. (2 marks)
  b Explain how a tor is formed. (4 marks)

Exam tip

- Look at the number of marks as a guide e.g. for number of labels needed in **1** and what might be needed in **(2a)**.
- When explaining landforms, always explain how the processes forming them operate. Which process needs to be explained in **(2b)**?

FURTHER RESEARCH

Find out more about the centre of Dartmoor by visiting the weblink www.contentextra.com/aqagcsegeog.

Carboniferous limestone

In what ways and why are limestone landscapes different from those of all other rocks? Why do limestone rocks have so many different uses?

Although outcrops of Carboniferous limestone occur widely in the uplands of England and Wales, the greatest number and variety of distinctive landforms can be seen in two of the English National Parks – the Yorkshire Dales around Malham and Ingleton, and the Peak District near Castleton. Outcrops of Carboniferous limestone cover much more extensive areas around the Mediterranean – you may have visited caves and caverns while on holiday in places such as Majorca.

Figure 1 *The swallow hole at Gaping Gill. Fell Beck is the surface stream seen disappearing underground. It drops 110m as a waterfall into a giant chamber more than 150m long and 30m high (i.e. large enough to fit a cathedral into). The stream flows several kilometres underground, through a complex system of caves, before it reappears on the surface through Ingleton Cave and forms Clapham Beck.*

Landscape features

Carboniferous limestone weathers by the solution process described on page 29 to produce distinctive landforms both above and below ground. **Limestone pavements** (page 25) are flat surfaces of bare rock broken up into separate blocks. The flat surfaces of the blocks are **clints** and the gaps are grykes. Rivers disappear underground either through small holes in the rock, called **sink holes**, or down larger holes with a funnel shape above, called **swallow holes** (**Figure 1**). Underground the limestone is full of holes: small passageways, or cave systems, which from time to time open out into large chambers, or caverns. Stalactites made of lime hang down from the roofs like long icicles, while stalagmites are the thicker columns built up from the floor. In places the two meet to form a pillar of limestone. Where water trickles out of the roof more or less continuously, a **curtain** or wavy screen of lime may grow across the cave to add to the weird and wonderful scenery underground. The **resurgence** of rivers from underground to surface occurs when they reach impermeable rocks. When many limestone landforms occur together in an area, they form **karst scenery** (**Figure 2**).

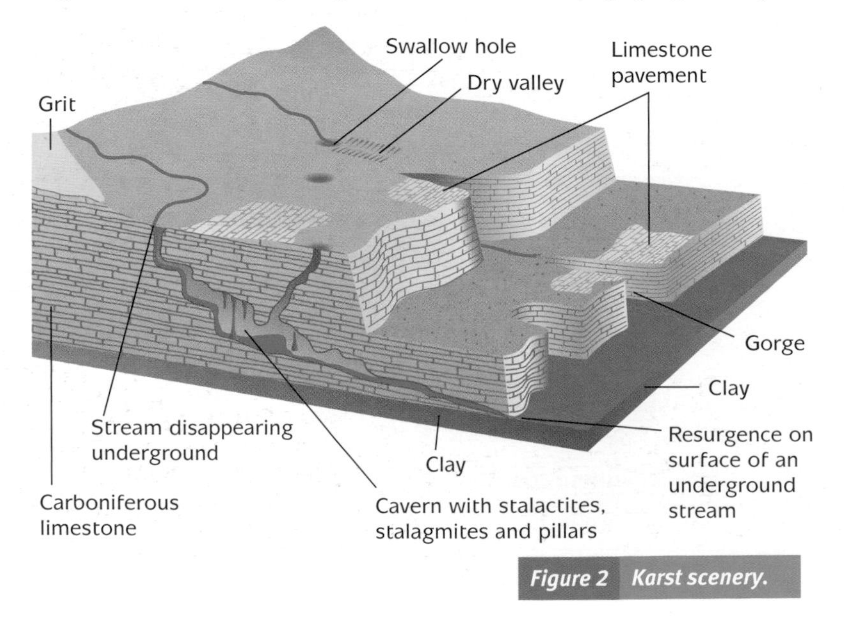

Figure 2 *Karst scenery.*

Figure 3 *The gorge at Gordale, a possible collapsed cavern.*

Occasionally the holes that are formed by solution become so large that the roof collapses. When the roof of a long underground passageway falls in, a deep steep-sided valley, or **limestone gorge**, forms with the river flowing at the bottom of it. A possible example is Gordale, where the blocks that may have formed the cavern roof can be seen as debris on the floor (**Figure 3**).

Land use and economic uses

Carboniferous limestone lies on or close to the ground surface, so the soil is too thin to be used for cultivation, and also dry. However, a turf-like grass covers the surface. This is good for sheep farming because the sheep graze short grass. Population density in limestone areas is low, but the limestone landforms are attractive to visitors. Service sector employment has been boosted in the villages and small towns, while some farmers earn a supplementary income from camping and caravan sites and bed and breakfast.

Limestone is of great economic importance. It is widely used as building stone. It is more easily worked than a hard rock such as granite. Limestone has been used in well-known buildings such as St Paul's Cathedral, the Houses of Parliament and the front of Buckingham Palace. When crushed up, it is used for chippings for drives or for making concrete and cement. Farmers spread lime on their fields as fertiliser. Limestone is also used as a cleanser in many industries such as steel smelting, to absorb harmful sulphur dioxide from coal-fired power stations, and to purify water. With so many varied uses, there are many quarries in limestone areas, and some visitors feel that these ruin the scenic beauty (pages 36–8).

EXAMPLE: Carboniferous limestone scenery near Malham

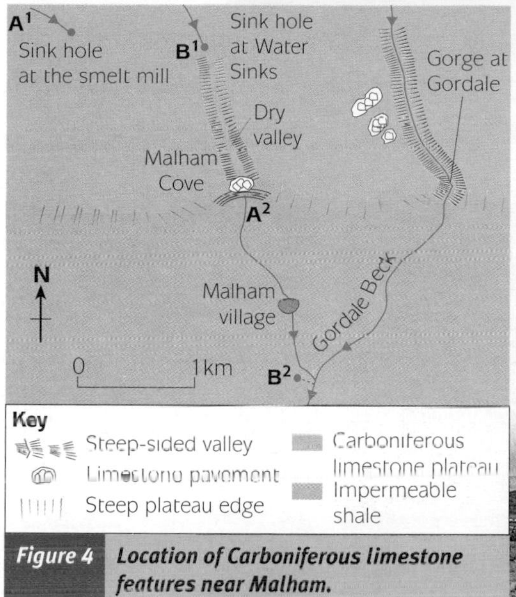

Figure 4 *Location of Carboniferous limestone features near Malham.*

The stream that disappears underground at A¹ in **Figure 4** reappears at A² at the bottom of Malham Cove. The stream that disappears underground at B¹ in **Figure 4** is shown in **Figure 5**. It reappears on the surface as a spring at B² in **Figure 4.** The network of underground cave systems is very complex.

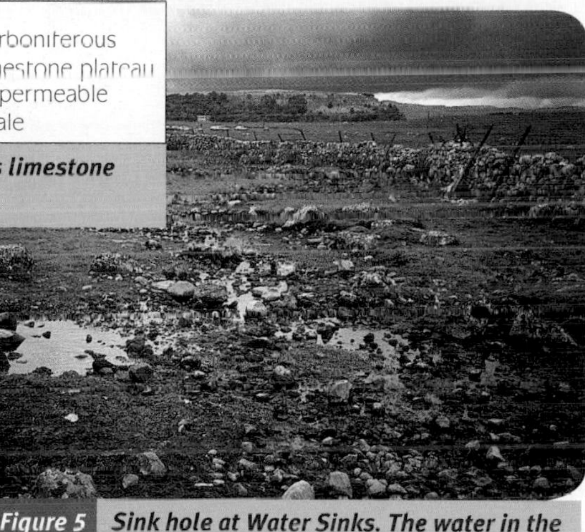

Figure 5 *Sink hole at Water Sinks. The water in the foreground was the last in the stream bed. Below this point, the valley was dry.*

Figure 6 *Dry-stone walls, made of limestone, on a farm in the Yorkshire Dales. Income from visitors supplements the low income from sheep farming.*

ACTIVITIES

1 What is similar and different about each of the following pairs of limestone features? Draw simple labelled sketches to illustrate your answers.
   (a) swallow hole and sink hole (b) clint and gryke
   (c) stalactite and stalagmite (d) gorge and dry valley

2 Name and locate an example of each of these limestone features from the Malham area.
   (a) limestone pavement (b) swallow hole (c) sink hole (d) cave (e) cavern (f) point of resurgence of an underground stream

3 What underground features are likely to be present between Water Sinks and the bottom of Malham Cove? Explain your answer with the help of diagrams.

4 (a) Draw a spider diagram to show the economic uses of limestone.
   (b) Why does stonework on buildings made of limestone 'rot away' over time?

Chalk and clay

Why do these two sedimentary rocks form landscapes that are so different? Why are water companies in south-east England in trouble when water levels in the chalk drop?

Figure 1 *Chalk and clay meet at the foot of the South Downs near Fulking. In many places in England chalk (on the right) and clay (on the left) outcrop next to one another; together they form a distinctive but contrasting landscape of chalk escarpment and clay vale. Chalk and clay are both sedimentary rocks and only outcrop in Lowland Britain; they have little else in common.*

Landscape features

The **chalk escarpment** (also known as a **cuesta**) is the most distinctive feature of chalk scenery in England. It consists of two parts – the **scarp slope**, which is steep, and the **dip slope**, on which the land falls away more gently. The top of the escarpment has gently rolling hills with rounded summits. There is little surface drainage and rivers are few and far between; however, in places the dip slope has been cut by deep, steep-sided, V-shaped **dry valleys**, which are marked landscape features (**Figure 2**). After spells of wet weather temporary streams may flow in the valleys; these are known as **bournes**, and this term is used in place names such as Bournemouth and Eastbourne.

Figure 2 *Devil's Dyke dry valley east of Fulking.*

Chalk outcrops along the coast often lead to high cliffs such as the famous 'white cliffs of Dover', and to prominent headlands, such as Beachy Head in Sussex and Flamborough Head in Yorkshire. Erosion around headlands can lead to the formation of caves, arches and stacks. The Needles off the north-west corner of the Isle of Wight are examples of stacks (page 121).

In contrast, the **clay vale** is a wide and often almost flat area of land. Surface drainage is abundant and the vale is crossed by meandering streams. At the coast, clay forms weak cliffs that slide and collapse.

Formation of the chalk escarpment

There are two requirements before an escarpment can be formed (**Figure 3A**):

1. Alternate outcrops of different types of rocks. One rock needs to be soft and the other needs to be more resistant to erosion.
2. Beds of rock that dip at an angle to the ground surface. Instead of being horizontal, the beds have been tilted by earth movements so that they lie at an angle to the surface.

These two requirements are commonly met in eastern and southern England. The clay is eroded more quickly than the chalk. As the clay is eroded down into a vale, the chalk is left standing up because of its greater resistance. The scarp slope forms a prominent feature where the layer of chalk reaches the surface. The dip slope is gentler, following the tilt of the beds of rock.

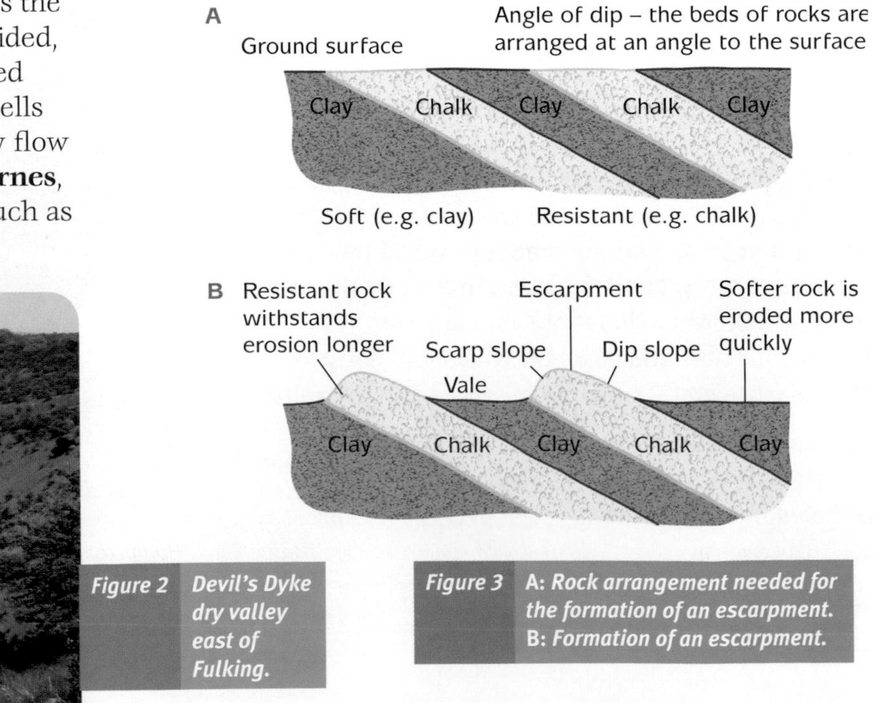

Figure 3 *A: Rock arrangement needed for the formation of an escarpment. B: Formation of an escarpment.*

Land use and economic uses

Chalk and clay offer very different opportunities for human activities. Chalk escarpments were some of the first places to be settled in the UK. They were drier than the wet clay vales and contained flint that could be used for tools and weapons. The main problem was shortage of water. Therefore, permanent village settlements grew along the **spring line** at the junction of the chalk and clay. Water seeping down through the spaces in the **porous** chalk meets the **impermeable** clay and reappears on the surface as a flow of water.

Land uses

Pastoral farming (keeping livestock) dominates on both the chalk and clay, but the focus is different. The short but rich turf on chalk is good for grazing sheep and training racehorses. In the wetter clay vales the grass is longer and better suited to dairy cows; in general, soils are too wet and heavy to plough. **Arable farming** (growing crops) is of increasing importance on the gentler, lower slopes of the chalk, where the soils are deeper. Some south-facing slopes are being used for vine cultivation. With their deep tap roots, vines do well in dry, stony chalk soils; and English wines are increasing in quality and reputation.

Economic uses

Chalk with flint provides a strong and attractive building material. Chalk has many of the same economic uses as limestone, such as in the manufacture of cement. Clay taken from pits is a raw material for brick making.

About 70 per cent of water supplies in south-east England come from underground stores (**aquifers**). Many would argue that water storage is the most important economic use of chalk.

In many places the geological structure (arrangement of the rocks) is really favourable for underground water storage; nowhere more so than in the London Basin (**Figure 4**). The chalk of the Chiltern Hills dips below London and reappears as the North Downs. Rain, falling on the chalk hills north and south of London, seeps down and fills the chalk layer below the city. The water cannot escape, trapped by the Gault clay below and the London clay above.

Two hundred years ago the chalk layer was saturated. When the fountains in Trafalgar Square were first built, the pressure on the trapped water was so great that water gushed out naturally. Since then, the amount used by people has been greater than that replaced by rainfall. Water levels have fallen, so water now has to be pumped out of boreholes drilled into the chalk.

EXAM PREPARATION

aquifer – underground store of water in permeable rock

groundwater – water stored underground in permeable rocks such as chalk

impermeable rock – rock that does not allow water to pass through it

permeable rock – rock with spaces and gaps so that water can pass through it or be trapped

porous rock – rock with pore spaces between the grains that allow it to absorb water

spring – place where water reaches the surface from underground, commonly at a junction of permeable and impermeable rocks

ACTIVITIES

1 Make two lists of the differences between areas of chalk and clay. Use these headings:
- landforms inland
- landforms at the coast
- drainage (surface and underground)
- farming types

Illustrate your answers with labelled sketches.

2 Refer back to pages 30–31.
 (a) Describe the pattern of reservoirs on Dartmoor (**Figure 3**).
 (b) Explain why reservoirs are the main source of water supply for people in south-west England but not south-east England.

3 Find out where your tap water comes from.
 (a) Is it from surface or underground supplies?
 (b) Explain this in relation to rock type and climate in your home area.

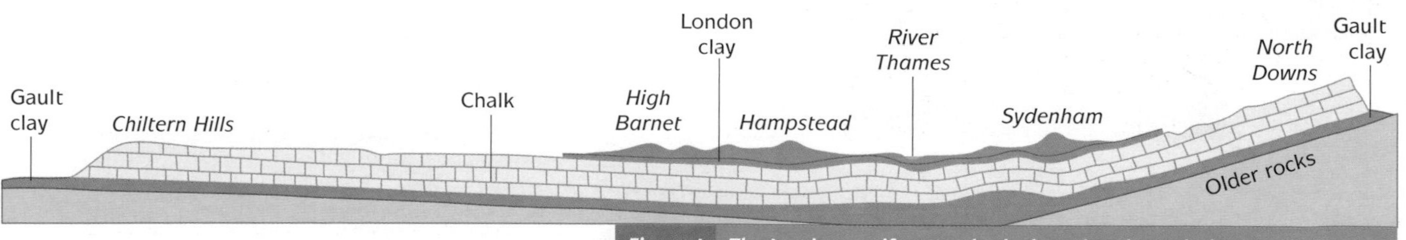

***Figure 4** The London aquifer – geological section through the London Basin.*

Quarrying

Why is quarrying rock an essential human activity? Why are some people much more in favour of it than others?

Quarrying is almost as old as settlement in the British Isles. Stone is a valuable natural resource. Early settlers used stone for building shelters, defensive works and burial mounds. Demand for stone increased over the centuries as populations grew and economic development occurred. Quarrying limestone, for example, is big business in the Yorkshire Dales today, because limestone has so many uses (page 33). About 5 million tonnes of the rock are extracted each year from quarries located within the Yorkshire Dales National Park (**Figure 1**).

Quarrying has two major advantages:

- It provides the raw materials for the huge demand for building, road construction, cement manufacture and sea defences.
- It is a major source of employment for people in upland areas where few other employment opportunities exist.

Unfortunately, quarrying is an example of a **non-sustainable** activity. It involves removing natural resources from the ground to be used up and not replaced; they will not be available for use by future generations. Quarrying has a mixture of environmental, economic and social disadvantages, which are illustrated in **Figure 2**. Can you distinguish the environmental disadvantages from the social and economic? Some negative impacts continue well after quarrying has ceased.

Figure 1 Horton-in-Ribblesdale quarry, in the Yorkshire Dales National Park, about 10 kilometres north of Settle.

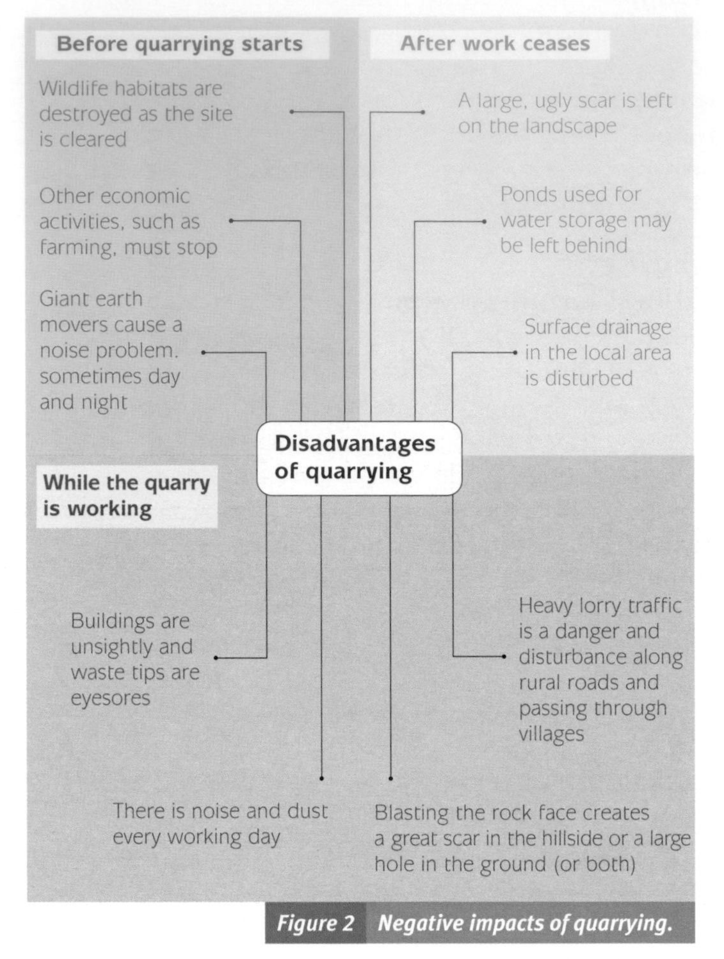
Before quarrying starts
Wildlife habitats are destroyed as the site is cleared
Other economic activities, such as farming, must stop
Giant earth movers cause a noise problem, sometimes day and night
After work ceases
A large, ugly scar is left on the landscape
Ponds used for water storage may be left behind
Surface drainage in the local area is disturbed
Disadvantages of quarrying
While the quarry is working
Buildings are unsightly and waste tips are eyesores
Heavy lorry traffic is a danger and disturbance along rural roads and passing through villages
There is noise and dust every working day
Blasting the rock face creates a great scar in the hillside or a large hole in the ground (or both)

Figure 2 Negative impacts of quarrying.

Reducing the negative effects of quarrying

Planners and local authorities have an important role to play in placing controls upon the commercial quarry companies, who are really only interested in making the largest profits. A full consideration of likely environmental effects is needed before planning permission is given and work begins. Planning authorities can restrict the size of the quarry, insist that buildings and waste tips are screened by trees in areas of great scenic beauty, limit noisy operations like blasting to certain times and impose binding commitments to clean up the site after work finishes. The Horton quarry is in the same area as the popular Three Peaks walk and close to the famous Ribblehead Viaduct on the Settle–Carlisle railway. The quarry company here is aware of its responsibilities. It sponsors conservation work around the industrial site, and encourages local people to become involved in community work for the benefit of birds and other wildlife.

Figure 3 *Quarry lorry in the centre of Horton-in-Ribblesdale, just below the train station. This was one of nine counted in a 20-minute period on a weekday. How could this negative impact be reduced?*

Alternatives after quarrying stops

The most environmentally friendly action is for the quarrying company to fill the hole and replace the topsoil – in order to leave the land looking similar to the way it looked before the work began. Trees, grass and shelter belts can be planted to landscape the land. This is an example of **reclamation**: the land has been reclaimed from quarrying to be used again for farming and other rural land uses. Usually this is only possible where the quarry is small and the company is forced to abide by strict planning rules. It is an expensive option.

Figure 4 *The Eden Centre near St Austell in Cornwall.*

Large holes and old quarries are convenient places for **landfill**, a cheap and easy way to dispose of waste. From time to time large machines are used to level it off and compact the waste that has been tipped into the hole. When full, the land can be reclaimed for other uses such as forestry, farming or recreation, although it is rarely very productive. The land never seems to look natural. Disposal of waste in landfill sites needs to be managed carefully; otherwise it may contaminate the land and water courses, and become a hazard to the health of people living in the area.

Two large quarries in the UK have attracted special uses. The glass domes of the Eden Centre, which house plants from different world environments, are in a china-clay quarry in Cornwall (**Figure 4**). The large Bluewater out-of-town shopping centre, east of London, is built in an old quarry, in a part of the UK where new building land is in short supply.

ACTIVITIES

1 (a) Rearrange the disadvantages (negative impacts) of quarrying from **Figure 2** in a table, using the headings: Environmental, Social and Economic.
(b) Choose one disadvantage from each heading. For each one:
 (i) explain why it is a disadvantage of quarrying
 (ii) describe ways to reduce its negative effects.

2 Write a case study of a quarry (Horton-in-Ribblesdale), using the guidance below.
(a) Location
(b) Labelled sketch from **Figure 1**
(c) Economic and social advantages
 (i) for people in local area
 (ii) general (usefulness of limestone).
(d) Environmental disadvantages
(e) Local issues and conflicts

3 It is usual to call a local meeting when an extraction company is seeking permission to increase the size of a quarry. Among the people likely to attend will be:
- retired residents
- young married couples
- the manager of the construction company
- activists from an environmental group
- chairman of the parish council
- an officer from the local planning authority
- the secretary of the local tourist board.

(a) Write down a list of points that one of these people might use if given the opportunity to speak at the meeting.
(b) In a debate, how would you expect some of the others to reply to your points? Make a brief list of their likely counter-arguments.

Tourism and quarrying in the Peak District National Park near Castleton

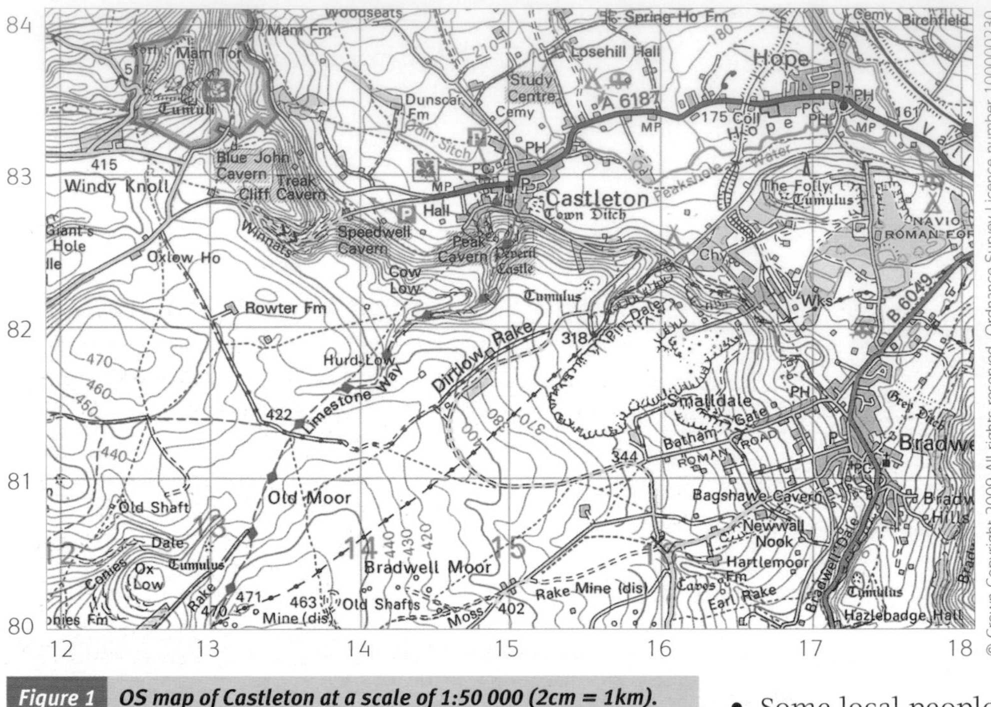

Figure 1 *OS map of Castleton at a scale of 1:50 000 (2cm = 1km).*

Figure 2 *Hope quarry and cement works.*

Castleton is a tourist **honeypot** – a magnet for thousands of visitors each year. For many the main attractions are the limestone caves and caverns, associated with the outcrops of Carboniferous limestone, which dominate the scenery of the surrounding area (**Figure 1**). Some of the popular caverns, such as Blue John and Speedwell Caverns, had previously been used by miners. There is a long history of lead mining here.

Nearby is the Hope limestone quarry and cement works (**Figure 2**). It is the largest single employer in the area; some 300 people are employed, nearly all of whom live locally. Without the cement works, many more people would be forced to commute by road to towns and cities outside the Park, particularly Manchester and Sheffield.

Conflicts of interest exist between local people and tourists:

- At peak times Castleton does get lots of visitors, since large population centres like Manchester, Stockport and Sheffield are within one hour's drive.
- Some local people are fed up with the numbers of cars and tourists. For other tourist issues in the UK National Parks, see Chapter 13 (page 239).
- For others, making at least a part-time living from tourism in hotels, B&Bs, camp sites, cafés, pubs and shops, it offers great commercial opportunities.

Then there is the quarry/cement works:

- You cannot have a working quarry without dust and noise. Notice the smoking chimney from the cement works in **Figure 2**. Both the quarry and cement works are eyesores within a scenic area.
- They provide work, save people from having to commute and keep the people in the area, where it is otherwise difficult to make a living. Farming is never very profitable in upland limestone areas.
- As long ago as 1943 the company that owns Hope quarry and cement works began to landscape the area to make it less of an eyesore. Some 2.5 million tonnes of stone were rearranged to conceal the quarry entrance and create a feature known as Jellicoe Brow. Over a 35-year period some 75 000 trees have been planted in stages to screen the site. At the same time, worked-out clay pits have been restored as fishing lakes and a golf course.

London aquifer: opportunities and limitations for water supply

Figure 3 is a summary sketch of the London aquifer. The water line indicates average water levels. So great is the present demand in London that the aquifer can supply only 20 per cent of the capital's water, but in surrounding areas aquifers provide a much higher proportion of water. For example, in mid-Kent aquifers supply 90 per cent of the water used.

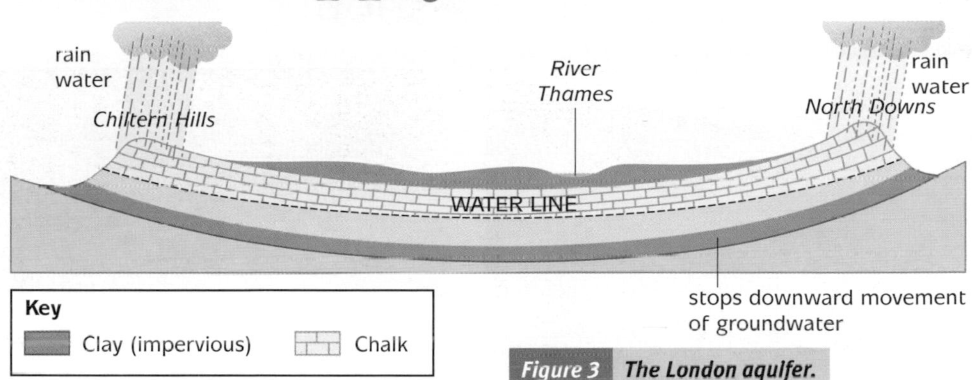

Figure 3 *The London aquifer.*

EXAM PREPARATION

Aquifers – opportunities and limitations for water supply

Opportunities
- clean (non-polluted) water
- long-term supplies; do not dry up in short-term droughts
- boreholes can be sunk close to where the water is needed

Limitations
- finite supply of water
- takes a long time for water used to be replaced
- people cannot increase storage capacity as happens with surface reservoirs

The limitations are under human control. To be sustainable, water use has to be kept to the amount that nature can replace. Although precipitation totals vary from year to year, over a period of years everything averages out; wet years will make up for dry years. By February 2006 water companies in south-east England were desperately worried about future water shortages, following two winters (2004/5 and 2005/6) of below average rainfall (**Figure 4**).

After many months with above average rainfall, these worries had gone away by the end of 2007, and were not mentioned in early 2008 after high rainfall in winter and spring. But 2006 did highlight the limits of sustainable water supply from the south-east's aquifers.

FURTHER RESEARCH

Check current water levels in the London aquifer on the weblink www.contentextra.com/aqagcsegeog.

Drought warning for the south-east

A drought in south-east England is almost certain this summer. Rainfall has been below average in eight of the last twelve months. Groundwater levels in the chalk are now at their lowest since records began over 300 years ago.

Demand for water is increasing all the time, particularly in the south-east, where the population is growing fast.

Underground aquifers rely upon rain in winter for replenishment, but December and January were very dry months, with only 40 per cent of normal rain falling. One water board official said 'Unless we get much more than average rainfall in the next three months, hose-pipe bans are inevitable. People must cut back on their water use. Water is in danger of becoming a scarce resource in the south-east.'

Figure 4 *Newspaper report, February 2006.*

ACTIVITIES

1 State the evidence from the OS map (**Figure 1**) showing that Castleton is a tourist honeypot.

2 In grid squares 1581 and 1682 in **Figure 1** there is a limestone quarry with the hope cement works to the north-east.
   (a) From the OS map, measure the length and width of the quarry in metres, and state the approximate area covered by it.
   (b) Use **Figures 1** and **2** to describe other ways in which the presence of the cement works is affecting the landscape.
   (c) Describe the land uses in square 1782 on the OS map (east of the quarry and works).
   (d) Explain how conflicts between the quarry and visitors have been reduced here.

3 Explain how the London aquifer is replenished with water.

4 Is the London aquifer an example of sustainable human use of a natural resource? Explain fully.

Practice GCSE Question

See a Foundation Tier Practice GCSE Question on the weblink www.contentextra.com/aqagcsegeog.

Photograph A Chalk and clay rocks in Sussex.

Photograph B Chalk soil.

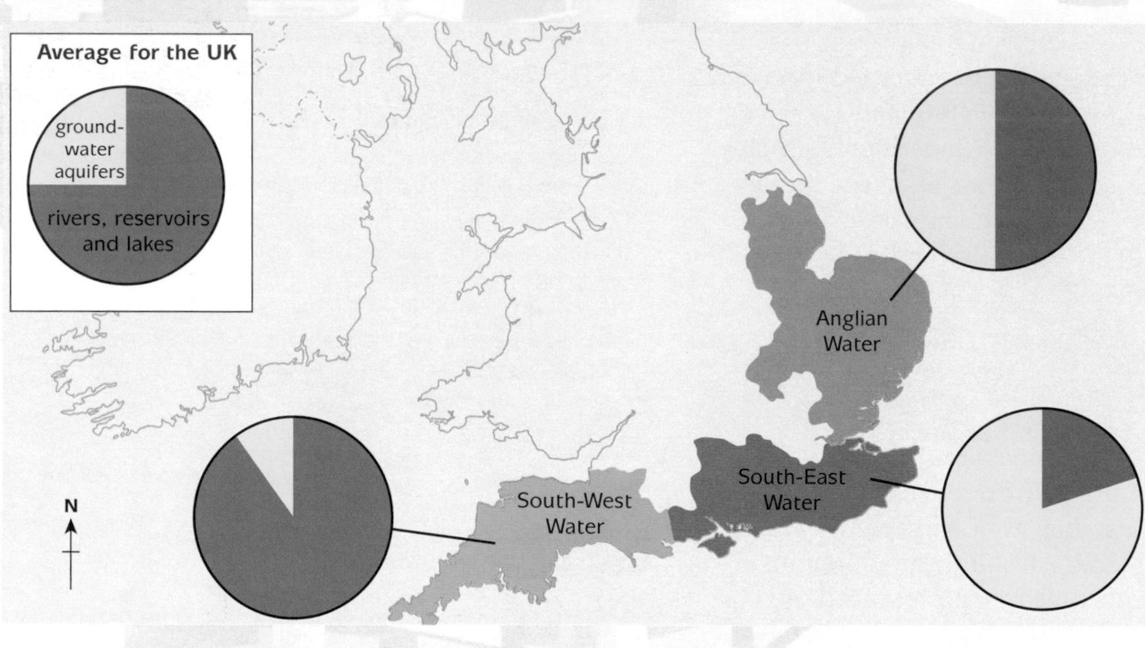

Figure 1 Sources of water supply.

2 (a) Study **Photograph A**, which shows a landscape of chalk and clay rocks in Sussex. Describe the differences in relief and land uses between the chalk and clay rocks. **(4 marks)**

(b) **Photograph B** shows a chalk soil.

(i) Draw a labelled sketch to show soil characteristics. **(3 marks)**

(ii) Why do many farmers with land on chalk not grow crops? **(2 marks)**

(c) Study **Figure 1**, which shows sources of water supply in the UK as a whole, and in three English water regions.

(i) Describe what **Figure 1** shows about sources of water supply. **(4 marks)**

(ii) Explain how rock type helps to explain differences in sources of water supply between the west and east of England. **(4 marks)**

(d) Using the case study of a quarry, explain the advantages and disadvantages of quarrying for local people and visitors. **(8 marks)**

Total: 25 marks

Exam tip

The question for this topic will be Question 2 in Paper 1.

Improve your GCSE answers

How to draw labelled sketches

A Exam question

Draw a labelled sketch of **Figure 2** to describe how it shows rock weathering. (4 marks)

Guidance

- Make a frame (whenever possible, of the same size).
- Draw in the skyline first (if there is one).
- Mark on the major features to provide the fixed points.
- Sketch in or highlight any minor features that are relevant.
- Label your sketch according to the needs of the question.
- Use at least as many labels as question marks.

***Figure 2** Area of rock weathering (past and present).*

B Example answer

Look at **Figure 3**.

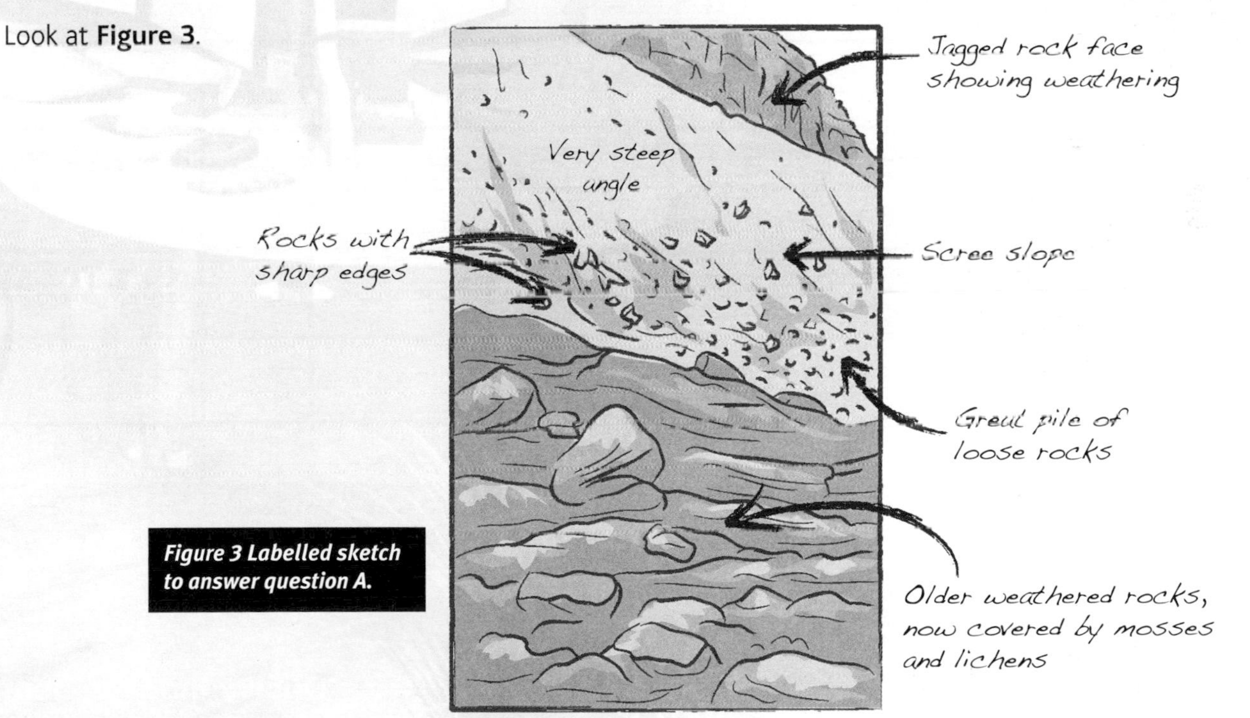

***Figure 3** Labelled sketch to answer question A.*

Remember

- Label only what can be seen and shown.
- Label what the question asks for.
- Take into account the number of marks – the more marks, the greater the number of labels.

ExamCafé

REVISION

Key terms from the specification

Exam Café

Conflict – opposing views about issues, leading to debate between people about them
Igneous rock – rock formed by volcanic activity, from magma that has cooled
Landscape – inland scenery, with varied landforms
Metamorphic rock – rock that has been changed by natural agencies
Rock cycle – rocks weathered, eroded, transported and used to form new mountains
Resource – something useful for human needs
Sedimentary rock – rock formed by sediments laid down in the sea bed
Sustainable management – planning ahead and controlling development for a long future
Weathering – breakdown of rock in the place where it outcrops (in situ)

Checklist

	Yes	If no – refer to
Can you name the three rock groups and examples of each type?		page 26
Do you know how the rocks differ between the north-west and south-east of Britain?		page 27
Can you explain the differences between mechanical, chemical and biological weathering?		pages 28–9
Do you understand how the rock cycle works?		page 29
Can you describe granite landscapes, and explain how tors are formed?		pages 30–31
Can you describe features of Carboniferous limestone scenery, and explain how limestone solution leads to their formation?		pages 29, 32–3
Can you describe the features of a chalk escarpment and explain why chalk and clay landscapes are so different?		pages 34–5
Do you know the main uses of granite, limestone and chalk rocks?		pages 30, 33, 35

Case study summaries of one rock type

Rock type for farming/ aquifer/tourism	Working quarry	Quarry management
Location in the UK	Location	Location
Opportunities	Advantages	Strategies during extraction
Limitations	Disadvantages	Strategies after finishing

Chapter 3

Challenge of Weather and Climate

A winter morning in the Lake District under anticyclonic conditions. Notice the local variations – sunny summits, foggy valleys and frost-covered slopes in the shade.

QUESTIONS

- **What is the UK's climate really like?**
- **Why do depressions and anticyclones bring weather that is so different?**
- **Is the UK's weather becoming more extreme?**
- **To what extent are people responsible for global warming and can it be stopped?**

What is the UK's climate like?

Can you describe the UK's climate? What is good and what is bad about it?

Climate is the summary of a place's weather conditions, such as temperature and precipitation, averaged out over a long period of time (usually at least 30 years). It includes the conditions that can normally be expected at a place, month by month during the year.

When asked to describe the climate of the UK, what did you say? Wet? Cold? Unpredictable? Unless you had time to research the answer, you would be unlikely to give the proper geographical description – *mild (or cool) wet winters* and *warm wet summers*. The name for this type of climate is **temperate maritime**.

- *Temperate* means that the place experiences neither the heat of the tropics, nor the coldness of the poles (**Figure 1**).
- *Maritime* means that the climate is strongly influenced by the sea. One effect of the sea is to reduce temperature differences between winter and summer. Another is increased precipitation.

A climate graph shows a summary of the climate of a place. Manchester is used as an example in **Figure 2**. Can you understand how it fits the UK summary of *mild wet winters* and *warm wet summers?* Manchester is famous for rain – is this reputation justified? Don't be put off by the amount of information in a climate graph; if you know what to look for (see the Exam Preparation Box), interpreting these graphs is easy.

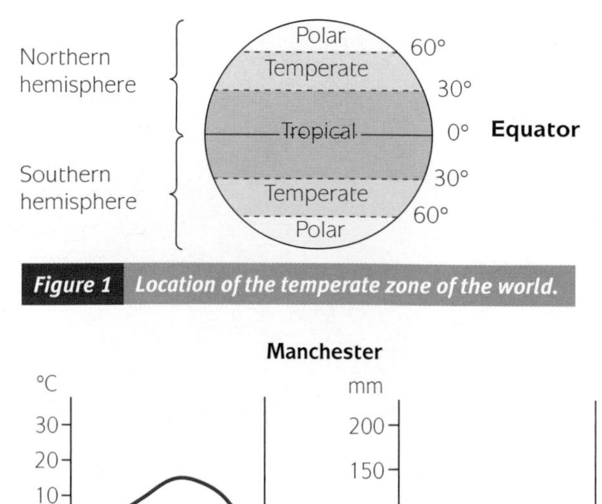

Figure 1 *Location of the temperate zone of the world.*

Figure 2 *Climate graph for Manchester.*

EXAM PREPARATION

Climate graphs – what to look for

A Temperature line

1. Highest (maximum) temperature – state value (°C) and month
2. Lowest (minimum) temperature – state value (°C) and month
3. Calculate annual range (°C) – highest minus lowest
4. Describe shape (gradient) of the line (e.g. steep, gentle change, level)

B Precipitation bars

1. Month with most precipitation – name and amount (mm)
2. Season with most precipitation – the answer can be 'all year'
3. Month with lowest precipitation – name and amount (mm)
4. Season(s) with less or little precipitation
5. Total annual precipitation – low, moderate, high or very high

Exam tip
Double-check answers from graphs for accuracy.
Always state the units for temperature and precipitation.

Global influences on the UK climate

Latitude is the most important. It greatly influences temperature, pressure and winds.

A Temperature

Temperatures are highest in low latitudes in the tropics and lowest in polar latitudes. The gradual decrease in temperature between Equator and poles is due to reduced insolation (**Figure 3**). The Sun shines from a high angle in the sky in the tropics all year. As a result, the Sun's light travels directly through the Earth's atmosphere so that less is lost by reflection. Also, because the Sun's rays are almost vertical, there is a smaller area of the Earth's surface for each ray to heat up. In contrast, near to the poles the Sun's rays approach the Earth's surface at an oblique angle. This means that each ray has a larger surface area to heat up. As the rays have had a longer journey, less sunlight remains to be absorbed by the Earth's surface.

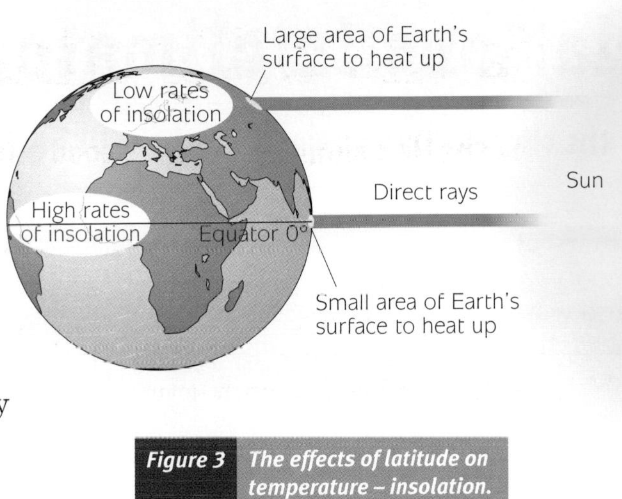

Figure 3 *The effects of latitude on temperature – insolation.*

B Pressure and winds

The UK lies in a zone of **low pressure** (**Figure 4**). Here the air is rising so that the weight of air at the surface becomes lower than average – hence there is low surface air pressure. Areas of low pressure are associated with cloud and rain. As air rises, it cools; moisture in the air condenses and forms clouds. Water droplets increase in size in the clouds until they become too heavy to be held and fall as rain.

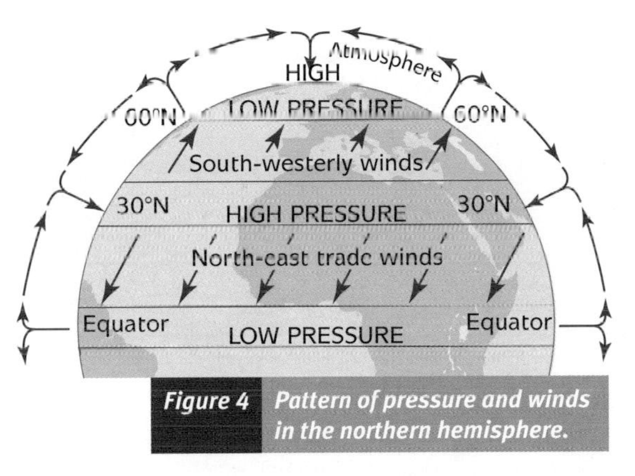

Figure 4 *Pattern of pressure and winds in the northern hemisphere.*

Hot air that rises at the Equator sinks around 30° north of the Equator to form a zone of **high pressure**, which includes the Azores High. Winds blow from high to low pressure, i.e. air is transferred from areas where it is sinking towards areas where it is rising. Southerly winds (winds blowing from south to north) are deflected by the Earth's rotation from west to east and reach the UK as south-westerly winds. South-westerly winds are the **prevailing winds** in the UK – the winds that blow most often. They have a long sea journey across the Atlantic Ocean before reaching the UK.

Distance from the sea

The **maritime influence** from the sea on the UK's climate is great, affecting both temperature and precipitation. As far as temperature is concerned, the sea has different thermal (heating) properties to the land.

- During *summer*, when rates of insolation are highest, the sea heats up less quickly than the land. The Sun's light penetrates below the water surface so that the Sun's rays have more than the surface to heat up, and constant movement mixes warm and cool water.
- In *winter*, the opposite happens. The sea retains its store of summer heat longer than the land.

The result is that places near the sea have less hot summers and less cold winters than those further inland, where the **continental** effect dominates. In the UK the winter warming effect of the sea is magnified by the presence of a warm ocean current, the North Atlantic Drift, over which south-westerly winds cross before reaching the UK. Winds that take long sea journeys pick up lots of moisture and are always likely to bring rain.

ACTIVITIES

1 Following the guidance in the Exam Preparation Box (page 44), describe the climate of Manchester.

2 Study **Figure 5**.

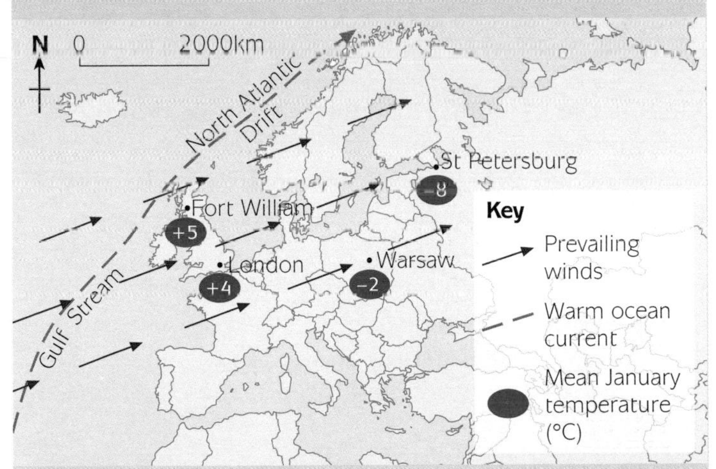

Figure 5

(a) Describe the pattern of winter temperatures shown in Europe.
(b) Give reasons for this pattern.
(c) How would you expect the temperature pattern to differ in summer?

Variations in climate within the UK

What is the climate like in your part of the UK? Is it better or worse than elsewhere in the UK?

The summary of mild (or cool) wet winters and warm wet summers applies to the whole country. However, there are quite marked differences in average temperature and precipitation from one part of the UK to another. **Figure 1** shows some of them.

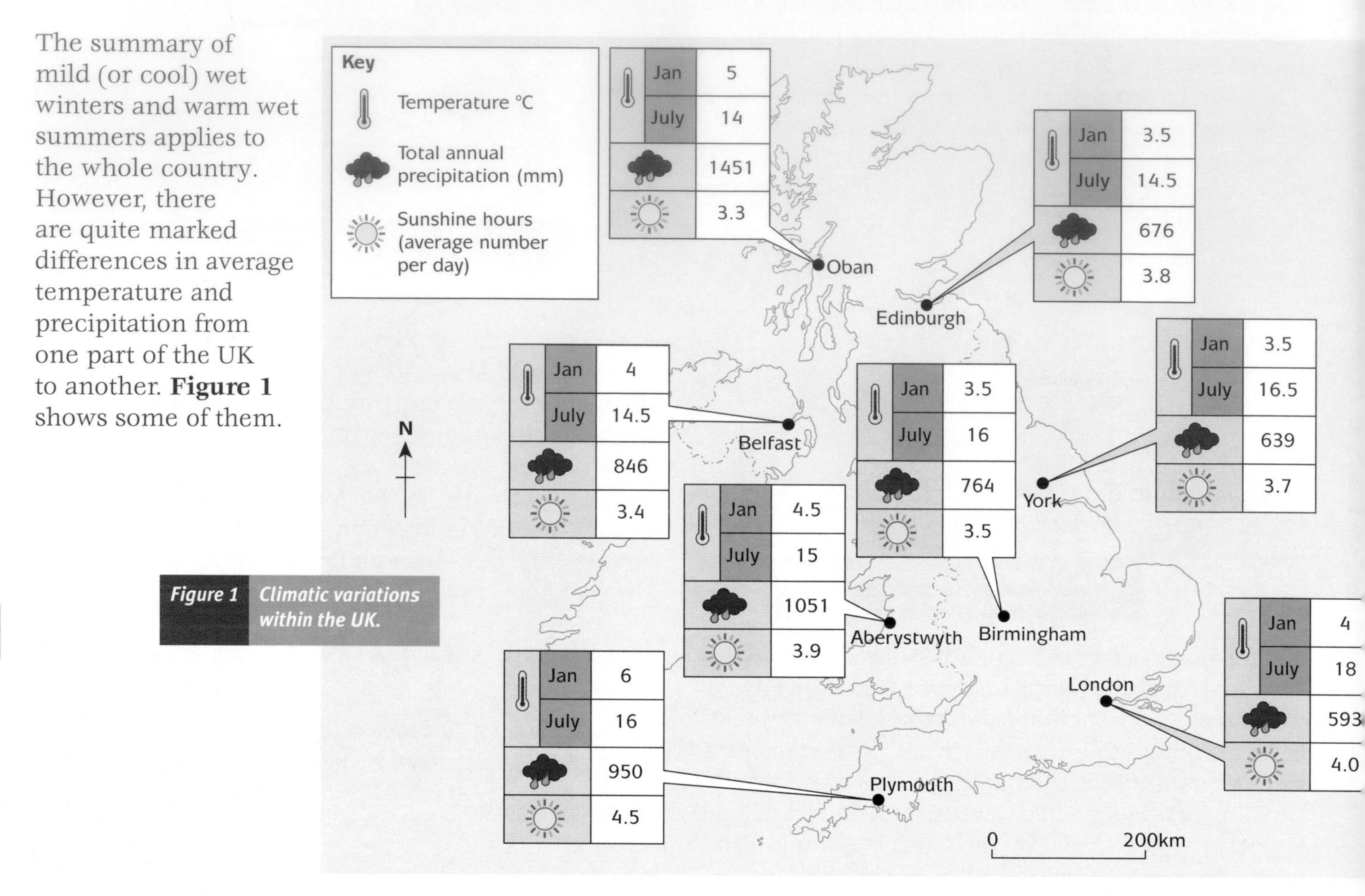

Figure 1 Climatic variations within the UK.

What is the evidence in **Figure 1** for each of these statements? How strong is the evidence?

A Winters are milder in the west than in the east.

B Summers are warmer in the south than in the north.

C Precipitation is higher in the west than in the east.

D There is more sunshine in the south than in the north, and more sunshine in the east than in the west.

Temperature

The temperature pattern changes between summer and winter (**Figure 2**). In *summer* (**Figure 2A**) the south is warmer than the north. The isotherms (lines linking places with the same average temperature) run mainly from west to east. The Sun shines from a higher angle in the sky in the south. This means that rates of insolation are higher here than in the north of Scotland, leading to a difference of about 4 °C in average temperatures. The built-up area around London, where the effects of the **urban heat island** are felt, is particularly warm. The lines bend southwards near to and over the sea, showing that the sea has a cooling effect in summer.

In *winter* (**Figure 2B**), in some places the isotherms run north to south, and the west of the country is generally warmer than the east. North-west Scotland is warmer than many parts of England. At this time of the year the sun has less influence on temperatures because it is at a low angle in the sky and there are fewer hours of daylight. The winter warmth of the sea and the North Atlantic Drift is of greater importance; their warmth is transferred onshore by prevailing westerly winds, which warm up the western side of the country first.

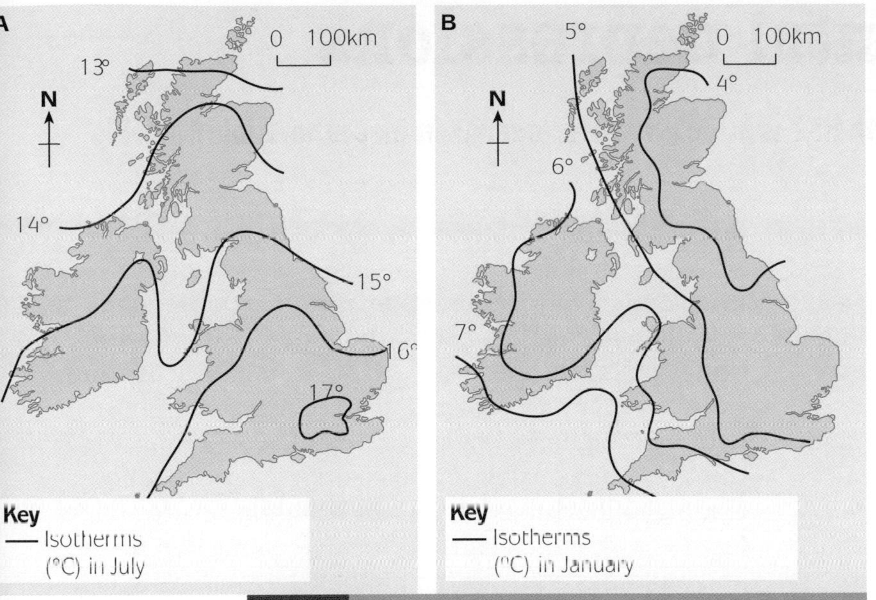

Figure 2 A: *Summer temperatures.* B: *Winter temperatures.*

INFORMATION

Urban heat island

Buildings and dark surfaces, such as tarmac roads, store heat. Further heat comes from car fumes, factories, lights and central heating systems. The larger the built-up area, the greater the heat. This is why the effect is so noticeable in London (**Figure 2A**), where the centre may be 4–5°C warmer than the suburbs.

Precipitation

Precipitation in the UK is generally highest in the west and lowest in the east (**Figure 3**). The factor most responsible for this is the *direction of the prevailing winds*. The south-westerly winds have had a long sea journey and are laden with moisture when they blow onshore. They release less moisture as they move east. Other factors contribute as well. *Frontal depressions* (areas of low pressure with warm and cold fronts) are also driven from west to east by the westerly circulation and they drop rainfall first in the west. What makes the difference in the amount of precipitation between east and west so large is *relief*. In the west are upland areas; in the highest of these, annual precipitation totals exceed 2000 millimetres. The amount of precipitation released increases when winds laden with moisture are cooled even more by being forced to rise over the uplands. In contrast, in some places in eastern England, such as East Anglia, the annual precipitation total barely reaches 500 millimetres. East Anglia is one of the lowest-lying parts of the country; winds and frontal depressions arriving from the west have crossed the widest area of land so the rain shadow effect is felt more strongly here than elsewhere (**Figure 4**).

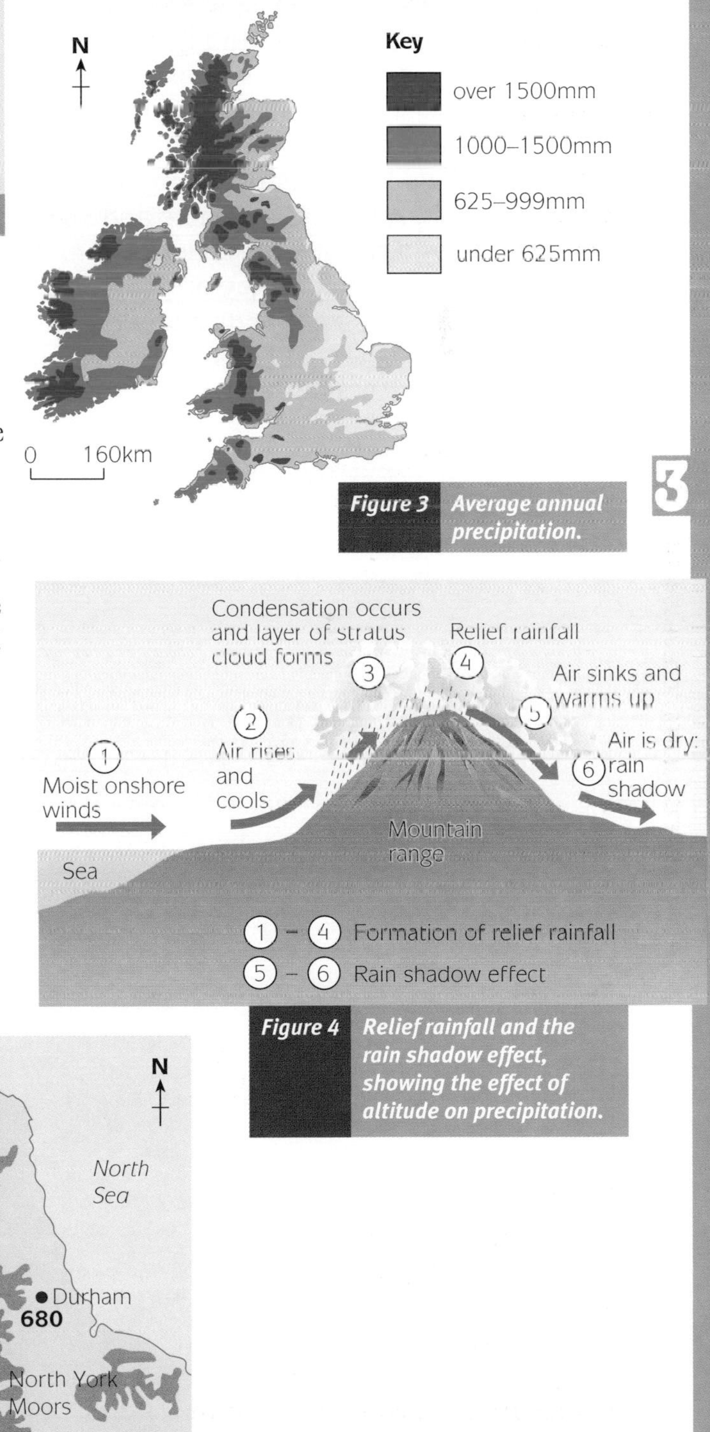

Figure 3 *Average annual precipitation.*

Figure 4 *Relief rainfall and the rain shadow effect, showing the effect of altitude on precipitation.*

ACTIVITIES

1. Study the relief map of part of the north of England (**Figure 5**).
   (a) Describe the pattern of precipitation shown.
   (b) Explain how the pattern of precipitation is influenced by (i) prevailing winds (ii) altitude.
2. Answer questions A–D on page 46 about **Figure 1**.

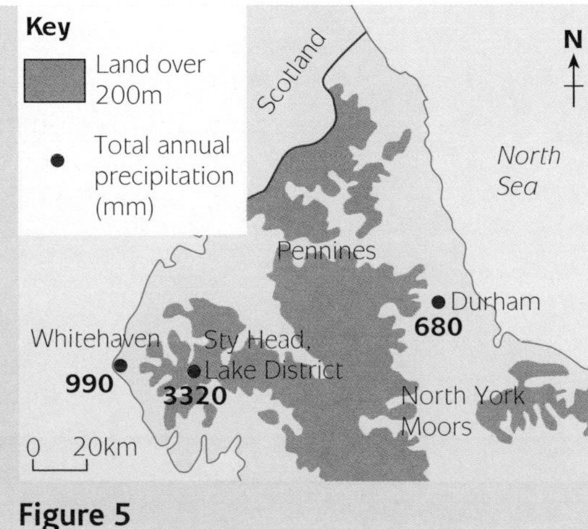

Figure 5

UK weather – frontal depressions

The UK is famous for wet and cloudy weather – why are frontal depressions responsible for this?

While climate is the long-term average, **weather** is the day-to-day conditions of temperature, precipitation, cloud, sunshine and wind. A typical weather forecast for the UK on a winter's day might be 'Rain in the morning with strong winds will be followed by sunshine and showers in the afternoon; overnight there will be fog and frost'. A cynic might say that the weather forecaster has no idea what the weather is going to be and has mentioned just about every type of weather that can occur in the UK! That would be a harsh judgement – British weather is notoriously changeable and usually no two days are alike. There has been a tremendous improvement in the accuracy of weather forecasts, which now rely on satellite data and pictures, radar showing precipitation and computer models for the behaviour of depressions and anticyclones.

Frontal depressions

Depressions are areas of **low pressure**. There is relatively low air pressure at the surface because the air is rising. Air, carried by surface winds, is drawn into the centre of the depression to replace the rising air. In the northern hemisphere, winds blow in an anti-clockwise direction around a depression, spiralling towards its centre. In general, depressions bring spells of unsettled weather with plenty of wind, cloud and rain. Occasional deep depressions (those with a particularly low pressure below 960 millibars (mb)) are responsible for releasing heavy downpours leading to local flooding, or for bringing severe storm-force winds, which cause extensive damage to trees and buildings, such as the famous 'hurricane', which hit south-east England in October 1987.

Formation

Frontal depressions form where warm air from the tropics meets – and is forced to rise above – cold air from the poles. The rising air creates the centre of low pressure. The line that separates the two air masses is called a **front**. As the depression develops and moves east, driven by the westerly circulation, two fronts can be recognized. The leading front is the **warm front**, so called because once the warm front passes over a place, the warm air of the tropical air

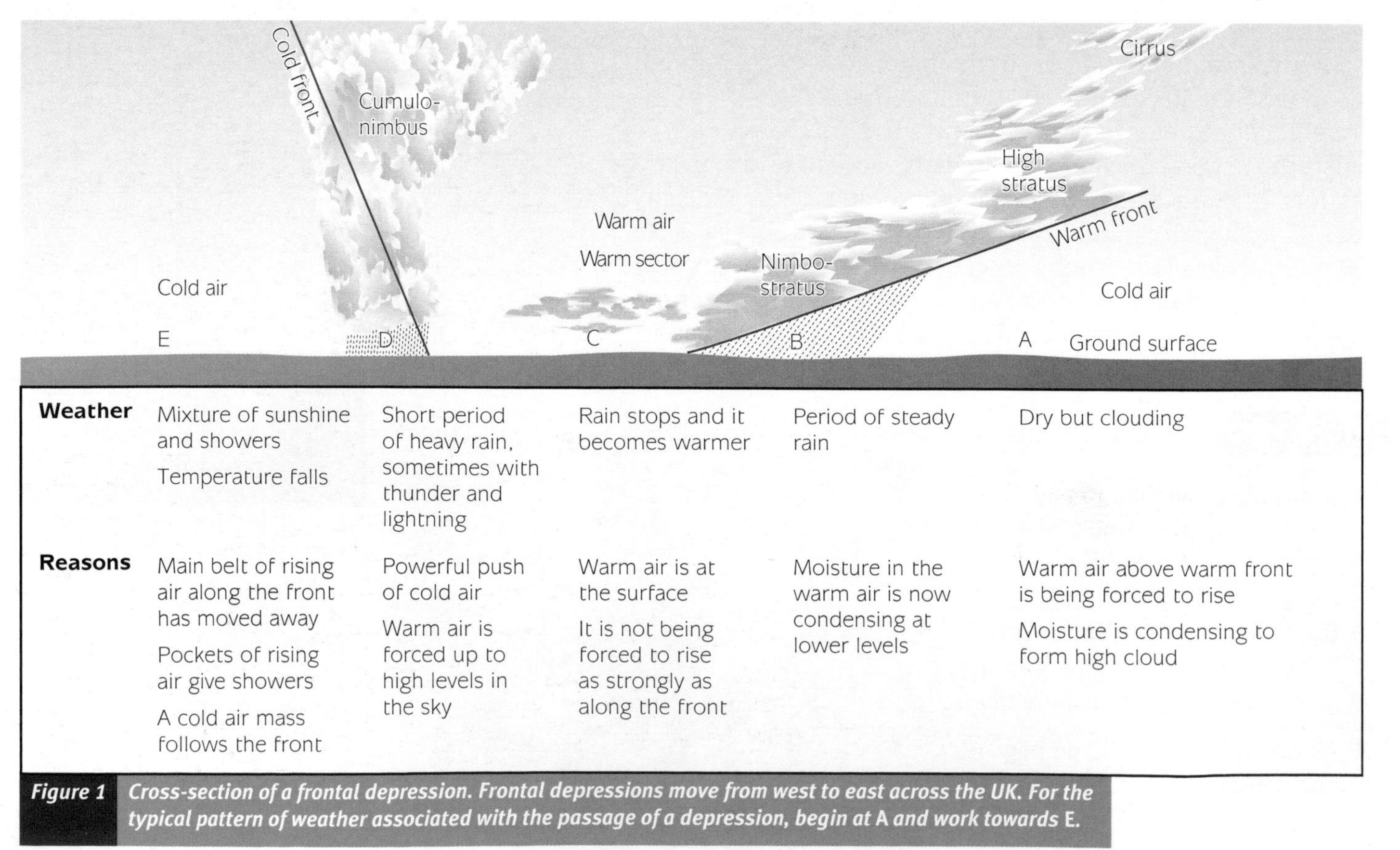

	E	D	C	B	A
Weather	Mixture of sunshine and showers Temperature falls	Short period of heavy rain, sometimes with thunder and lightning	Rain stops and it becomes warmer	Period of steady rain	Dry but clouding
Reasons	Main belt of rising air along the front has moved away Pockets of rising air give showers A cold air mass follows the front	Powerful push of cold air Warm air is forced up to high levels in the sky	Warm air is at the surface It is not being forced to rise as strongly as along the front	Moisture in the warm air is now condensing at lower levels	Warm air above warm front is being forced to rise Moisture is condensing to form high cloud

Figure 1 *Cross-section of a frontal depression. Frontal depressions move from west to east across the UK. For the typical pattern of weather associated with the passage of a depression, begin at A and work towards E.*

mass brings warm weather. The front at the rear of the depression is the **cold front**, so called because after it passes the cold air of the polar air mass brings colder weather. In well-developed depressions, like the one in **Figure 2B**, the cold front catches up with the warm front and forms an **occluded front**.

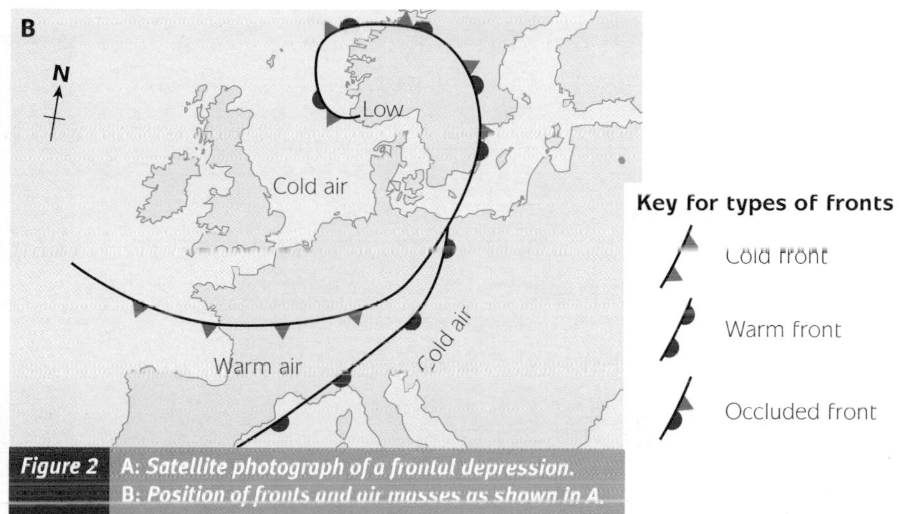

Figure 2 A: Satellite photograph of a frontal depression. B: Position of fronts and air masses as shown in A.

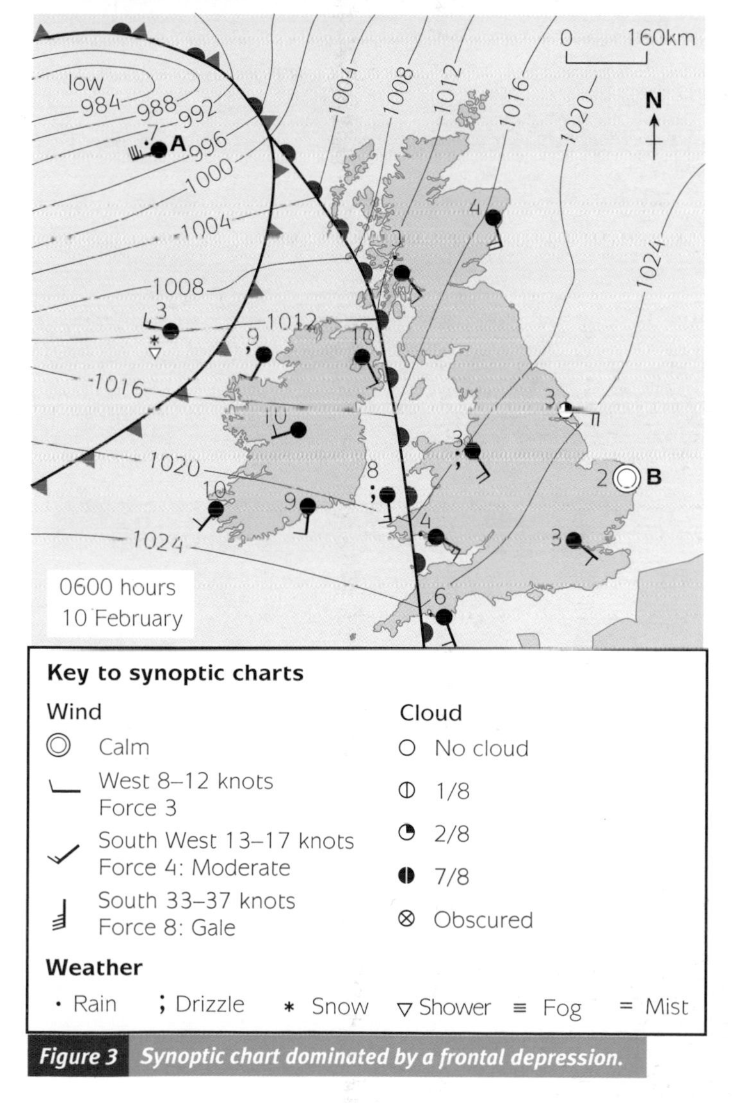

Figure 3 Synoptic chart dominated by a frontal depression.

Frontal depressions on satellite photographs and synoptic charts

In **Figure 2A** the centre of the low pressure can be detected by the swirl of cloud. This is where the wind is being sucked into the centre from surrounding areas of high pressure. The swirl is caused by the rotation of the Earth. The fronts show up as trailing lines of cloud on the southern sides of the depression. It is along the fronts that most activity is occurring as air is being pushed upwards, leading to condensation of its moisture into cloud. The white speckles of cloud in the area behind the cold front indicate shower clouds. Cold air is being warmed up by its journey over the sea, encouraging air to rise and to form the cumulo-nimbus shower clouds shown.

GradeStudio

1 Study **Figure 3**.
   a State the:
      (i) pressure in the centre of the depression. (1 mark)
      (ii) wind speed and wind direction at station **A**. (2 marks)
   b (i) Describe the differences in temperatures between England and Ireland. (2 marks)
      (ii) Give reasons for these differences. (2 marks)

2 Explain why the western areas of Scotland, Wales and England are being affected by a belt of rain. (4 marks)

3 The weather system is moving eastwards. Describe how the weather at station **B** is likely to change during the rest of the day. (4 marks)

UK weather – anticyclones

Anticyclones bring dry weather to the UK – but is it always good weather?

These are areas of **high pressure** in which the air is descending to the Earth's surface. The air 'piles up' near the surface to create a higher than average air pressure. The air drifts outwards from the centre of the anticyclone and the winds blow in a clockwise direction around the centre in the northern hemisphere. Only light winds are caused by the gentle **pressure gradient** (small differences in pressure over wide areas, which are shown by wide spacing of the isobars on synoptic charts).

A strong anticyclone, above 1030 millibars, may establish itself over the British Isles and stay for several days, or even weeks, blocking out the frontal depressions. At any time of year, dry and settled weather with little wind is expected when an anticyclone is dominant. In *summer*, temperatures are likely to be higher than average as strong sunshine from clear cloudless skies heats up the land. This leads to a *heatwave* if the anticyclone does not move and release its grip on the weather. In *winter*, temperatures are lower than average. Sunny but cool days are followed by long cold nights with frost and fog, both of which are most likely on valley floors (page 43). The intensity of the frost can build up night after night – leading to a *big freeze*, which brings problems such as burst pipes.

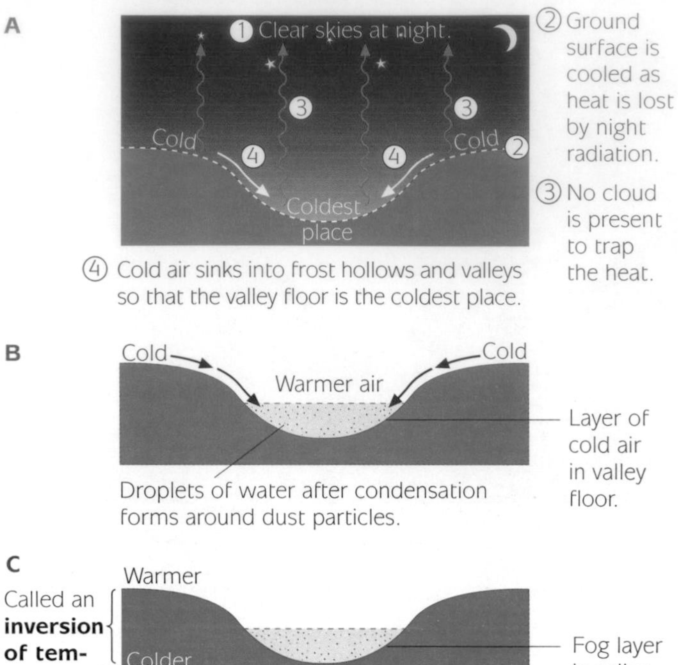

A
① Clear skies at night.
② Ground surface is cooled as heat is lost by night radiation.
③ No cloud is present to trap the heat.
Cold
Cold
Coldest place
④ Cold air sinks into frost hollows and valleys so that the valley floor is the coldest place.
B
Cold
Cold
Warmer air
Layer of cold air in valley floor.
Droplets of water after condensation forms around dust particles.
C
Warmer
Colder
Called an **inversion of temperature** because the temperature rises as you go higher up out of the valley.
Fog layer in valley floor.
Called **radiation fog** because it is the night radiation of warm air that leads to surface cooling.

Figure 1 *Formation of frost (A) and radiation fog (A, B and C).*

Anticyclonic weather and the reasons for it

Dry weather all year	Little wind at any time	Hot and sunny in summer
• Sinking air is warmed up by increased air pressure near the surface • Warm air can hold more moisture	• Gentle pressure gradient • Little difference in pressure over a wide area	• Greater insolation from the sun at a high angle in the sky • Long hours of daylight and cloudless skies for much sunshine
Cold weather in winter	**Frost mainly in winter**	**Fog mainly in winter**
• Sun shines from a low angle in the sky (low insolation) • Short hours of daylight from weak winter sun • Long hours of darkness for heat loss from clear skies	• Clear skies allow heat loss from ground surface at night • Ground temperatures can fall below freezing point, especially inland and in valleys • Moisture in contact with the cold ground condenses into ice	• Visibility is less than 1km for fog and 2km for mist • Radiation fog forms on cold nights with little wind (**Figure 1**) • A strong breeze is often needed to clear the fog or it can last all day

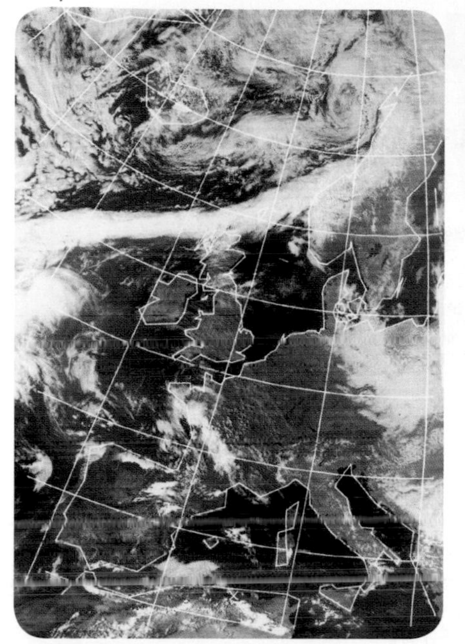

Figure 2 *Satellite photograph taken when an anticyclone is dominating the weather over most of the British Isles. Cloud associated with the fronts of the depression centred off the coast of Norway can be seen touching the north of Scotland. Its progress southwards over the rest of the UK has been blocked by the anticyclone, which has given largely cloudless skies to the rest of the UK and much of central Europe.*

Summer heatwaves and droughts in the UK

These occur when a summer anticyclone sits over the UK for several weeks, and refuses to be pushed out of the way by depressions forming in the Atlantic Ocean. These depressions are forced to go around the anticyclone, often to the north of Scotland, leaving most of the UK cloudless and dry in the sinking air of the high pressure. High rates of solar insolation and long hours of daylight mean that the heat builds up, especially inland and in the big cities. The last nationwide drought in the UK was in 1995, when reservoirs even in the usually very wet Lake District (**Figure 5**, page 47) dried up (**Figure 4** below).

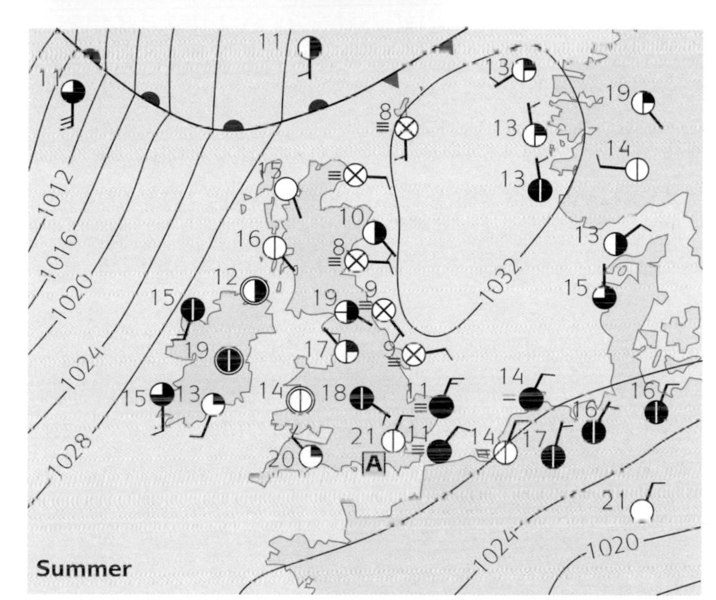

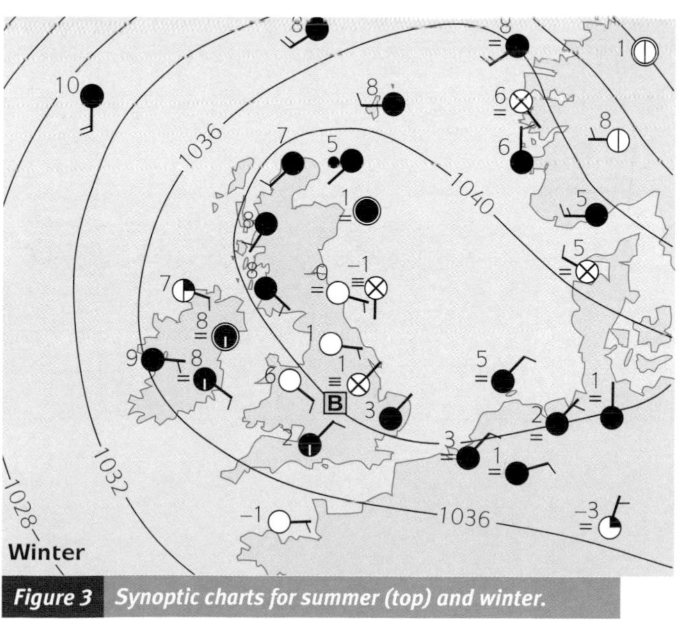

Figure 3 *Synoptic charts for summer (top) and winter.*

Figure 4 *Haweswater in the Lake District, August 1995. Instead of a reservoir full of water to supply Manchester, this became a tourist attraction.*

ACTIVITIES

1 Make a table to summarise the differences between depressions and anticyclones using these headings: Typical pressure (mb); Wind speed; Precipitation; Weather in summer; Weather in winter.

2 (a) Describe the differences in weather between stations **A** and **B** in **Figure 3**.
  (b) Give reasons for temperature differences between **A** and **B**.

3 Study the photograph on page 43.
  (a) Draw a labelled sketch to show how weather conditions vary from place to place.
  (b) Explain why these weather conditions are typical of winter anticyclonic weather in the UK.
  (c) Suggest why local weather variations like these occur with anticyclonic weather.

Extreme weather events in the UK

What do we regard as extreme or freak weather in the UK? How bad are the consequences for people?

UK weather is described as **extreme** when it is more severe than what is normally expected. Examples include:

- very cold in winter – 'long cold snap'/'big freeze'
- very hot in summer – 'heatwave'/sunny summers like the summer of 1995
- heavy rain – 'cloudburst leading to a flash flood'/'one month's rain falling in just a few hours'
- drought – 'many months without significant rainfall'
- gales – 'gale and storm-force winds'/'the hurricane of October 1987'
- snow – 'blizzards'/winters famous for heavy snowfalls like the winters of 1947 and 1962/3
- fog – thick fog night and day/'fog causes motorway chaos'.

Figure 1 *Motorway in fog. 'Pile-ups' occur because drivers are slow to adapt to poor weather conditions.*

People in other parts of the world sometimes laugh at what we call *extreme weather* in Britain. The UK cannot experience a real hurricane (it lies too far from the tropics, page 56), nor a winter deep freeze as in Canada and Siberia (it is too close to the warm Atlantic Ocean, page 45). However, people and businesses adapt their activities and operations to average conditions, with only slight variations. Radiation fog can be expected in winter with anticyclonic weather. However, when this type of fog persists all day, day after day, it often causes major traffic disruption on the motorways (**Figure 1**) and at the airports.

Most of the seven extreme events listed above are caused by depression or anticyclonic weather, which is either more intense or lasts longer than usual. The variety, which is such a characteristic feature of the UK's weather, is temporarily lost.

A Which extreme events are caused by high pressure dominating the weather for weeks and months?

B Which ones result from low pressure and fronts sitting over the UK for several days instead of moving on?

C Which one is caused by a very deep depression, with a very low pressure and steep pressure gradient?

Some people believe that extreme weather events in the UK are becoming more frequent, particularly droughts and floods. Look at **Figure 2**, which shows the pattern of monthly rain from 2004 to August 2007. The driest year was 2005, when south-east England received only 66 to 80 per cent of average rainfall. By January 2006, alarm bells were ringing here about future water shortages. When did the alarm bells stop ringing?

There are good physical and social reasons why south-east England is always going to be the area at greatest risk of drought.

- Average rainfall is low in the south-east (page 47).
- It is the last part of the UK to be reached by Atlantic fronts.
- It is mostly lowland (page 27), with few good natural sites for reservoirs.
- Underground water stores are being used up faster than they are being replenished (page 39).
- It is experiencing population growth and there are plans for a lot of new houses.

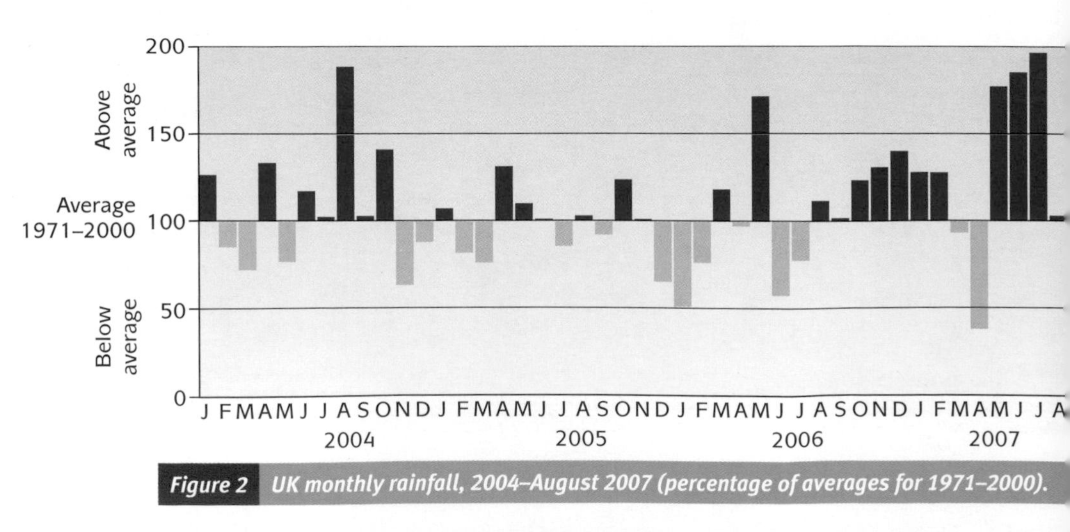

Figure 2 *UK monthly rainfall, 2004–August 2007 (percentage of averages for 1971–2000).*

The summer floods of 2007

Summer 2007 was the wettest on record (since 1766). After a dry April (**Figure 2**), over 400 millimetres of rain fell across England and Wales in May, June and July. Wet weather in May and early June meant that water levels were already high. By the end of June the saturated ground could take no more rain and all the water ran off into rivers. Large parts of England were affected by flooding, with some areas – such as those around Hull, Doncaster, Sheffield and Tewkesbury – severely affected (**Figure 3**). Weather experts referred to it as 'freak weather' rather than the result of climate change.

Only 13 deaths were recorded (mainly people swept away by fast-running flood waters); however, many more people needed to be rescued by the emergency services. The RAF airlifted some from homes, cars and boats; at one time thousands were stuck on the M5. In addition, there was flood damage to properties and businesses. A total of 56 000 homes and nearly 7000 businesses were flooded. Many more were left without power or drinking water. Around 350 000 people in Gloucestershire were left without tap water after water treatment works were submerged, and they had to rely upon bottled water and water tanks. Fields of crops, ready to harvest, were ruined.

Update 2008

One year later, 5500 families had not been able to return to their homes; of these, 1400 were still living in caravans. Those who have returned are finding that their home insurance premiums have increased by three or four times; that is if they can still get insurance.

Figure 3 ***Tewkesbury cut off by flood waters of the River Severn in July, after 78 millimetres of rain fell in a 12-hour period, is one of the most memorable images of the floods of 2007.***

INFORMATION

The Environment Agency has:

- signed up 73 000 more people to its free flood warning scheme
- completed 34 000 new flood defences, protecting about 30 000 homes
- planned to spend another £1 billion to protect 145 000 more homes over the next few years.

INFORMATION

A BBC Yorkshire survey of South Yorkshire people from streets worst affected by flooding revealed:

- average time out of the home was nine months
- two out of three had suffered physical ill health (e.g. chest infections, stomach upsets, skin complaints)
- one out of four had suffered significant mental health problems (e.g. depression requiring medication)
- one out of three reported damaging effects on family life and relationships (e.g. worse-behaved children, young children afraid of heavy rain)
- communities had become united as people worked together in difficult times.

ACTIVITIES

1 People directly affected by the weather include farmers, fishermen, airline pilots, owners of beach-front shops, cricketers, skiers. Choose three of them and for each one:
   (a) describe what is good and bad weather
   (b) explain when and why this good and bad weather is likely to occur.

2 Write notes for a case study of an extreme weather event in the UK using these headings:
   (a) Causes (b) Areas affected and damage
   (c) Economic impacts (d) Social impacts (people's homes, lives, health) (e) Immediate responses to the event (f) Longer-term responses for the future.

The global warming debate

Can we be certain that global warming is happening? If it is, can people do anything about it?

The Earth's climate is *dynamic* – climate is always changing. Only 10 000 years ago snow and ice covered the whole of the UK, except for the area south of the line between Bristol and London. Then the world started warming up. Around AD 1300 the UK climate was warmer than today and vines were being grown as far north as York. Temperatures fell away from this peak. The middle of the nineteenth century was a time of cold winters. The River Thames froze regularly and Londoners skated on it, producing scenes like the ones still shown on many Christmas cards. During the 1960s, after cold winters such as 1962/3, newspaper journalists speculated about the return of an ice age. In contrast, during the last ten years winter snowfalls have become rare events in many parts of the UK. Today media comment is solely focused on global warming.

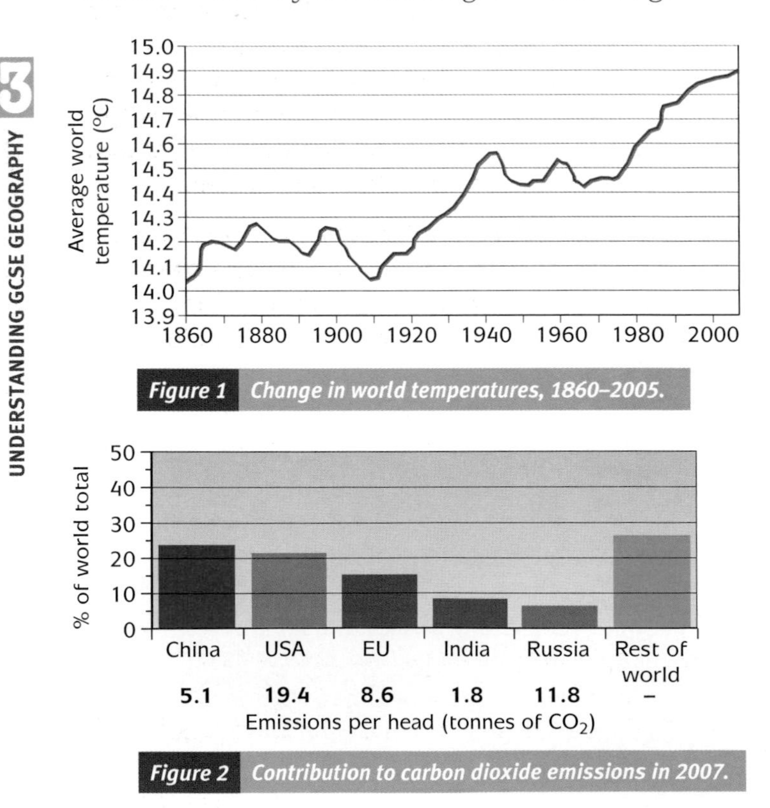

Figure 1 Change in world temperatures, 1860–2005.

Figure 2 Contribution to carbon dioxide emissions in 2007.

You are growing up through a time of increasing average world temperatures, of **global warming** (**Figure 1**). Despite the fluctuations, the general trend is upwards, most notably during the last 30 years. Meteorologists take great care when measuring the weather and their results can be trusted. Further evidence comes from melting ice sheets and retreating glaciers (see below). Yet forces exist that are capable of causing global cooling, such as a supervolcano erupting (page 13).

Possible causes of global warming

The greenhouse effect is a natural process. Water vapour and carbon dioxide in the atmosphere absorb long-wave (heat) radiation from the land surface. This delays radiation loss from the Earth into space and keeps the Earth warmer (by some 30°C) than it would otherwise be. This is what makes life on Earth possible.

General world temperatures have been rising since the end of the Ice Age 10 000 years ago, clearly as a result of natural causes, because the human population was small and levels of technology were low. Although some scientists and American politicians remain unconvinced, the majority have come to the view that the current rate of warming is too rapid to be due to natural causes alone. Most agree that people are to blame, at least in part. Human activities from the Industrial Revolution onwards have sent levels of greenhouse gases in the atmosphere soaring, making it likely that they are causing the 'enhanced' or 'accelerated' greenhouse effect. Burning fossil fuels and deforestation are most responsible. However, the blame is not shared out equally (**Figure 2**). Note the US contribution per head and compare it with that of other countries.

Consequences of global warming

Global warming is already having impacts that can be observed and measured. The Arctic ice cap is thinning by 10 centimetres a year; icebergs of record sizes are breaking off the Antarctic ice sheet during summer. Glaciers in the Alps and other high mountain areas are retreating to higher levels. Sea levels are already 18 centimetres higher than they were 100 years ago. The sequence of events shown in **Figure 3** is already happening and is expected to continue and become more widespread in its effects. Island countries – like the Maldives and the Seychelles in the middle of the Indian Ocean – are particularly alarmed at the prospect of further global warming, as are delta countries such as the Netherlands and Bangladesh. One gloomy prediction is that a complete melting of the Greenland ice sheet

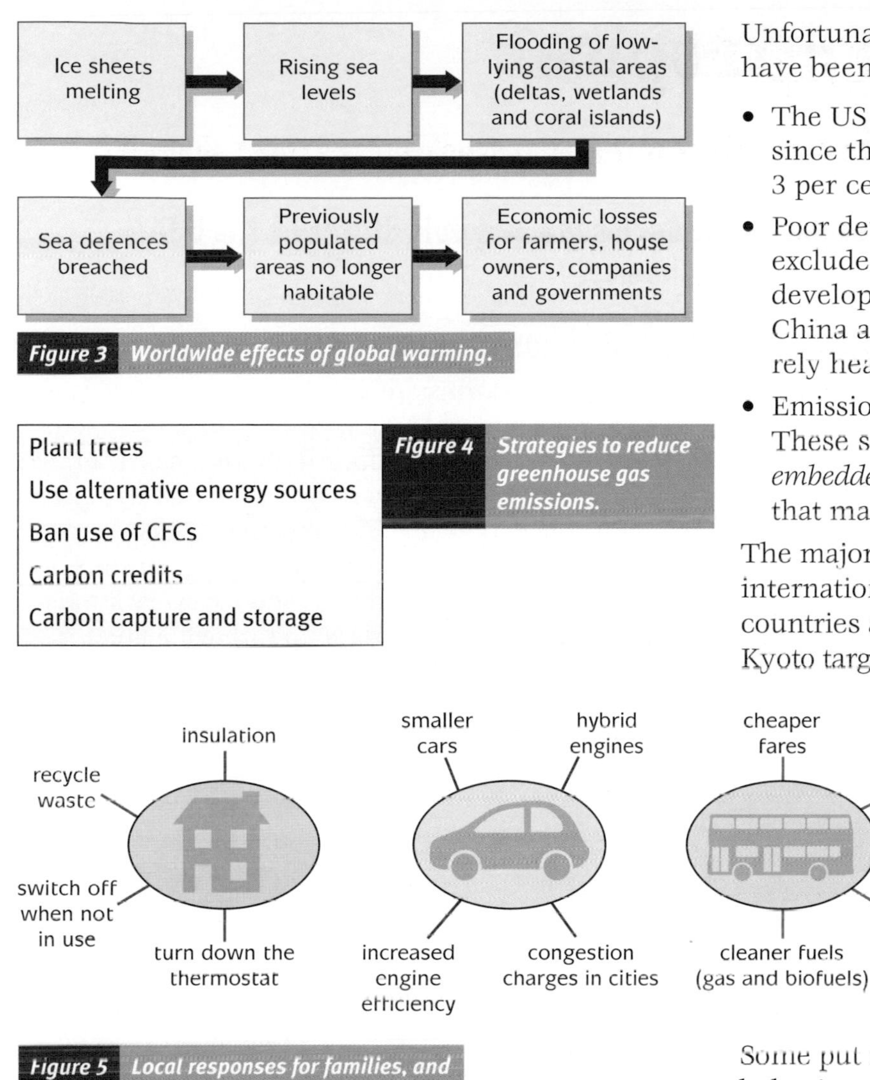

Figure 3 *Worldwide effects of global warming.*

Figure 4 *Strategies to reduce greenhouse gas emissions.*

Figure 5 *Local responses for families, and private and public transport.*

would result in a sea level rise of 7 metres, which would be enough to drown many of the world's big cities (including London). What is more, some scientists believe that global warming is already causing major changes in atmospheric circulation, to the point where world weather patterns are being affected. Extreme weather events, such as hurricanes, storms and droughts, appear to be happening more frequently and to be more intense, with serious consequences for people in affected areas.

Responses and strategies for the future

Given that we are facing a global problem, the responses need to be international. The Kyoto climate change conference in 1997 was a pioneering attempt to find an international solution, at a time when agreement on the existence of global warming was not as great as it is now. The Kyoto conference proposed compulsory reductions of 5 per cent in the carbon dioxide emissions of all rich developed countries by 2010, compared with levels in 1990.

Unfortunately the benefits of the Kyoto proposals have been limited for the following reasons:

- The US Senate refused to ratify the agreement; since then US emissions have increased by 3 per cent per year.
- Poor developing countries were deliberately excluded so as not to hinder their economic development, although some of them, such as China and India, are industrialising rapidly and rely heavily upon their own large coal deposits.
- Emissions from some sources were not covered. These sources included aviation, shipping, and *embedded emissions* (those added by the distance that manufactured goods and food have to travel).

The major benefit of Kyoto has been to increase international awareness of the issue. Many EU countries are struggling to meet even the modest Kyoto target, at a time when some environmentalists suggest that a worldwide reduction of 60 per cent is the minimum needed to have an impact. A list of strategies is given in **Figure 4**. Stricter than Kyoto, the EU target for the UK is 15 per cent of energy from renewables by 2020, which the government will struggle to meet.

Some put more faith in local responses and people behaving responsibly, summed up by the slogan **'think globally, act locally'**. Each individual contributing a small amount of carbon dioxide means a large total world contribution of greenhouse gas emissions (**Figure 5**). To what extent can government and local authorities influence individual responses? Public awareness is greater than ever, but what will make people change their lifestyles?

ACTIVITIES

1 Give supporting information for each of these three statements.
   (a) The greenhouse effect is a natural process.
   (b) Global warming already exists.
   (c) Not all countries are equally responsible for greenhouse gas emissions.

2 (a) Why is international action needed to reduce global warming?
   (b) How can individuals help?
   (c) Will it ever be possible to stop global warming from increasing? Explain your views on this.

Climatic hazards – tropical revolving storms

Why is the hurricane season feared by people in the Caribbean and Gulf States of the USA? What can people do to protect themselves and their homes?

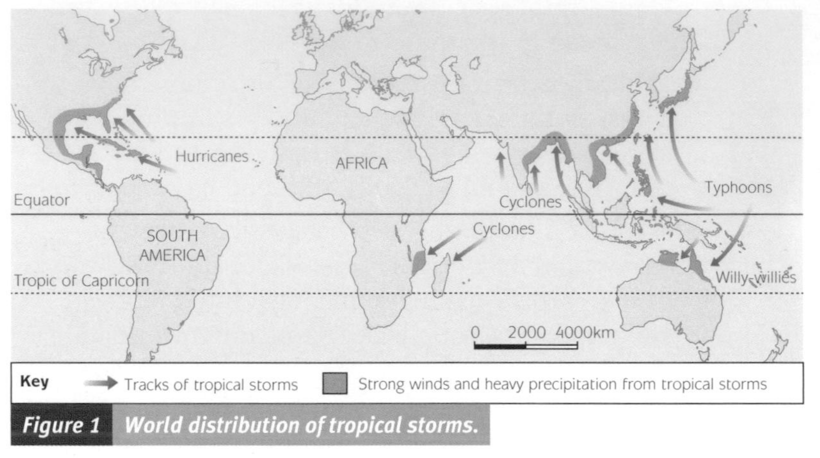

Figure 1 *World distribution of tropical storms.*

A **climatic hazard** is a short-term weather event that is a threat to life and property. People living in the tropics are extremely fearful of tropical storms. These storms go under a variety of local names, including hurricanes, cyclones, typhoons and willy-willies (**Figure 1**), but all are formed in the same way.

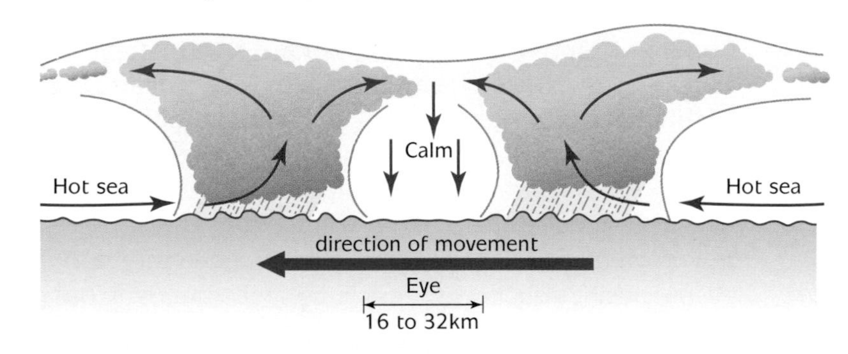

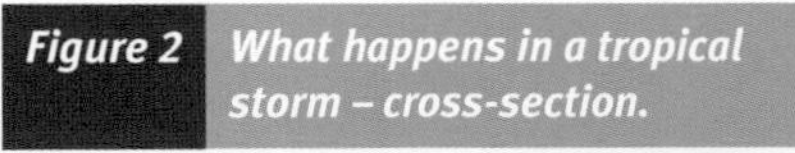
Figure 2 *What happens in a tropical storm – cross-section.*

Formation and characteristics

Tropical storms are formed when the seawater is at its hottest (27°C or more), usually in late summer and autumn. Air above the sea surface is heated and the warm moist air starts to rise in convection currents, creating a very deep centre of low pressure. To feed the currents, air is sucked in, and this produces very strong winds. The hot air rises up through a great swirl of towering cumulo-nimbus clouds, which cause torrential rain. These clouds form the **eye wall**. The **eye** of the storm is different; here the weather is calm and dry, because it is the only place for hundreds of kilometres around where air is sinking (**Figure 2**). Since the energy driving the system comes from continuous supplies of heat and moisture from ocean surfaces, tropical storms soon lose their power when they cross land surfaces.

Effects

Tropical cyclones cause immediate loss of life and a great deal of damage. Winds speeds of 250 kilometres per hour or more can destroy everything in their path; even strong buildings can be flattened by the strength and swirl of the winds. Torrential rains (500 millimetres or more rainfall in 24 hours is not uncommon) lead to flooding. In low-lying coastal areas this is made worse by storm surges, whipped up by the winds, which can cause walls of water up to 10 metres high to crash down on the shore. On steep slopes, landslides of soil, stones and rocks are set off. There can be almost total devastation in coastal regions.

After the storm dies down, people are in a state of shock from *social* losses (deaths of relatives and friends) and *economic* losses (damage to homes, possessions and businesses, loss of crops and animals on farms). Public utilities are badly disrupted: life can be very difficult without access to electricity, telephones, transport and fresh water. Disruption to fresh water supplies, sewage treatment and waste disposal can lead to serious health problems and the spread of diseases such as cholera and typhoid. There are often adverse *political* consequences for governments that fail to respond well to the disaster.

Figure 3 *Satellite view of Hurricane Wilma. Notice the great swirling mass of cloud around the eye, as air is sucked in. This explains why hurricanes are called revolving storms.*

	A Rich world: Hurricane Floyd, North Carolina, USA (September 1999)	B Poor world: Hurricane Mitch, Honduras and Nicaragua, Central America (October 1998)
Wind speed	up to 250kph	up to 260kph
People affected	3 million	4 million
Loss of life	7	18 000 dead or missing
Damage America (October 1998)	scores of houses destroyed roads washed away	over 1 million homeless landslides washed away whole villages
Estimated losses	US$ 16 billion	US$ 7 billion
Percentage of losses insured	75%	2%

Figure 4 *The impact of two tropical storms compared.*

As with other natural hazards (e.g. volcanoes and earthquakes – Chapter 1), the poor are affected more severely than the rich. Look at the summary of the impact of two tropical storms in **Figure 4**, of similar intensity and only a year apart. How much worse were the effects in the poor countries of Central America than in the rich USA?

Responses

Without pre-evacuation, the lives of thousands of people are at risk. Meteorologists can watch tropical storms continuously thanks to weather satellites; warnings can be given to evacuate areas. Police and fire services can practise emergency drills; people can be educated in advance about emergency procedures; and emergency shelters can be prepared with rations of food and water. However, the behaviour of tropical storms is notoriously unpredictable; they can suddenly become stronger and change direction. Some people are always unwilling to leave their homes and property unattended. Others are too poor or do not have the means of transport to move inland out of the storm's reach. Theory and reality can be two different things.

The long-term goal is for profitable economic activities (farming, industry and services) to be resumed as soon as possible. For this to happen, outside help from all sources (government, companies, charities and aid organisations) is essential. There is an urgent need for infrastructure (power supplies, transport and public services) to be repaired. Until power lines are restored, and roads and bridges are back in use, offices and businesses cannot communicate with the outside world and are therefore unable to operate effectively. This can take weeks, or even months.

For farmers, it might take years for livestock numbers to increase and for newly planted bush and tree crops to yield fruit. The economies of many small Caribbean island countries are dependent on tourism; not only do hotels have to be rebuilt, but massive advertising campaigns are also needed to convince visitors that it is safe to return. It is important that governments and local authorities get the message and plan better for the next storm – by improving warning systems, building shelters, training emergency teams and educating people about what to do. 'Be prepared' is the best motto.

GradeStudio

1 Study **Figure 1**.
   a State one similarity between different tropical storms for (i) source (ii) track followed (iii) type of area affected. (3 marks)
   b For one type of tropical storm, describe the location of its source area and areas affected. (2 marks)

2 Draw a sketch of **Figure 3**. Label it to show the features of this tropical storm. (3 marks)

3 Describe and explain the sequence of weather associated with the passage of a tropical storm. (4 marks)

4 a Describe how **Figure 4** shows that the poor suffer more than the rich in tropical storms. (3 marks)
   b Explain why strategies for preventing loss of life in tropical storms are usually more successful in rich developed countries than in poor developing countries. (5 marks)

Exam tip

Make sure you know the names of the seven continents and five oceans.
Use them when describing from world maps.

FURTHER RESEARCH

Investigate Hurricane Katrina 2005 or a more recent tropical storm on the weblink www.contentextra.com/aqagcsegeog.

Practice GCSE question

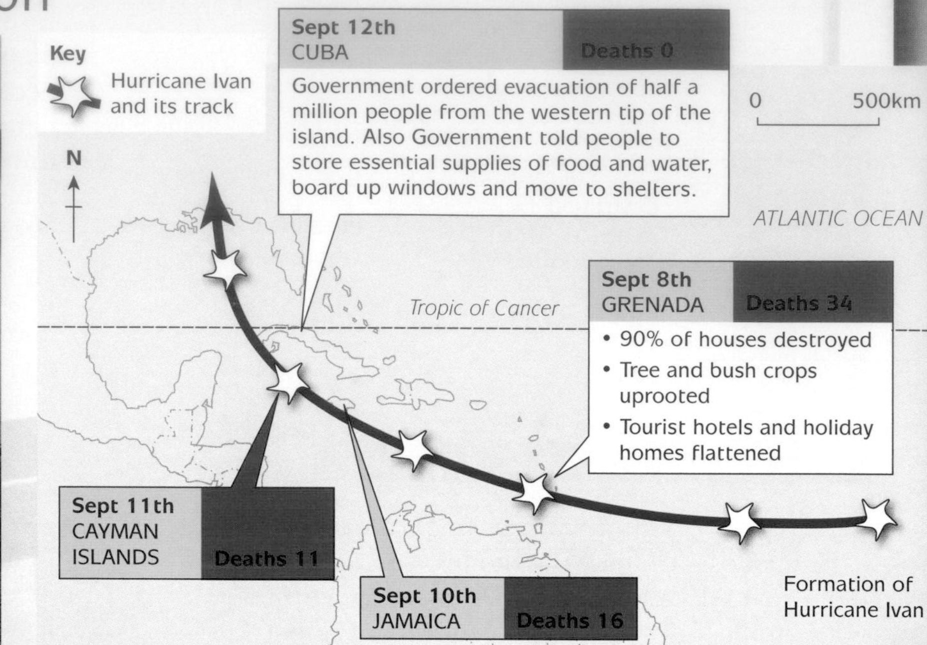
Key
Hurricane Ivan and its track
N
0 500km
ATLANTIC OCEAN
Tropic of Cancer
Sept 12th CUBA Deaths 0
Government ordered evacuation of half a million people from the western tip of the island. Also Government told people to store essential supplies of food and water, board up windows and move to shelters.
Sept 8th GRENADA Deaths 34
• 90% of houses destroyed
• Tree and bush crops uprooted
• Tourist hotels and holiday homes flattened
Sept 11th CAYMAN ISLANDS Deaths 11
Sept 10th JAMAICA Deaths 16
Formation of Hurricane Ivan

Figure 1 Anticyclonic weather in the UK.

Figure 2 Track of Hurricane Ivan, September 2004.

3 (a) **Figure 1** shows anticyclonic weather in the UK.

(i) Describe the weather in the area shown in **Figure 1**. **(3 marks)**

(ii) Choose one of the weather conditions shown. Explain why it occurs in an anticyclone. **(3 marks)**

(b) Study **Figure 2**, which shows the track of Hurricane Ivan through the Caribbean in September 2004 and some of its effects.

(i) State the trend in the pattern of deaths and give two reasons for it. **(3 marks)**

(ii) Explain why revolving tropical storms form in places like this in the northern hemisphere in September. **(4 marks)**

(iii) Before Hurricane Ivan, the economy of Grenada depended on the export of crops (mainly spices from bushes and trees) and tourism. State and explain the possible responses of the government and people of Grenada in the first few years after Hurricane Ivan. **(4 marks)**

(c) **Figure 3** shows some of the possible consequences of global warming.
Explain the possible consequences of global climate change for the world and the UK. **(8 marks)**

Total: 25 marks

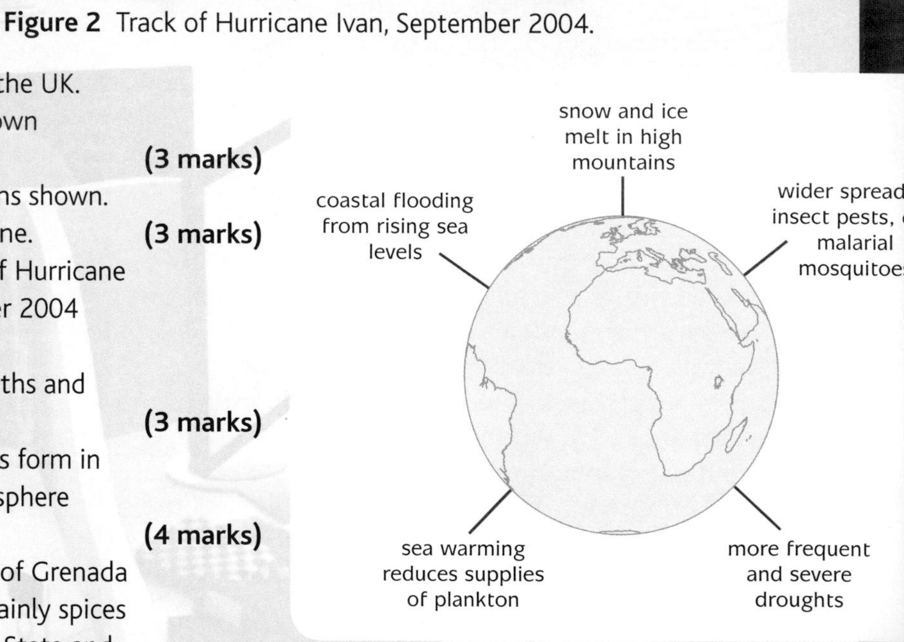
snow and ice melt in high mountains
coastal flooding from rising sea levels
wider spread
insect pests,
malarial
sea warming reduces supplies of plankton
more frequent and severe droughts

Figure 3 Possible consequences of global warming.

See a Foundation Tier Practice GCSE Question on the weblink www.contentextra.com/aqagcsegeog.

Exam tip
The question for this topic will be Question 3 in Paper 1.

Improve your GCSE answers

Know the differences between labels for describe and explain – social, economic, political and environmental impacts

1 Choosing between different labels for 'describe' and 'explain' used on photographs and diagrams.

Question A
Add labels to *describe* the weather shown.

Question B
Add labels to *explain* how convectional rain is formed

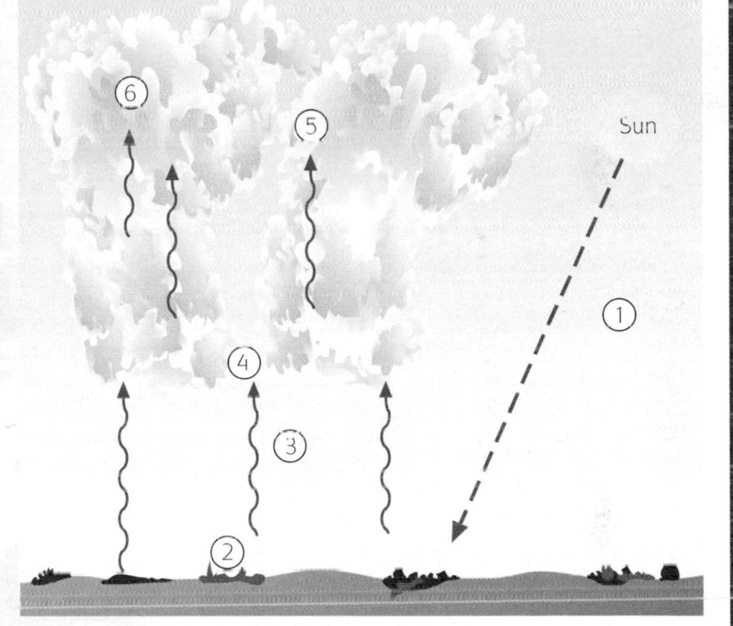

Labels useful for either question A or question B

- towering cumulo-nimbus clouds form
- cumulus clouds in the sky
- water droplets increase in size in the cloud
- moisture condenses
- hot air rises and cools
- great heating of ground surface
- direct rays of sun
- heavy rain falling
- looks like a thunderstorm
- some blue sky shows storm is passing

(a) Choose the four mark-earning labels for question A.
(b) For the other six labels, give the label number from the diagram in question B.
(c) Why were the six labels numbered 1–6 in that order for answering question B?

2 Knowing the differences between impacts (effects) that are social, economic, environmental and political.

Show that you can identify the different types of impacts by reorganising them under the headings: Social, Economic, Environmental, Political. For the best answer, you need to list three social impacts, four economic impacts, two environmental impacts and one political impact.

Examples of impacts from a climatic hazard/extreme event (such as a tropical storm)

fishing boats washed 100 metres inland by storm surges

heavy rains set off landslides on steep slopes

children and old people suffer health problems from contaminated water supplies

hotels and factories damaged and destroyed

sandy beaches eroded and destroyed

transport disrupted by broken roads and bridges

government blamed for being unprepared and voted out at the next election

homeless people forced to migrate

families broken up by deaths

crops flattened and livestock washed away

ExamCafé

REVISION

Key terms from the specification

Exam Café

Anticyclone – area of high pressure
Climate – average weather conditions recorded at a place over many years
Continentality – influence of land surface on weather and climate
Depression – area of low pressure
Extreme weather – weather event more severe than normally expected
Global climate change – variations in temperature and rainfall affecting the whole world
Hazard (climatic) – short-term weather event that threatens lives and property
Maritime influence – influence of the sea on weather and climate
Precipitation – all moisture that reaches the Earth's surface from the atmosphere
Responses – actions immediately after the event or in the long-term
Tropical revolving storm – area of very low pressure in low latitudes, with strong winds and heavy rain
Weather – condition of the atmosphere at any one time, day-to-day variations

Checklist

	Yes	If no – refer to
Can you state the main characteristics of the UK climate?		page 44
Can you describe variations in climate in the UK?		page 46
Do you know how and why temperatures and precipitation vary within the UK?		pages 46–7
Can you describe and explain the weather changes as a frontal depression moves over the UK?		pages 48–9
Do you understand how and why anticyclonic weather in the UK differs between summer and winter?		pages 50–51
Can you write about an example of an extreme weather event in the UK, like a drought, a flood or a great storm?		page 53
Do you know some of the possible causes and consequences of global warming?		pages 54–5
Do you understand why tropical storms can cause so much damage and destruction?		page 56–7

Case study summaries

Extreme weather event in the UK	Tropical storm
Areas affected	Name and location
Causes	Causes
Economic impacts	Effects (economic, social and environmental)
Social impacts	
Responses	Short-term responses
Plans for the future	Long-term responses

Chapter 4
Living World

A young boy in Nigeria takes the measure of an old and once mighty ironwood tree, which has just been felled by clearance of the tropical rainforest.

QUESTIONS

- **What is an ecosystem?**
- **Why is the natural balance of ecosystems easily disrupted?**
- **Why do some ecosystems offer more economic opportunities for people than others?**
- **How can these ecosystems be used more sustainably in the future than in the past?**

Ecosystems

What is an ecosystem? Why are they called systems? What happens when they are disturbed?

An **ecosystem** is a living community of plants and animals, which is linked to the natural environment where they live. Each element in the system, whether living or natural, depends upon, and influences, others. They are interrelated. This is why the diagram summarising the ecosystem (**Figure 1**) shows some two-way relationships. Ecosystems can be studied at many different scales, from small-scale such as a single oak tree or a pond, to large-scale, such as **deciduous** woodland or tropical rainforest. Ecosystems tend to be named according to the vegetation cover.

Climate
Animals
Soil
Vegetation

Figure 1 **Ecosystem – systems diagram.**

A small-scale ecosystem

Figure 2 illustrates a hedgerow ecosystem in England.

- The UK's temperate maritime climate (page 44) allows certain types of plants to grow, many of them deciduous (losing their leaves in winter).
- The vegetation in this hedgerow is varied and of different heights. The richness of the vegetation is partly due to the fertile soils found in many parts of lowland Britain, known as **brown earths** (page 68). These soils benefit, of course, from the plentiful organic material returned to the soil when plants drop their leaves and eventually die.
- The wealth of vegetation cover provides a good habitat for wildlife, for food, shelter and breeding. Many of the birds that regularly visit urban gardens are typical hedgerow species. Animals help to fertilise the soil and to spread seeds.
- And then there are people, who – like other animals – sometimes help natural ecosystems, but more often than not destroy them, intentionally or otherwise.

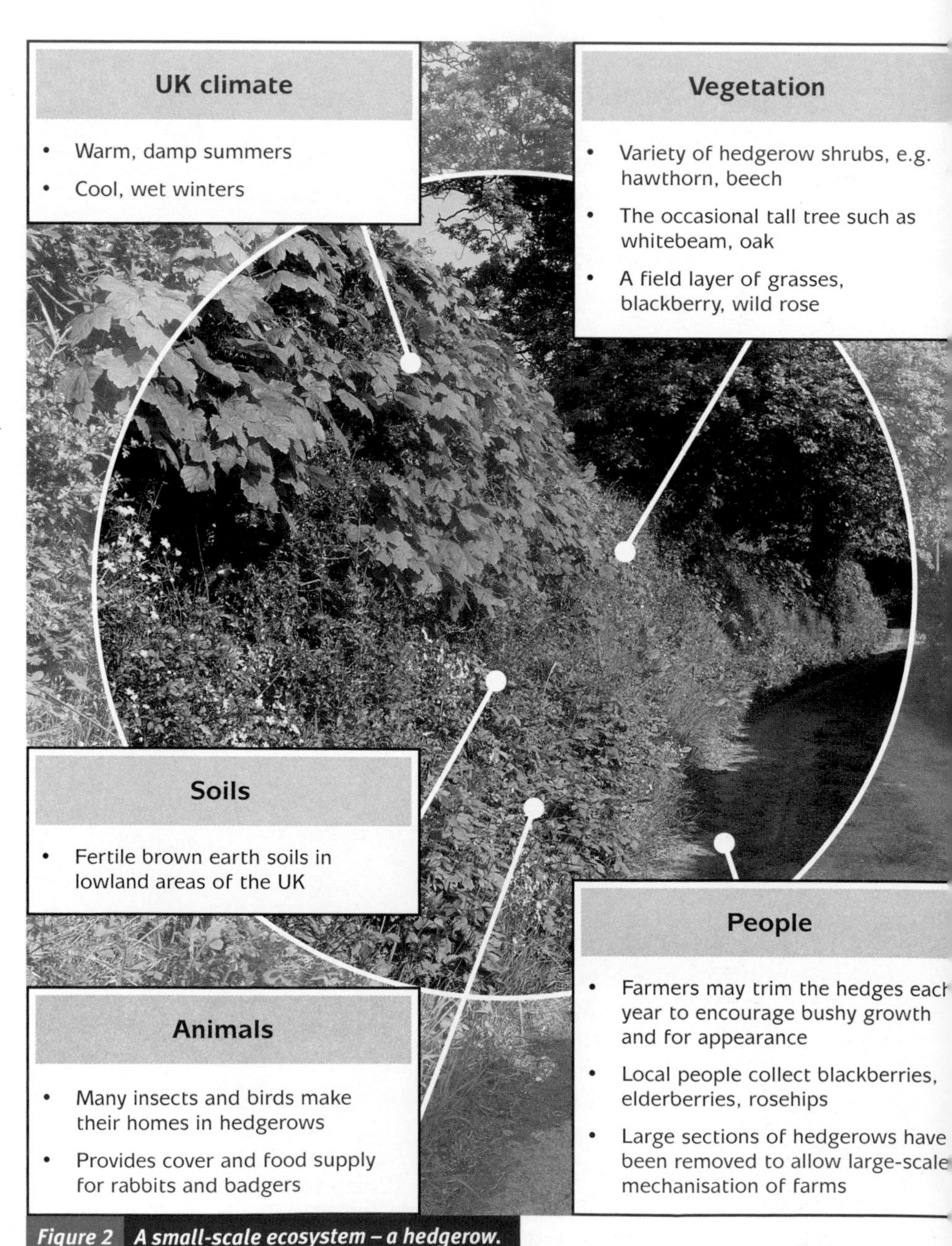

UK climate
• Warm, damp summers
• Cool, wet winters
Vegetation
• Variety of hedgerow shrubs, e.g. hawthorn, beech
• The occasional tall tree such as whitebeam, oak
• A field layer of grasses, blackberry, wild rose
Soils
• Fertile brown earth soils in lowland areas of the UK
Animals
• Many insects and birds make their homes in hedgerows
• Provides cover and food supply for rabbits and badgers
People
• Farmers may trim the hedges each year to encourage bushy growth and for appearance
• Local people collect blackberries, elderberries, rosehips
• Large sections of hedgerows have been removed to allow large-scale mechanisation of farms

Figure 2 **A small-scale ecosystem – a hedgerow.**

FURTHER RESEARCH

Undertake an investigation of a small-scale ecosystem in your local area.

The ecosystem in balance

In an ecosystem the flow of energy and the cycling of nutrients need to be constant to maintain the balance between the different elements. Green plants take substances from water, air and weathered rock in the soil and trap the Sun's light energy to create living matter. The name of this process is photosynthesis.

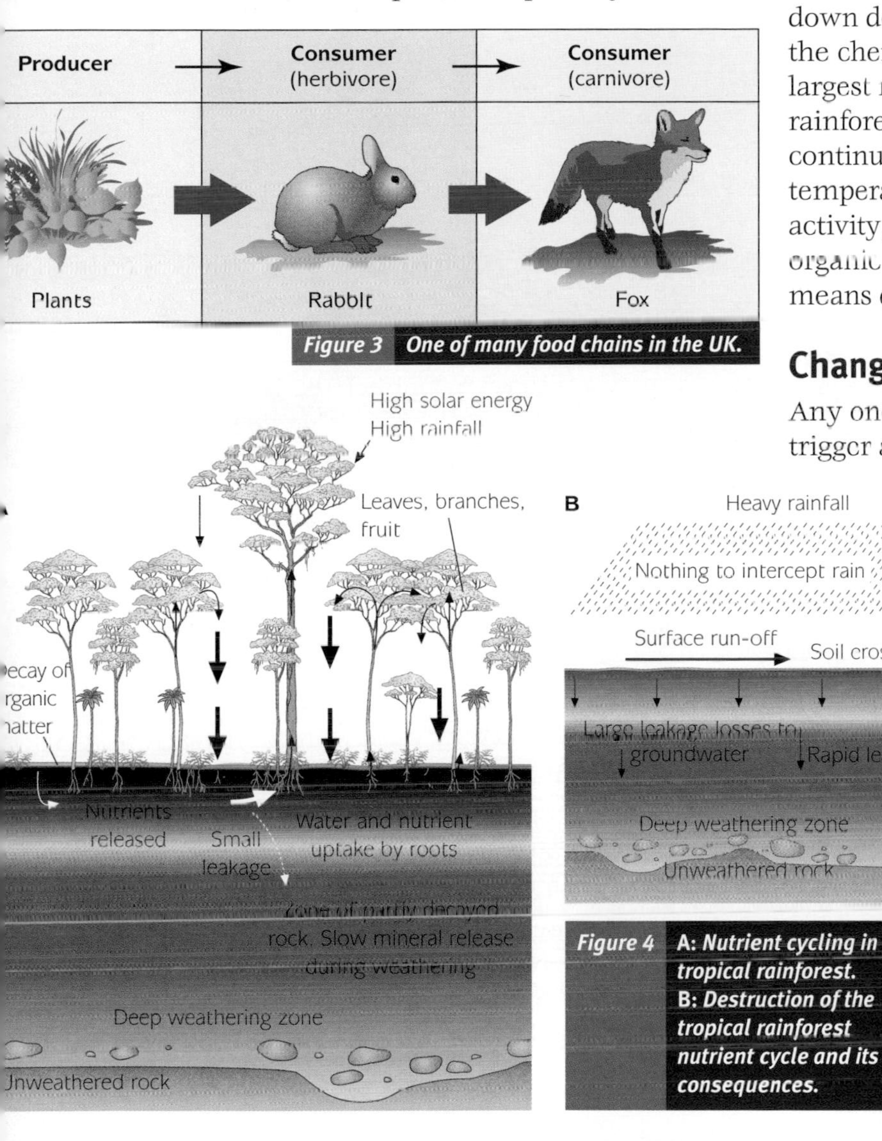

Figure 3 *One of many food chains in the UK.*

Figure 4 ***A: Nutrient cycling in tropical rainforest. B: Destruction of the tropical rainforest nutrient cycle and its consequences.***

Plants are **primary producers** – they make the living matter that can then be used by animals (people included). The nutrients and energy absorbed by plants are passed along a **food chain** of living things, first to herbivores (plant eaters) and then to carnivores (meat eaters). These are the **consumers**. The food chain shown in **Figure 3** has only one producer, but two levels of consumer. Each level in the food chain is called a **trophic** or feeding level. Between each level there is a great energy loss, about 90 per cent, mostly in the form of heat. This leaves only 10 per cent available for growth.

Nature recycles everything. Nothing is wasted. Humans are only beginning to get the message that we should do the same! Bacteria and fungi, especially those in the soil, are the **decomposers**. They break down dead remains of plants and animals, and release the chemicals for plants to use again. The fastest and largest nutrient recycling system is found in tropical rainforests. Leaf fall and falling branches provide a continuous supply of litter to the forest floor. High temperatures and rainfall favour intensive biotic activity, which leads to rapid decomposition of the organic material (**Figure 4A**). All-year plant growth means quick reuse of nutrients released.

Changes to the ecosystem

Any one alteration in the balance of nature can trigger a chain reaction throughout the whole system. Sometimes the reasons for change are natural (physical); examples include climate change (increases or decreases in rainfall and water availability, rising or falling temperatures), and plant and animal diseases. More often human activities are the cause. In trying to use and control habitats, humans have been ignorant of, or have ignored, the complex natural structures in ecosystems – sometimes with disastrous results. A natural forest is a closed system with little leakage of nutrients. After rainforest clearance, the cycle shown in **Figure 4A** is broken. Nutrients are washed out with nothing to replace them, leading to devastating effects on soils (**Figure 4B**).

ACTIVITIES

1 Show that you understand the difference between
   (a) the natural and living elements in an ecosystem
   (b) producers and consumers.

2 For the hedgerow in **Figure 2**:
   (a) draw a larger and more detailed version of **Figure 1** to summarise its main features
   (b) describe how the ecosystem shown appears to be in balance
   (c) suggest ways in which this balance might be upset.

3 Draw another diagram similar to **Figure 1** to show changes in each of the four elements that can upset the balance of the system.

4 Study **Figures 4A** and **B.** Explain the effects of rainforest clearance on the soils.

Global distribution of large-scale ecosystems

How does natural vegetation change from the Equator to the poles? Why can plants grow even in the world's driest places?

Tropical rainforest
Climatic summary
- Hot all year
  Average temperature 27–30°C
- Wet all year
  Annual precipitation 2000–3000mm

Deciduous forest
Climatic summary
- Warm summers
  Average temperature 16–20°C
- Mild/cool winters
  Average temperature 3–8°C
- Precipitation all year
- Annual rainfall 550–1500mm

Hot desert
Climatic summary
- Very hot in summer
  Average temperature 35–45°C
- Hot in winter (20–30°C)
- Dry all year
- Annual rainfall under 250mm

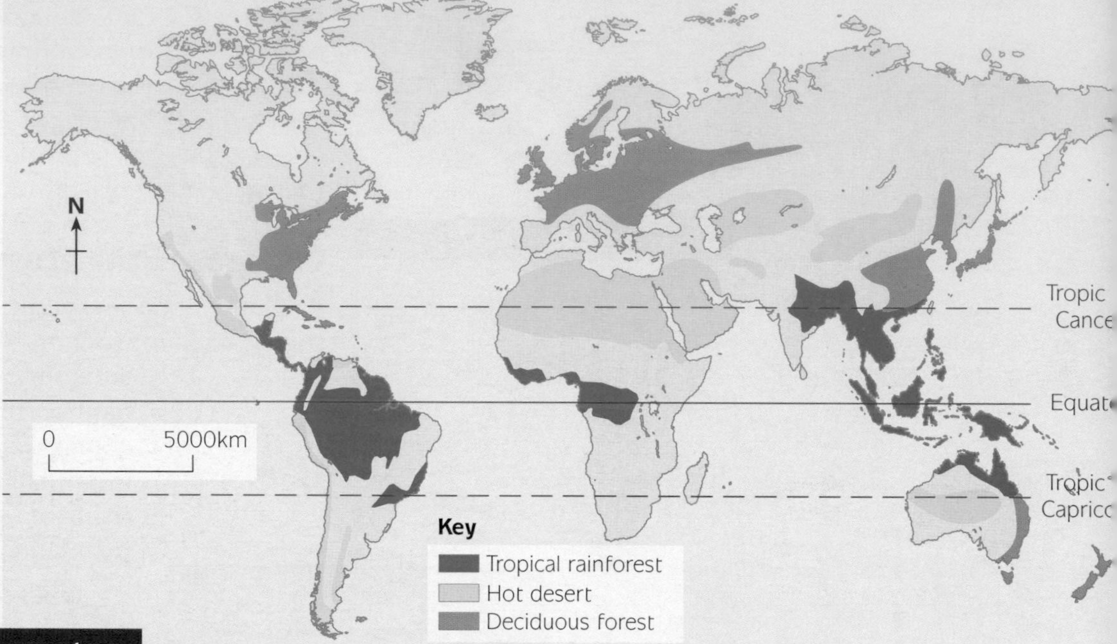

Figure 1 *World distribution of three ecosystems.*

On a world scale, *climate* is the main factor that controls the characteristics and extent of the natural vegetation cover. **Figure 1** shows the distribution of the three ecosystems to be studied in more detail. Outline information about the climate of each ecosystem is given down the side. Notice how they stretch across continents from east to west (in general), parallel to the Equator.

- Tropical rainforests are close to the Equator, and located down the wet eastern sides of the continents in the tropics.
- Hot desert ecosystems are closer to the Tropics of Cancer and Capricorn. The greatest extent is in the Sahara, where the African continent is widest.
- Deciduous forests are in temperate latitudes; they are located further north on the western sides of continents where winters are warmer than on the eastern sides. They form the natural vegetation cover not only for the UK, but also over much of western and central Europe.

Figure 2 *The changing pattern of vegetation cover from the Equator to the polar regions in the northern hemisphere.*

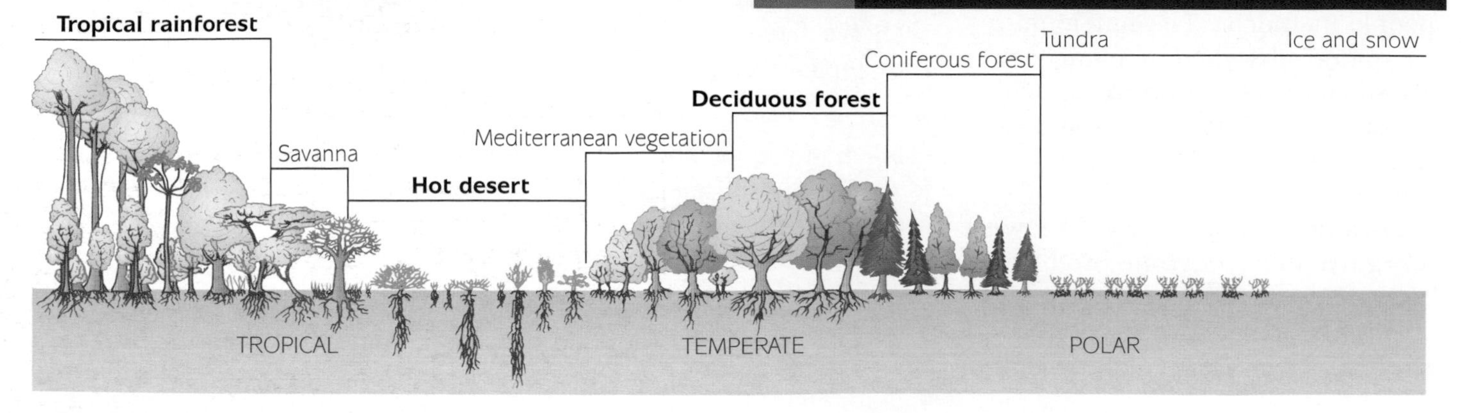

Figure 2 allows comparisons between ecosystems. The hot wet Equatorial climate supports the most abundant vegetation cover on Earth, whereas the hot desert climate cannot support continuous vegetation cover. In tropical latitudes, since it is always hot enough for plant growth, *rainfall* is the most important climatic factor explaining the vegetation differences shown. *Temperature* gradually becomes more important and dominant in higher latitudes. Vegetation cover decreases in height and variety as latitude increases.

Example of a large-scale ecosystem – hot deserts

Hot deserts are examples of extreme environments caused by great daytime heat under a baking sun and absence of rainfall. Although surface vegetation is sparse, small and separated by large areas of bare ground, it does exist, even in the driest place on Earth (**Figure 3A**). Plants need to grow far apart so that they are not competing for scarce resources, principally water. Vegetation with adaptations for living in a dry environment is called **xerophytic**.

Figure 3 A: Plant living in the Atacama desert, the world's driest desert. How much more of the plant is underground? It can be up to 20 times. B: Cactus.

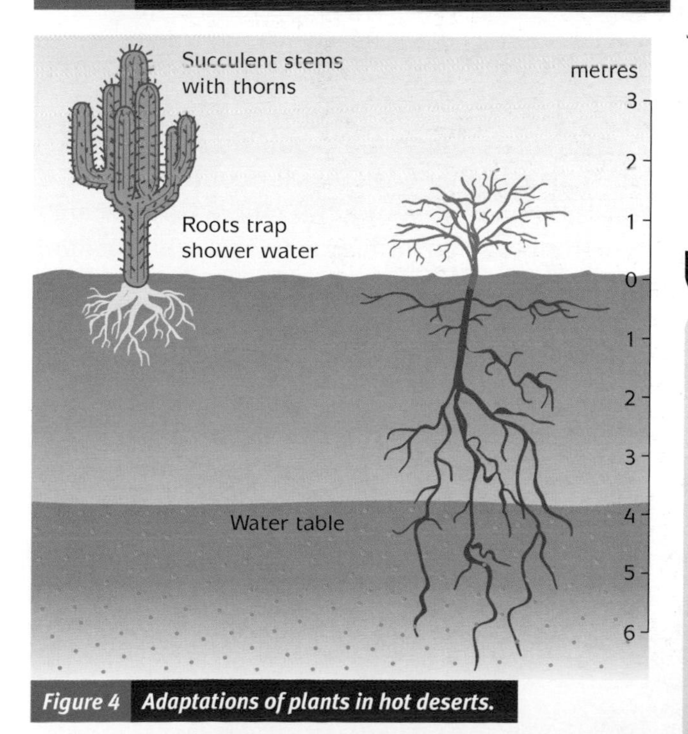

Figure 4 Adaptations of plants in hot deserts.

Figure 5 An irrigated oasis in Egypt. The date palm, which has a deep tap root, is one of the symbols of the deserts of North Africa and the Middle East, along with the camel.

Figure 4 shows two contrasting types of xerophytic adaptation. One type is a **succulent** – a plant that stores water in its stem for use during the long periods of drought. Cactus is the best-known example. It has shallow roots to trap water from infrequent rains. Thorns help reduce water loss, as well as protecting the plant from desert animals. The other type of desert plant has very deep roots to enable it to tap underground water supplies. Only a small part of the plant is above the surface, where it is exposed to the scorching heat. Plants like this are woody and often thorny to reduce water loss to a minimum. Any leaves they have are tough and leathery, reducing transpiration further by means of dense hairs that cover their surfaces.

Plants have adapted to survive in hot deserts in other ways as well. Some have incredibly strong seeds, which lie dormant for years, until it rains. Then the desert blooms. The appearance of bare rock and sand surfaces is transformed by the abundance of flowering plants. These can only bloom for as long as water is available; they have no xerophytic properties of their own. It is their seeds that are adapted to the dry desert climate.

Most desert soils have no organic content; they are just made up of pieces of rock and sand. In low-lying areas, where water is closer to the surface, desert soils are often saline (salty). One of the few plants that has the ability to grow in saline groundwater is the date palm (**Figure 5**). This view shows how people can transform the desert with water.

ACTIVITIES

1 (a) Areas of tropical rainforest in 2005 – Africa 19 per cent, Asia and Oceania 36 per cent, Central and South America 45 per cent. Draw a pie graph to show these percentages.
(b) Name one country with large regions of rainforest in each of these three areas.
(c) Looking at **Figure 1**, describe the distribution of tropical rainforest in Africa.

2 (a) State the main characteristics of a hot desert climate.
(b) Explain the different ways in which the desert plants shown in **Figures 3A**, **3B** and **5** are adapted to this climate.
(c) Describe what desert soils are like.

A rich part of the world – the desert west of the USA

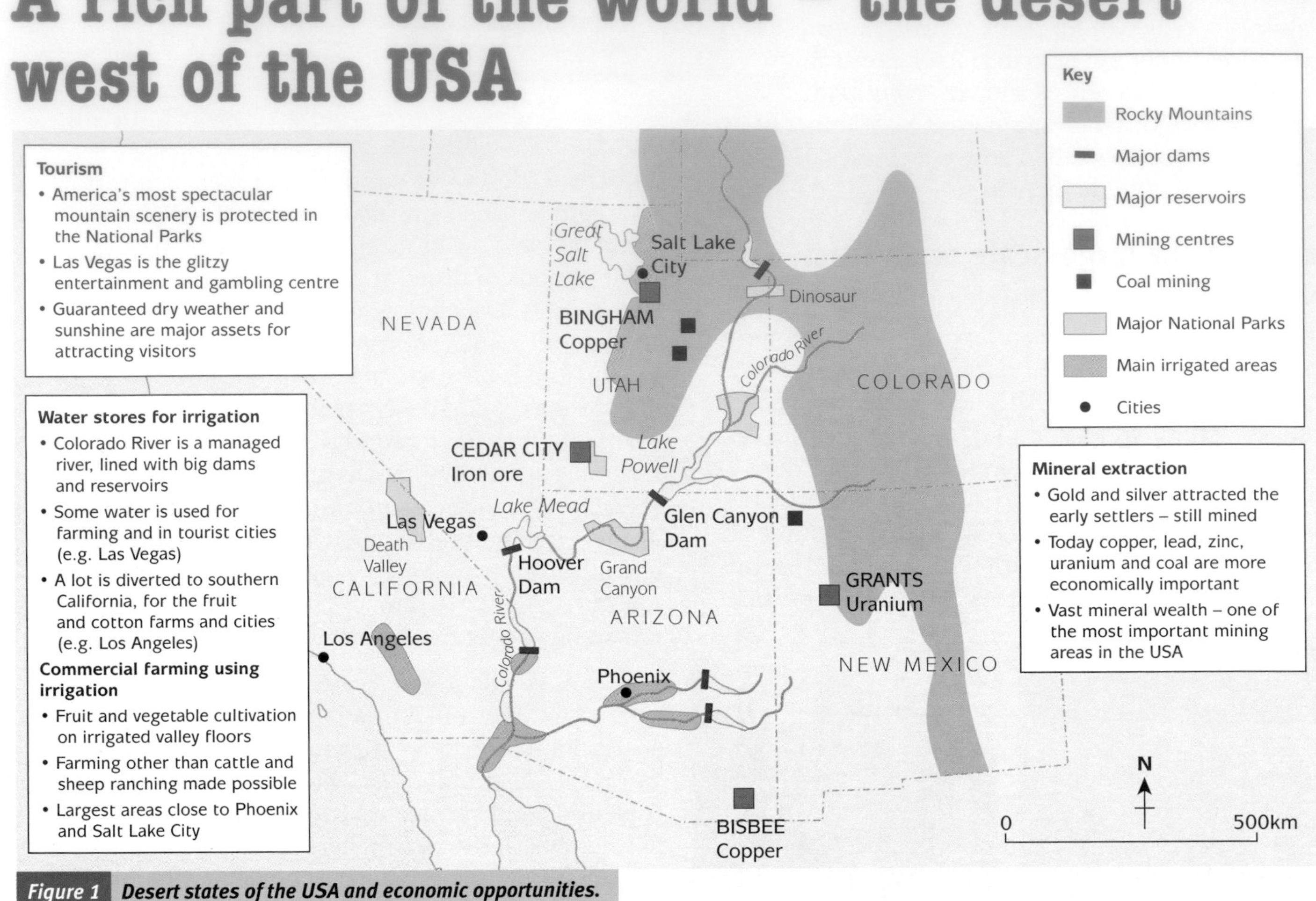
Key
Rocky Mountains
Major dams
Major reservoirs
Mining centres
Coal mining
Major National Parks
Main irrigated areas
Cities
Tourism
• America's most spectacular mountain scenery is protected in the National Parks
• Las Vegas is the glitzy entertainment and gambling centre
• Guaranteed dry weather and sunshine are major assets for attracting visitors
Water stores for irrigation
• Colorado River is a managed river, lined with big dams and reservoirs
• Some water is used for farming and in tourist cities (e.g. Las Vegas)
• A lot is diverted to southern California, for the fruit and cotton farms and cities (e.g. Los Angeles)
Commercial farming using irrigation
• Fruit and vegetable cultivation on irrigated valley floors
• Farming other than cattle and sheep ranching made possible
• Largest areas close to Phoenix and Salt Lake City
Mineral extraction
• Gold and silver attracted the early settlers – still mined
• Today copper, lead, zinc, uranium and coal are more economically important
• Vast mineral wealth – one of the most important mining areas in the USA
Great Salt Lake
Salt Lake City
Dinosaur
BINGHAM Copper
NEVADA
UTAH
COLORADO
Colorado River
CEDAR CITY Iron ore
Lake Powell
Lake Mead
Las Vegas
Glen Canyon Dam
Death Valley
CALIFORNIA
Hoover Dam
Grand Canyon
GRANTS Uranium
ARIZONA
Los Angeles
Colorado River
Phoenix
NEW MEXICO
BISBEE Copper
N
0
500km

Figure 1 ***Desert states of the USA and economic opportunities.***

As in most hot deserts, overall densities of population in the desert states of the USA (**Figure 1**) are low. People are highly concentrated in just a few locations where economic activities are profitable – mining, commercial farming and tourism. However, annual population growth rates are currently the highest in the USA (2–3 per cent), fuelled by the inward migration of retired people. Arizona and Nevada are the new 'Sunbelt'.

What are the advantages for old people of a warm, sunny climate with low humidity? Many new housing estates in suburban towns around Phoenix are restricted to retired people and marketed as safe locations in the Arizona sun.

Figure 2 ***Part of the great open pit of the Bingham copper mine, near Salt Lake City.***

Challenges and issues

As in all deserts, the main issue is water. The region's largest river, the Colorado, can no longer meet all the demands placed upon it. It often reaches the sea in Mexico as a mere trickle, to the annoyance and cost of the Mexican people. Las Vegas is using up the underground aquifer, on which it sits, much more quickly than nature can replenish it. The demand is there for continuing growth in tourism and retirement, but where is the water going to come from? Are current growth rates sustainable?

Visitors to the National Parks need to be managed – by providing visitor facilities such as car parks, camp and camper sites, maintaining walking tracks, providing visitor information and controlling numbers at peak times. Park rangers supervise the parks to maintain the wildlife and natural beauty that visitors go to see, but which could be destroyed by those same visitors.

Figure 3 ***Monument Valley, made famous in 'Westerns' (films set in the Wild West); is any economic use other than tourism possible in desert areas like this?***

A poor part of the world – the desert in southern Pakistan

INFORMATION

	USA	Pakistan
Total population (millions)	300	160
Birth rate (per 1000)	14	35
Income per head ($US)	40 000	610
Employed in agriculture (%)	2	42
Living in cities (%)	80	35

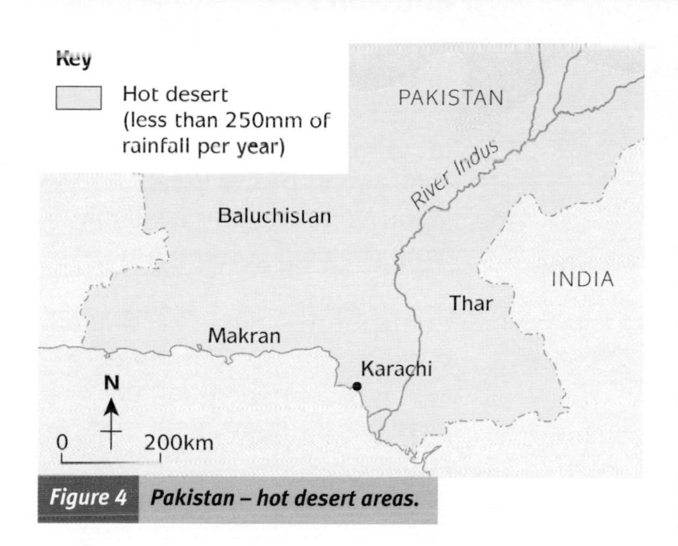

Figure 4 ***Pakistan – hot desert areas.***

Desert covers most of the southern half of Pakistan (**Figure 4**). Most desert surfaces consist of bare rock and stones. Vegetation at best is tough grass and scrub; soil formation is minimal because of the lack of organic content. Despite this, most inhabitants are subsistence farmers. Population densities are mostly low (under 10 people per square kilometre), but in the east densities are high by desert standards (sometimes over 50 people per square kilometre), which places great pressure on natural resources.

Traditional ways of life

Tribes such as the Baluchis are **nomadic pastoralists**. They have to keep moving, in search of fresh grazing for their mixed livestock groups of sheep, goats and camels, because there is nothing better than poor coarse grasses and tough shrubs for the animals to eat. All three are highly adaptable animals, able to survive on little water and dry vegetation (in areas where cattle cannot). Animals supply almost all the needs of the tribes – milk, meat, wool for weaving blankets, and skins for making leather goods. The people gather fuelwood for cooking. Where water can be obtained from underground using methods that have been used for centuries, such as the Persian wheel, people are able to grow staple food crops.

Challenges and issues

Sheep and goats, already found in abundance in the tribal areas in the deserts, are increasing in numbers. This is partly the result of high birth rates among the local people, leading to a rising population, requiring more livestock to supply their needs. Goats are particularly destructive because they eat every green thing in their path. This is leading to severe problems with overgrazing and consequent soil erosion, made worse by clearances of vegetation for firewood. In the crop-growing areas, fallow periods to allow the soil to recover are being reduced – again due to increasing population pressure. Therefore human activities are exposing the soil to strong wind erosion and sand invasions. Over-watering of crops results in salinisation: as water evaporates in the heat, surface concentrations of salt are left behind.

The desert tribes are too poor to undertake initiatives that would increase sustainability, such as reducing herd numbers but increasing animal quality, planting fast-growing trees to stabilise the dunes while providing fodder and fuel wood, and building small dams. Only in the few areas where aid agencies are active are small communities feeling the benefits of being introduced to water saving and more sustainable techniques such as building small earth dams.

ACTIVITIES

1 (a) Describe the differences between the two desert areas for (i) water supply for irrigation (ii) farming (iii) range of economic opportunities.
(b) Explain how the wealth gap between the USA and Pakistan helps to explain these differences.

2 (a) State what is (i) similar (ii) different about the challenges and issues between these two desert areas.
(b) In your view, which area faces the greater challenges for the future? Explain your choice.

3 Make case study notes for exam use for both hot desert areas using these headings:
(a) Sketch map for location (b) Economic activities
(c) Challenges for the future.

Deciduous forests

Why did deciduous forests cover most of the UK in the past? Why are there so few left? What are we doing about it?

The trees that dominate these forests in the UK, such as oak, beech and ash, are deciduous. They lose their leaves for several months of the year during winter. A definite forest structure can be recognised (**Figure 1**), but it is less complex and varied than in tropical rainforests (**Figure 1**, page 70).

This structure reflects the availability of, and competition for, light. Small plants on the forest floor, such as bluebells and wild garlic, come into flower early; in spring more light reaches them, because the canopy trees have not yet come back into leaf (**Figure 2**).

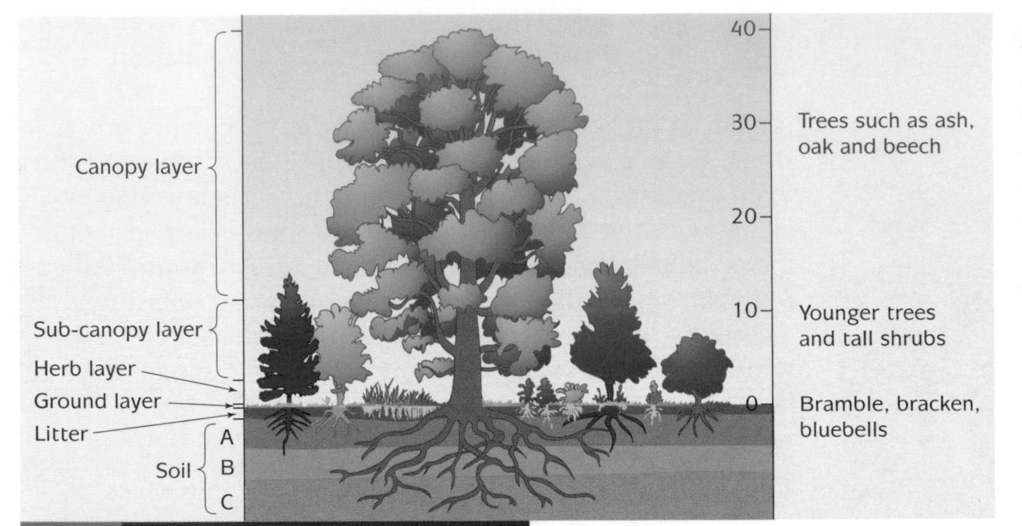

Figure 1 Deciduous forest – structure and plants.

- Canopy layer of dominant deciduous trees, between 20 and 30 metres above the ground.
- Sub-canopy layer of young trees or larger shrubs, in spaces between the tall trees.
- Herb layer of non-woody plants, such as bluebells, wild garlic, brambles and ivy.
- Ground layer close to the soil, mainly mosses.

Figure 2 Ash woodland in mid-summer. How will its appearance change in other seasons?

Adaptations to climate and soils

The trees shed their leaves in winter as a response to reductions in light and heat. A lot of moisture is lost through their broad fleshy leaves. Water is not readily available from the cold soil in winter. In summer, however, warm temperatures above 14°C allow rapid growth, as their large soft leaves intercept all the available sunlight. Rain falls regularly so the leaves do not need special protection against water loss, as in hot deserts, which is why they are neither thick nor waxy. The growing season, with mean daily temperatures comfortably above 6°C, lasts from four to seven months. Where it is less than four months, as in the uplands of the UK, deciduous trees are replaced by conifers.

Figure 3 Soil profile of a brown earth.

Brown soils, usually known as brown earths, are the typical soils (**Figure 3**). They are reddish-brown throughout, about 30 centimetres deep. Often they are slightly darker in the narrow 'A horizon' below the surface, where there is most organic material and biological activity. No sharp boundary between A and B horizons can be recognised. **Leaching** (the washing out of minerals from the top soil layer) is taking place in the wet climate, but only slowly. This means that the soils have a high pH value (above 5) and make fertile agricultural soils.

The National Forest

The few remaining ancient woodlands in the UK total well under 5 per cent of their original extent. Over the centuries, they were cleared for farmland, settlement and timber.

To help reverse this trend, the National Forest was created in 1990. It covers 520 square kilometres of the English Midlands, spanning parts of Derbyshire, Leicestershire and Staffordshire, an area that includes the towns of Burton upon Trent and Ashby de la Zouch. The original 6 per cent wooded cover in the area has been increased three-fold by the planting of over 7 million trees.

This is a *multi-purpose* forest for the nation, with ecological, commercial and recreational benefits.

- Conservation and community involvement – preservation of ancient woodlands with their unique habitats for woodland birds; involvement of the local community through educational and recreational visits, and by sponsored tree planting from individuals and companies.
- Environmental improvement of the old mining and clay working sites; reclamation and landscaping.
- Commercial forestry – old fully mature trees selectively felled from the ancient woodlands; younger trees also felled and saplings cut where their density is too high (both part of good forest management); a woodland economy is created, leading to local employment in forestry and a forest income; helps satisfy the plentiful demand for timber (the UK imports 85 per cent of the timber it needs).
- Recreational needs of walkers, cyclists and horse riders are met by a network of country lanes, walks and cycle trails; tourism supports over 4000 jobs; visitor numbers have reached almost 8 million per year; Conkers is the family visitor attraction near Ashby de la Zouch and there are craft workshops nearby; 90 per cent of woodland areas have public access.

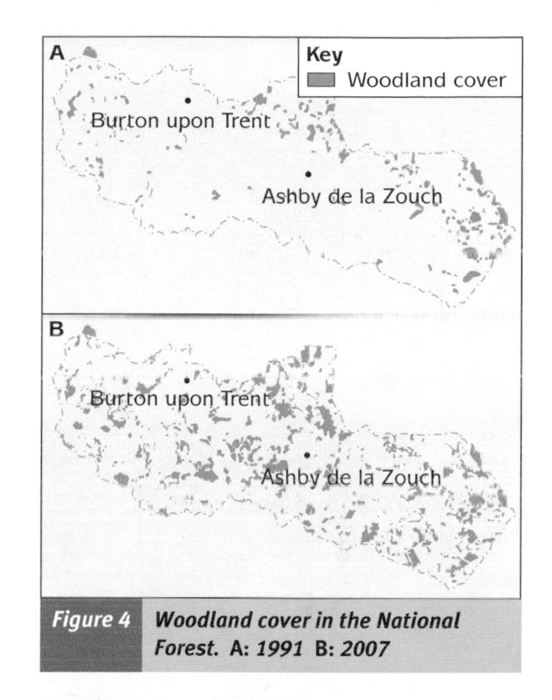

Figure 4 ***Woodland cover in the National Forest. A: 1991 B: 2007***

The overall aim is to increase woodland cover to about one-third of the area, with 30 million trees planted, most often deciduous. The emphasis is upon *sustainability* – in landscape change, wildlife habitats, local employment, economic activities and recreational use.

FURTHER RESEARCH

Read more about the National Forest on the weblink www.contentextra.com/aqagcsegeog.

GradeStudio

1 Study **Figures 4A** and **B**.
  a Describe the distribution of woodland in the National Forest in 1991. (2 marks)
  b Describe how the distribution and extent are different in 2007. (3 marks)

2 a State one benefit of the National Forest under each heading (i) Environmental (ii) Economic (iii) Social. (3 marks)
  b Explain how the National Forest provides both economic and environmental benefits at the same time. (4 marks)

Exam tip

Soil profiles – user guide for what to look for

- humus layer at the top
- partly weathered rock at bottom
- changes from A to B horizons

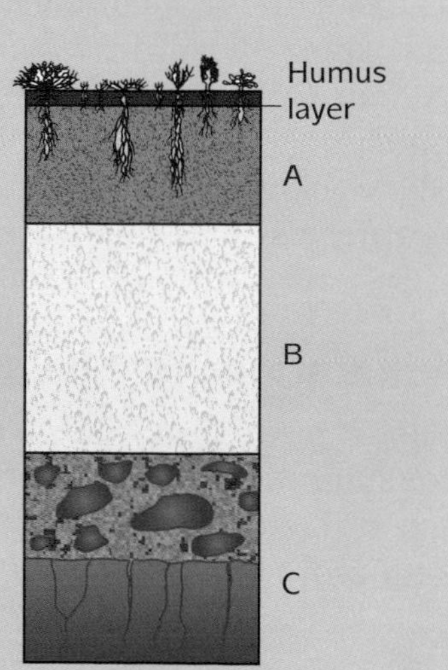

Tropical rainforests

Which features make tropical rainforests unique among ecosystems? Why are most tropical soils red?

Figure 1 ***Tropical rainforest – forest structure.***

Characteristic features

The vegetation mass is the greatest of all ecosystems. Despite the sheer amount of vegetation present, and the way in which climbing plants and creepers, known as **lianas**, run from tree to tree in a chaotic manner, it is possible to recognise five distinctive **forest layers** (**Figure 1**). The **canopy** provides a habitat for monkeys and numerous birds such as macaws; on the ground are some larger animals such as jaguars, tapirs and anteaters in the Amazon Basin.

The tall trees are deciduous, but they shed their leaves at different times and for only six to eight weeks each year, so that the forests always look green. The **emergents** (tallest trees), by reaching up to 50 metres high, stand head and shoulders above the forest canopy; these are hardwoods and include types such as mahogany and ironwood. They have long trunks without any branches until their rounded crowns extend out over the canopy. Their leaves are oval in shape with extended points known as **drip tips**, and they have dark green and leathery upper surfaces. The smooth bark is thin. Their shallow roots, which mainly extend sideways below the ground surface, extend above the ground as **buttress roots**. **Epiphytes**, parasitic plants growing on trees and tree branches, increase the abundance of vegetation.

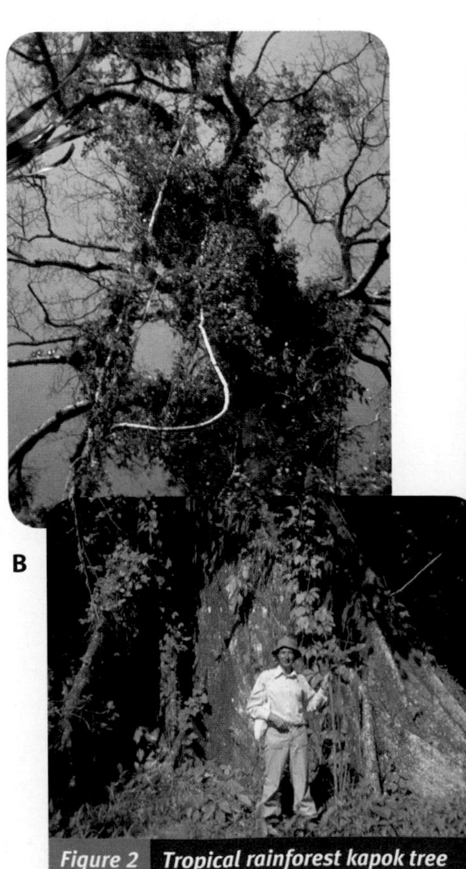

Figure 2 ***Tropical rainforest kapok tree A: Above the canopy with many epiphytes and lianas. B: Forest floor showing buttress roots.***

Adaptations to climate

The tropical rainforest's biodiversity is a response to climate. There are constant high temperatures, with a mean monthly average above 27 °C, accompanied by high solar light intensity. Rainfall is regular and high, with more than 2000 millimetres falling during the year, which creates humid conditions. There is no more favourable climate on Earth than this for plant growth.

Plant communities are fiercely competitive. There is 'survival of the tallest' as the tall trees are drawn upwards by the heat and light, which is why leaf growth is concentrated in the canopy. The leathery upper surfaces of the trees' leaves are necessary to withstand the great power of the Sun's rays. The drip tips help the leaves to shed water during the heavy rains.

In the lower layers of the forest, sunlight is in short supply. Ferns are adapted to life on the forest floor by having leaves that intercept a high proportion of the light that reaches them. The shrub layer is sparse because of lack of light, although shrubs quickly take advantage of any gap in the forest canopy.

Soils

Figure 3 shows a latosol, which is the name given to soils that form under tropical rainforest. They are red or yellowish-red in colour throughout. They are very deep soils, often 20–30 metres deep, compared with the 2–3 metres for brown earths (page 68). The black humus layer at the top is a narrow horizon of organic material. The red and yellow colours below it come from the oxides of iron and aluminium, which remain in the soil after other minerals have been washed out by leaching.

Figure 3 A latosol.

Figure 4 Some species of flowering plants even grow 40 metres above the ground in the crowns of emergent trees.

Looking at the density and diversity of the vegetation cover, you might think that it is growing from the world's most fertile soil. Nothing could be further from the truth. Rainfall is much higher than evapotranspiration; even with the protection of the forest canopy, there is the downward movement of rainwater through the soil. Leaching washes organic material and silica downwards and then out of the soil. The only fertile part of a latosol is the narrow organic layer, which is why the trees have shallow roots.

INFORMATION

Biodiversity in Ecuador

A typical 5-hectare patch of rainforest contains:

- 750 species of trees
- 1500 species of flowering plants
- 400 species of birds
- 150 species of butterflies
- 100 species of reptiles

The importance of tropical rainforests

It is their **biodiversity** that makes tropical rainforests unique among ecosystems. This term refers to the great number and variety of living species, plants and animals; it is estimated that 50 per cent of the world's 10 million species live in tropical rainforests (**Figure 4**). The Earth's genes, species and ecosystems have evolved through 3000 million years and form the basis for human survival. Wild varieties of plants are the basis for new seeds for farmers to use; many of today's drugs, such as aspirin, are derived from plants. When new combinations of genes are sought in the future for new food crops or cures for diseases, without the rainforests and their biodiversity, the number of potentially useful species from which to choose will be reduced. Some environmentalists emphasise the importance of tropical rainforests as suppliers of oxygen (the 'lungs of the world') and carbon dioxide stores (against the enhanced 'greenhouse effect').

ACTIVITIES

1 (a) Draw a labelled diagram to show the main characteristics of tropical rainforest vegetation.
(b) Explain the great importance of climate in determining these characteristics.

2 (a) Looking at **Figure 3**, draw a sketch of a latosol and label three main characteristics.
(b) Are latosols fertile soils? Explain your answer.

Tropical rainforests – sustainable uses and management

Which groups make a living from the rainforests? Which groups want to destroy the forests? How can the rainforests be conserved?

While natural deciduous forests have been cleared almost to extinction in Europe and North America, large areas of tropical rainforest remain in other parts of the world, mainly because of obstacles such as difficult access and infertile soils. Rainforests are mainly located in the world's poorer, less developed countries (**Figure 1**, page 64). However, many of these forests are now under threat as governments view their untapped resources as a passport to economic development and a way to reduce their international debts.

Traditional uses

Human settlement in rainforests is long established, but often of low density. Indigenous (native) peoples, living in tribes or groups, either collected and hunted to use the food naturally available in the forest, or practised 'slash and burn' and grew crops such as manioc, or did both. With slash and burn, the clearing was small and the group cultivated only for as long as the soil retained its fertility (probably two or three years). They then moved to another part of the forest, hence the alternative label of 'shifting cultivation'. Indigenous tribes, low both in numbers and levels of technology, barely left a mark on the forest, which re-invaded within a few years as if there had been no human presence. This is an example of **sustainable** human use of rainforests.

Uses after forest clearance

Indigenous peoples are pushed back into smaller and smaller areas of forest by the advances of loggers, miners and farmers, each group supported by superior modern technology. Direct contact results in the spread of diseases to which they have no resistance, and in the destruction of their traditional culture and ways of life. Unlike them, outsiders engage in activities that involve forest destruction.

The first forest clearances are usually associated with road building. Roads attract farmers, loggers and miners, enabling them to open up wider areas of forest away from the roadsides. Farming by outsiders is more likely to be cattle ranching (**Figure 2**). The ranchers are interested in only one thing – replacing forest with pastures. Often, they do not even save and sell the valuable hardwood timber and it is just burned (**Figure 3**).

Logging companies are only interested in certain types of tree, but one of the characteristics of rainforests is that individual species of tree are widely dispersed. In order to reach the trees they want, all the other trees are felled and cleared. Oil, gas, iron ore, bauxite (for aluminium), nickel and gold are just a few of the natural resources that have attracted mining companies to rainforest regions. Despite all the problems with access, once a mineral deposit of commercial size is discovered, then roads, railways and pipelines are built with little or no thought for the forests or their inhabitants. In remote locations, without government supervision or environmental controls, mining operations, disposal of untreated waste, leaks and spillages cause land and water pollution.

Figure 1 **Banana plants being grown in an area cleared by indigenous people. How great is the risk to the environment?**

Figure 2 **Farming on land previously covered by rainforest in Costa Rica (Central America). How great is the risk to the environment?**

Figure 3 **Timber from rainforest clearances piled up and burned in Brazil. How great is the risk to the environment?**

Sustainable uses and management methods

Human activities are sustainable when they have a long future because people are working with nature and the environment upon which they depend. An estimate made of net potential value of a Rainforest National Park in Africa showed that the long-term value of benefits from reusable forest products, tourism and absence of environmental damage (soil protection and flood control estimated at US$ 90m) dwarfed the more short-term gains from forest clearance and farming (US$ 30m). Selling the valuable hardwood timber only provided one-off income; returns from farming on the cleared land reduce with time, as the soil loses its fertility. This calculation might underestimate gains from conservation. It was made before the idea of 'debt for nature' swaps was first mooted; poor countries can trade conservation of rainforests against their financial debts to rich countries. Also, rainforests now have money value as carbon dioxide stores: rich countries buy carbon credits to offset their own carbon dioxide emissions. In other words, rich countries pay poor tropical countries to keep their untouched rainforests.

Logging does not have to be a destructive activity. Malaysia is a rare example of a tropical country with a Forest Management Plan for its rainforests. Its aim is to achieve sustainable forestry instead of predatory logging by:

A Dividing the forests into two *groups*:

- Protection and conservation forests: these include National Parks and wildlife and bird sanctuaries.
- Production forests: in these forests logging takes place but it is carefully planned and controlled.

B Making a *survey* of the area to be logged and its resources.

C Using *selective logging*. Only between seven and twelve mature and fully grown trees per hectare are cut down in each logging cycle. This allows the logged area to regain full maturity after 30–50 years. The forest recovers because the younger trees and saplings are given more space and sunlight to grow.

D *Monitoring* what happens. At all stages it is necessary to check that the work being done conforms to the plan. Selective logging is more expensive than total logging. Where large gaps already exist due to over-logging, the only management option may be reforestation using plantation types of trees, such as acacia and eucalyptus – poor substitutes for the original forest.

Education and training are needed to make local people, company bosses and politicians more aware of how forests work and to increase their appreciation of potential uses. Protecting rainforests by creating National Parks or nature reserves attracts visitors from overseas and the home country. After experiencing the wonders of nature, both groups are likely to be more conservation minded. The push is towards **ecotourism**. Many of the rainforest lodges in tropical countries that receive lots of tourists, such as Costa Rica and Ecuador, are of this type (**Figure 4**). Lodge visitors spend their time taking guided treks through the forest and riverboat trips to view colourful birds and hundreds of different types of butterflies. The lodge provides employment for local Indians as guides and boatmen; it also buys surplus produce from them.

Figure 4 ***Jungle lodge in Ecuador in the Amazon Basin.***

EXAM PREPARATION

Ecotourism – tourism that
- is environmentally sound
- safeguards wildlife and habitats
- protects natural resources in a sustainable manner
- causes no damage to local communities.

ACTIVITIES

1 Study **Figures 1** and **2.** For each one, **(a)** describe the land uses shown **(b)** explain whether the risk to the environment is high, medium or low.

2 Methods for sustainable management of tropical rainforests include selective logging, forest protection, ecotourism, reducing debt, replanting, education, and reducing demand for tropical hardwoods.

(a) Briefly describe how each method of management for conserving the rainforests works.

(b) Choose two of them. For each method, **(i)** explain more fully how it works **(ii)** comment on its strengths and weaknesses as a method of forest management.

3 Explain why international cooperation is needed to protect the tropical rainforests.

Tropical rainforest in the Amazon Basin of Brazil

The country with the largest area of untouched rainforest is Brazil (**Figure 1**, page 64), the world's fifth largest country by area. Tropical rainforest is the natural vegetation cover in 60 per cent of Brazil; despite clearances since 1960, rainforest still covers almost half its total area. Even in areas close to Manaus, the largest town in the Amazon Basin with over 2 million people, large areas of untouched forest remain (**Figure 1**).

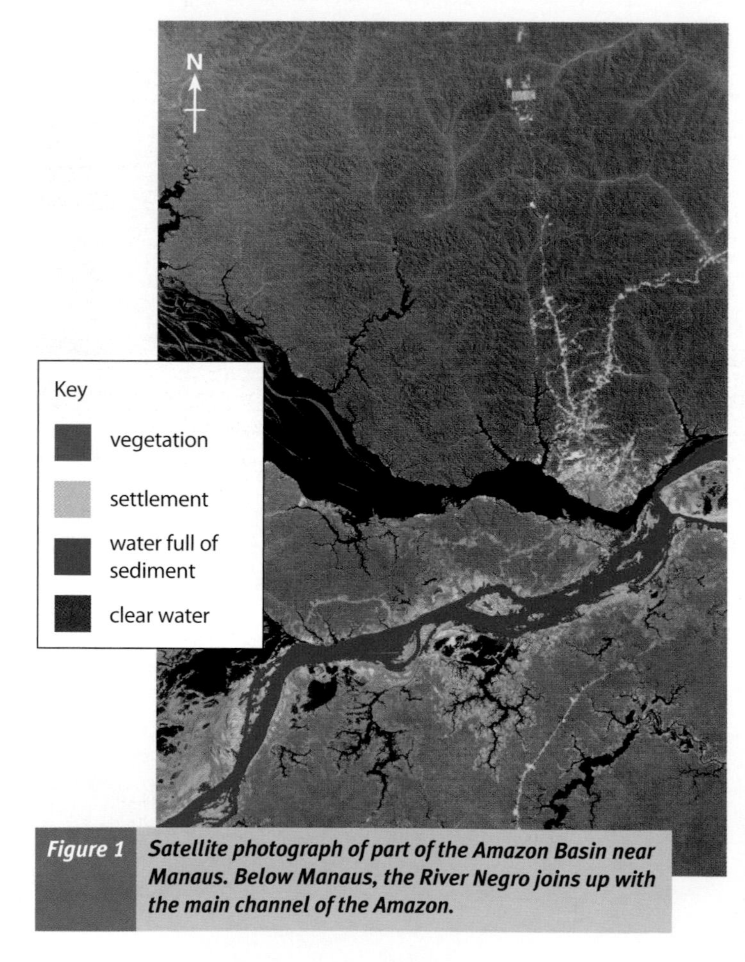

Figure 1 *Satellite photograph of part of the Amazon Basin near Manaus. Below Manaus, the River Negro joins up with the main channel of the Amazon.*

Forest destruction and the reasons for it

What alarms environmentalists and others living outside Brazil is the persistent and continuous forest destruction, which in some years has seen an area the size of Wales (20 754 square kilometres) disappear (**Figure 2**). Although there is still a lot of rainforest left in Brazil, environmental groups are dismayed by the figures, which generally show an increase in the rate of clearance since the low point in 1997. They argue that it cannot keep on increasing for ever. However, the figures suggest that powerful economic, social and political reasons lie behind the forest losses. These are summarised in the Exam Preparation Box.

Year	km²
1996	30 000
1997	12 600
1998	16 200
1999	16 700
2000	19 200
2001	17 600
2002	24 600
2003	23 400
2004	25 200
2005	17 800
2006	11 200
2007	16 800

Figure 2 *Estimated rainforest losses in the Amazon Basin of Brazil 1996–2007.*

EXAM PREPARATION

Development of the Amazon Basin

The following are some reasons why Brazilian governments have encouraged and supported development in the Amazon region since the 1960s.

1 *Economic*
- to export minerals, gain foreign exchange and pay off international debts
- to use the minerals and other raw materials in growing industries
- to extend areas of agriculture and export beef and soybeans; soybeans were Brazil's leading export crop in 2004 (up 50 per cent from 2001)
- to become a more economically developed country.

2 *Social*
- large total population (180 million and growing)
- to relieve population pressure along the coast
- to give landless peasants the chance to own land for the first time.

3 *Political*
- since 1960 it has been government policy to open up the interior ('march to the west')
- to take people's minds off problems such as poverty and landlessness.

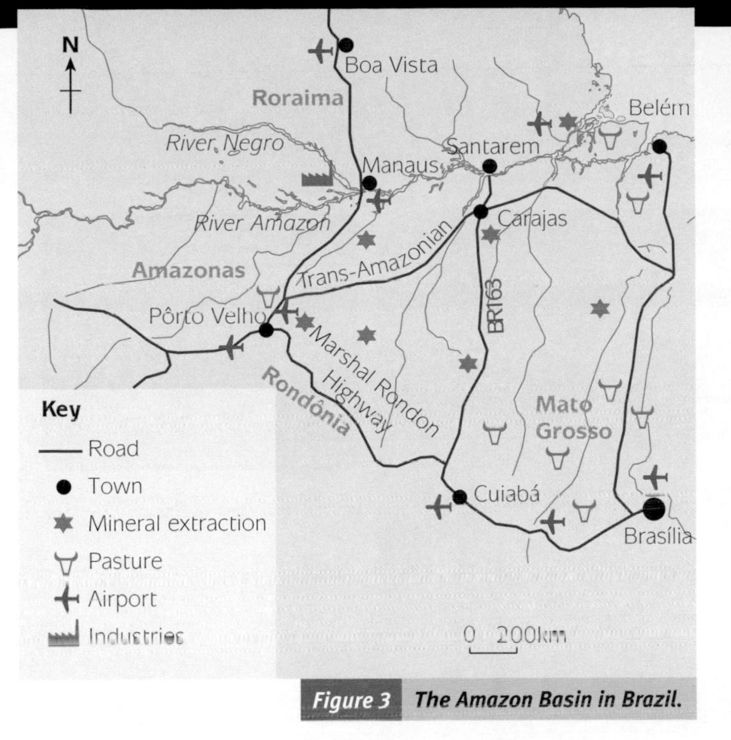
N
Boa Vista
Roraima
Belém
River Negro
Santarem
Manaus
Carajas
River Amazon
Amazonas
Trans-Amazonian
BR163
Pôrto Velho
Marshal Rondon Highway
Rondônia
Mato Grosso
Cuiabá
Brasília
Key
Road
Town
Mineral extraction
Pasture
Airport
Industries
0 200km

Figure 3 *The Amazon Basin in Brazil.*

The roads shown in **Figure 3** provide the framework for access into the Amazon rainforest. They are in the process of being improved and extended, although not without controversy (**Figure 4**). Brazil's big share of the Amazon Basin is rich in natural resources. The largest concentration of mineral resources is at Carajas, where there are major deposits of iron ore, bauxite, nickel, copper and manganese. Most of the cleared land is used for farming. Cattle ranching is everywhere on the cleared land. Much of the recent advance of agriculture westwards and northwards from Brasilia, has been by farmers growing soybeans. This crop offers farmers large profits. Most of the crop is exported to Europe, where soybeans are in high demand as animal feed. Good market prices for soybeans and for raw materials in general are boosting Brazil's export earnings, allowing the government to pay off some of the country's massive international debts.

The government of Brazil has designed sensible policies to combat deforestation including:

- increasing protected areas
- hiring more forest patrols
- supporting forest-based activities.

However, in practice their effectiveness is reduced by shortage of resources, corruption, red tape and opposition from powerful business interests. The indigenous tribes of Brazil are gradually being wiped out. They are increasingly confined to reserves, such as the Xingu reserve. Even here, their culture and way of life are not fully protected, as their territory is invaded illegally by miners and loggers.

Under pressure from the agri-business lobby, paving BR163 was planned to begin in 2005.

- Roads open up areas over which the government, police and army have little control.
- Land is invaded on either side to clear trees and raise cattle.
- Loggers build unofficial networks of roads through the forest from the main roads.
- The prospects of paving a road lead to land rushes by soybean farmers.
- More and more of the natural forest will be cleared and lost.

Figure 4 ***Arguments against building and paving roads in the Amazon Basin.***

GradeStudio

1 a Use the values from **Figure 2** to draw a bar graph of forest losses in Brazil. (3 marks)

b Describe the pattern of forest losses from 1996 to 2007. (3 marks)

2 Refer to **Figure 5**.

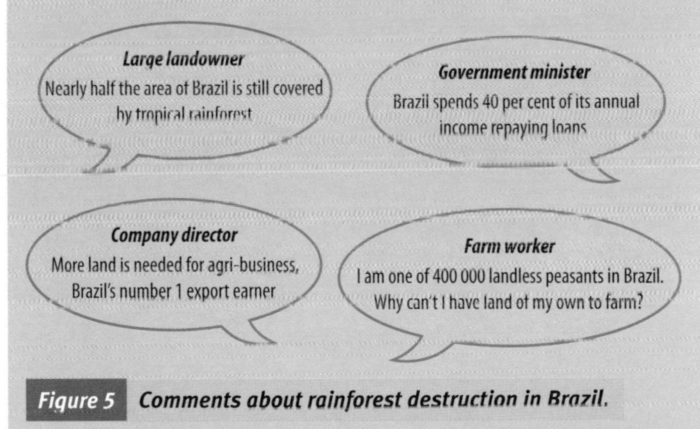
Large landowner
Nearly half the area of Brazil is still covered by tropical rainforest
Government minister
Brazil spends 40 per cent of its annual income repaying loans
Company director
More land is needed for agri-business, Brazil's number 1 export earner
Farm worker
I am one of 400 000 landless peasants in Brazil. Why can't I have land of my own to farm?

Figure 5 ***Comments about rainforest destruction in Brazil.***

a Identify one comment that is social and another one that is economic. (2 marks)

b Choose two of the four comments and explain their importance for rainforest destruction in Brazil. (4 marks)

c Name one group of people likely to be against paving the BR163. Explain their objections. (3 marks)

Exam tip

Make sure that you understand the differences between economic, social, political and environmental factors.

Practice GCSE Question

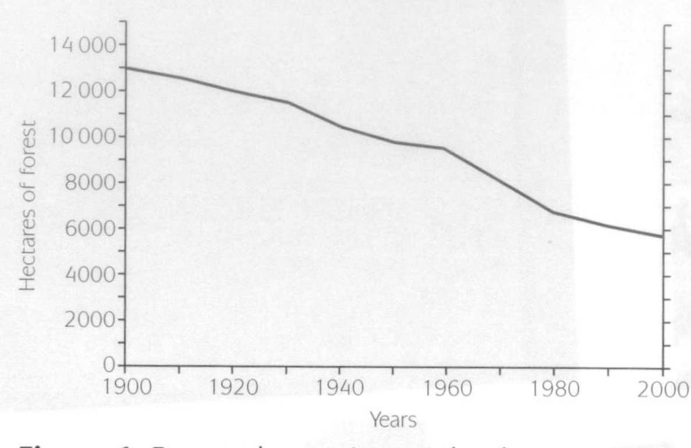

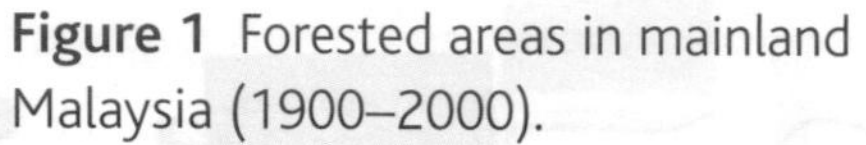

Figure 1 Forested areas in mainland Malaysia (1900–2000).

Figure 2 Clearing in a Malaysian rainforest.

See a Foundation Tier Practice GCSE Question on the weblink www.contentextra.com/aqagcsegeog.

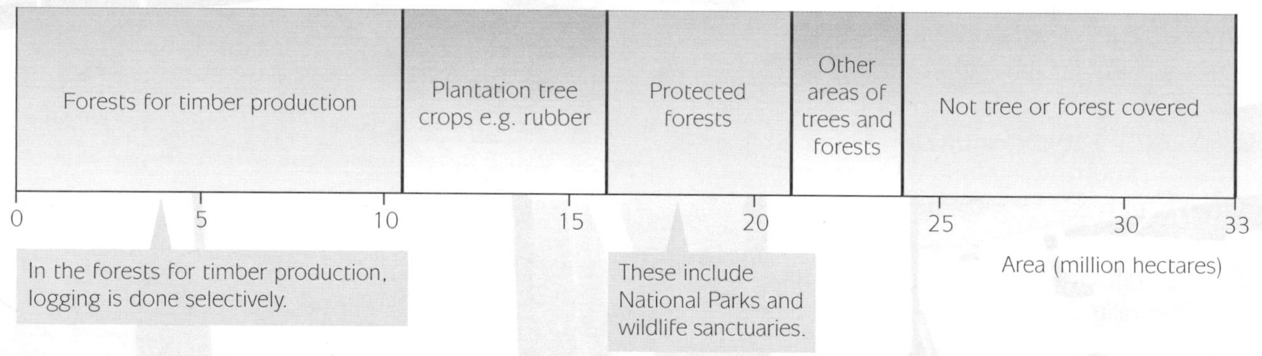

Figure 3 Land uses in Malaysia 2004.

4 (a) **Figure 1** shows changes in the forested area of mainland Malaysia, where most Malaysians live. Most of the natural forest was tropical rainforest.

(i) Describe the changes shown in **Figure 1**. **(3 marks)**

(ii) State the economic reasons why tropical rainforests are being cleared in many countries. **(4 marks)**

(b) (i) Study **Figure 2** and describe what it shows. **(2 marks)**

(ii) Explain why this happens once rainforests are cleared. **(4 marks)**

(c) **Figure 3** shows land uses in Malaysia in 2004.

(i) What percentage of the total area of Malaysia has some kind of forest and tree cover on it? Show your working. **(2 marks)**

(ii) Explain why many people believe that plantation tree crop farming is more sustainable in rainforest regions than other types of farming such as cattle ranching. **(2 marks)**

(d) Sustainable logging and conservation are possible in tropical rainforests with good management; however, predatory uses of tropical rainforest are more common. Explain this statement. **(8 marks)**

Total: 25 marks

Exam tip

The question for this topic will be Question 4 in Paper 1.

Improve your GCSE answers

Know and understand physical and human impacts of deforestation – sustainable management strategies for ecosystems

A Impacts of forest destruction

Rearrange the following impacts, dividing them between **physical (environmental)** and **human**, as well as the different elements in the ecosystem (see below). There is a minimum of three for each heading, with most for soils.

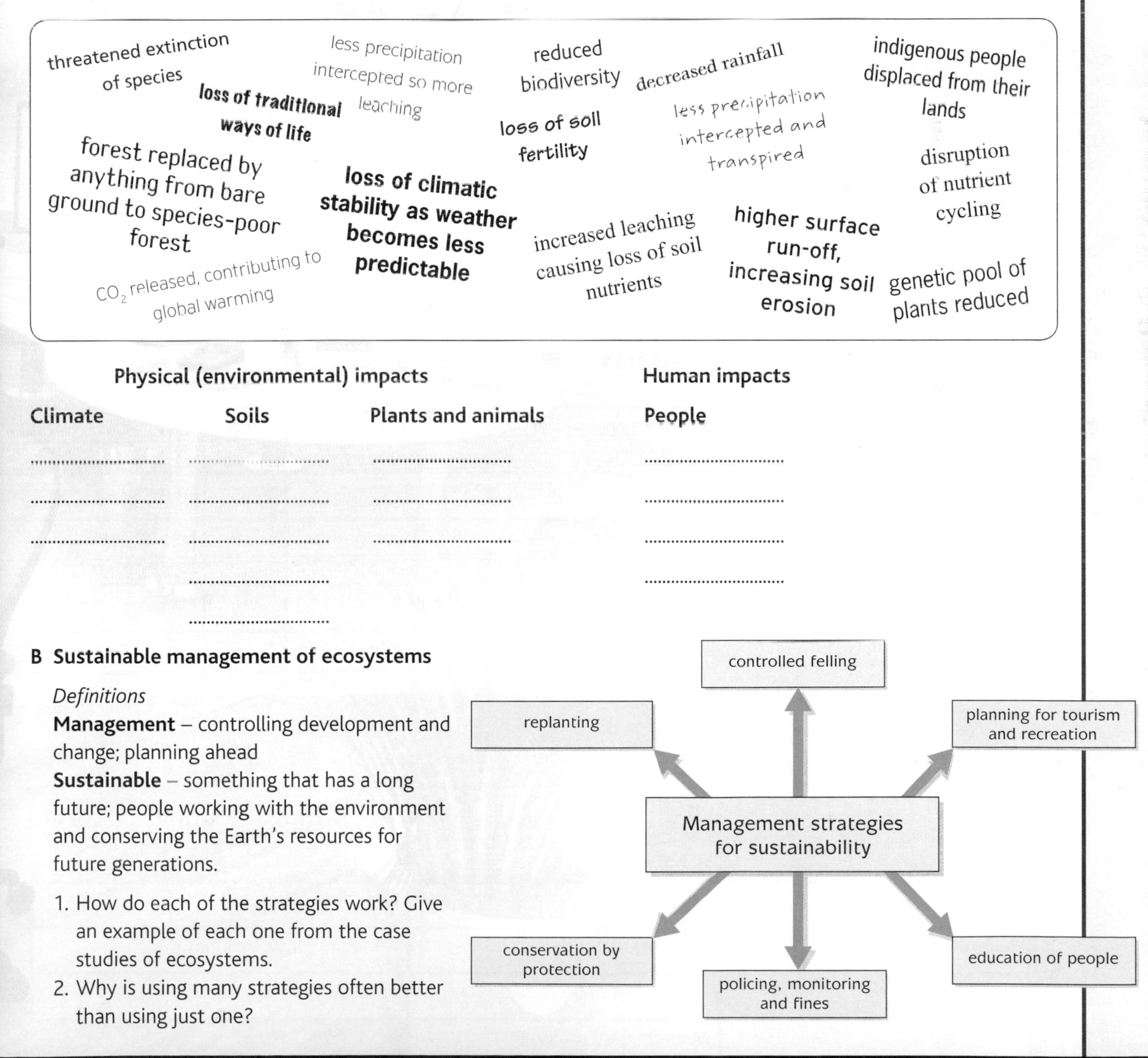

threatened extinction of species

less precipitation intercepted so more leaching

reduced biodiversity

decreased rainfall

indigenous people displaced from their lands

loss of traditional ways of life

loss of soil fertility

less precipitation intercepted and transpired

forest replaced by anything from bare ground to species-poor forest

loss of climatic stability as weather becomes less predictable

disruption of nutrient cycling

increased leaching causing loss of soil nutrients

higher surface run-off, increasing soil erosion

CO_2 released, contributing to global warming

genetic pool of plants reduced

Physical (environmental) impacts | **Human impacts**

Climate	Soils	Plants and animals	People
....	....	....	....
....	....	....	....
....	....	....	....
	....		....
	....		

B Sustainable management of ecosystems

Definitions

Management – controlling development and change; planning ahead

Sustainable – something that has a long future; people working with the environment and conserving the Earth's resources for future generations.

1. How do each of the strategies work? Give an example of each one from the case studies of ecosystems.
2. Why is using many strategies often better than using just one?

ExamCafé

REVISION

Key terms from the specification

Exam Café

Biodiversity – level of plant and animal variety in an ecosystem
Brown earth – uniform brown-coloured soil that forms under deciduous woodland
Ecosystem – system in which living things (plants and animals) and physical factors (climate and soils) are linked
Food chain/web – nutrients and energy absorbed by plants are passed along a line of living things
Latosol – deep soil, red or yellow in colour, which forms under tropical rainforest
Leaching – downward movement of minerals through soil
Nutrient cycling – dead remains of plants and animals are decomposed and used again
Salinisation – increasing concentrations of salt in the topsoil where evaporation rates are high
Soil erosion – loss of fertile top soil by action of wind and water
Succulent – plant that stores water in a fleshy stem to survive drought
Xerophytic – adaptations in plants that allow them to survive in a dry climate

Checklist

	Yes	If no – refer to
Do you know what an ecosystem is?		page 62
Can you name examples of large areas in the world covered by tropical rainforest, hot desert and deciduous forest ecosystems?		page 64
Do you know the different adaptations of hot desert vegetation for survival in the climate?		page 65
Can you describe the characteristics of the vegetation in deciduous forests and explain how it is adapted to climate and soils?		page 68
Can you describe the forest structure in a tropical rainforest?		page 70
Do you know the reasons why biodiversity in tropical rainforests is much greater than in other ecosystems?		pages 70–1
Can you state differences between desert soils, brown earths and latosols?		pages 65, 68, 71
How many methods of management for conservation and future sustainability of natural ecosystems can you name?		pages 69, 73, 77

Case study summaries

Hot desert areas (rich and poor – opportunities for economic development)	**Temperate deciduous woodland** (uses and management)	**Tropical rainforest** (destruction and impacts)
Location	Location	Location
Activities	Uses of the timber	Reasons for destruction
Challenges faced	Other uses of the forest	Impacts from destruction
Management	Management methods	

Chapter 5
Water on the Land

The Iguaçu Falls on the border between Brazil and Argentina are a stunning example of the erosive power of rivers. They make humans look small.

QUESTIONS

- **Rivers erode, transport and deposit – how and where?**
- **How do rivers and river valleys change their characteristics travel from mountainous terrain to sea level?**
- **Why do rivers flood and can they be controlled?**
- **Why are surface water supplies more abundant in some parts of the UK than in others?**

River basins and discharge

Where do rivers begin? In what ways do river channels and valleys change downstream? How is river discharge affected by drainage basin characteristics?

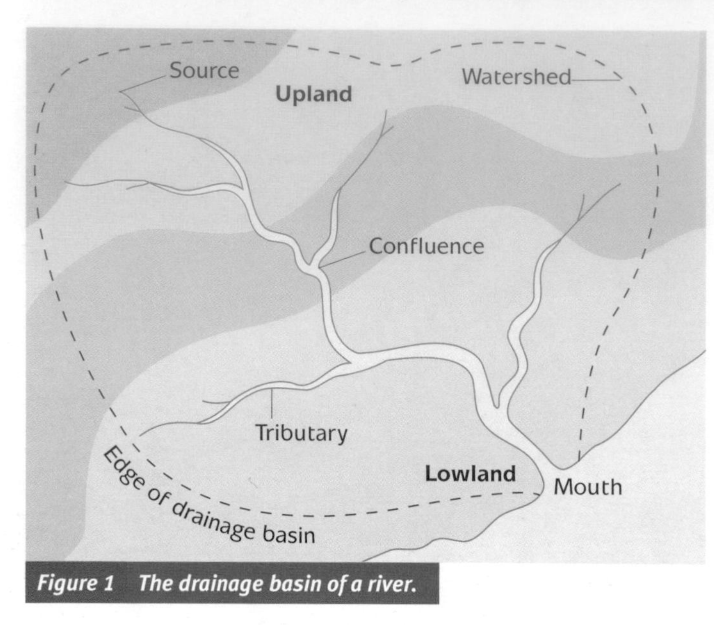

Figure 1 The drainage basin of a river.

Traction – large boulders roll along the river bed

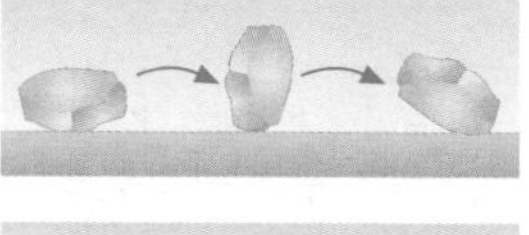

Saltation – smaller pebbles are bounced along the river bed, picked up and then dropped as the flow of the river changes

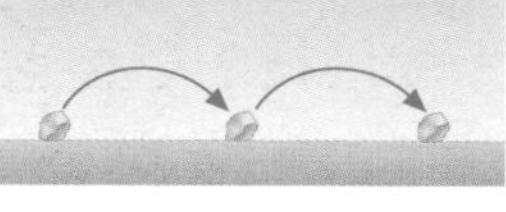

Suspension – the finer sand and silt-sized particles are carried along in the flow, giving the river a brown appearance

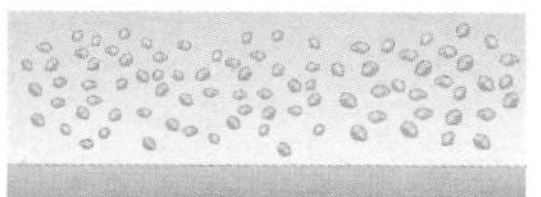

Solution – minerals, such as limestone and chalk, are dissolved in the water and carried along in the flow, although they cannot be seen

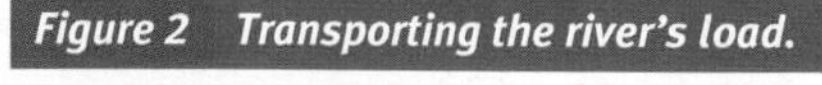

Figure 2 Transporting the river's load.

Rivers begin in upland areas and flow downhill, becoming wider and deeper, until they enter the sea. Where a river begins is called the **source** and where it ends is the **mouth**. Along a river's journey to the sea, other smaller rivers called **tributaries** may join the main river at a **confluence**. A river and its tributaries obtain their water from the surrounding land. The area drained by a river and its tributaries is called the **drainage basin** (**Figure 1**). The boundary of the drainage basin is called the **watershed** and it is usually a ridge of high land.

Some of the river's energy is used in transporting loose material downstream. The material is transported in one of four ways (**Figure 2**). The amount of **load** being carried depends on:

- volume of water – the greater the volume, the more load it can carry
- velocity – a fast-flowing river has more energy to transport and can move larger particles
- local rock types – some rocks, e.g. shales, are more easily eroded than others, e.g. granite.

Figure 3 shows some of the changes that take place downstream in a river valley. The water in a river flows within a **channel** unless the river floods and spills out onto the surrounding land. The size and shape of the channel changes as the river flows downstream, becoming wider and deeper. A river also flows within a valley; the size and shape of the valley changes downstream from a steep **V-shaped valley** to a broad, almost flat V-shaped valley.

Many of these changes are caused by changes in the river's energy. In the uplands, close to the source, the river is high above its base level (usually sea level). This gives the river a lot of potential energy. The river is also trying to reach its base level, so the main processes at work are erosional. The river mainly erodes in a downwards direction (**vertical erosion**) to try to reach its base level. This helps to create V-shaped river valleys in upland areas. As the river moves downstream, it uses a lot of energy to transport the material or load it has eroded. Surplus energy is now used to erode sideways (**lateral erosion**) because the river is much closer to its base level, and so the river valley becomes wider and flatter.

The **long profile** of a river shows an irregular steep gradient at the source, gradually becoming lower and less steep until the gradient is almost nil when the river gets near sea level (**Figure 3**). The normal profile is smooth and concave. The changes in the river valley and features along the profile allow the river to be subdivided into three sections – the upper, middle and lower courses.

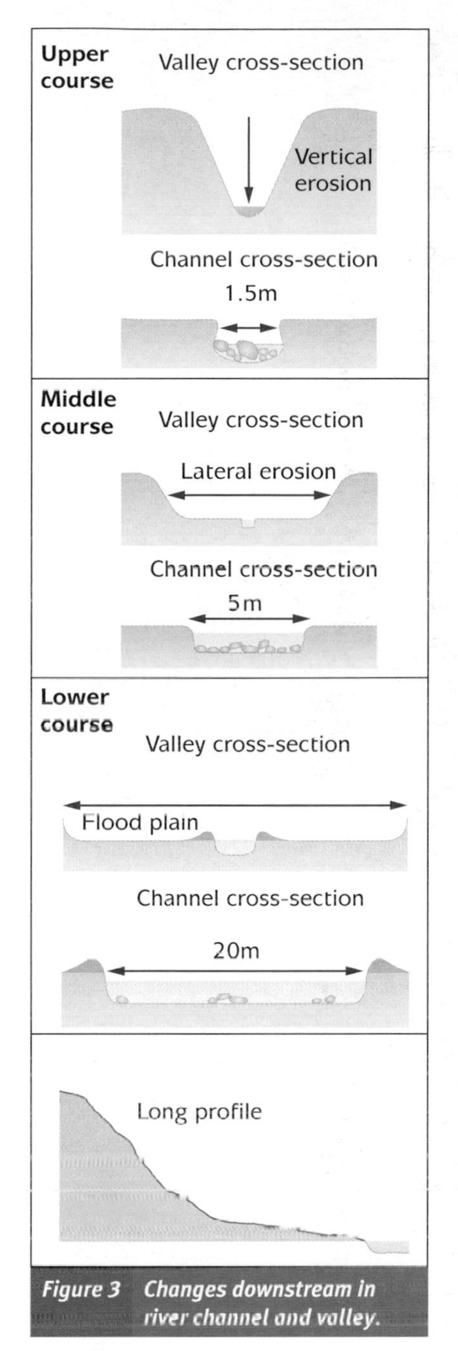

Figure 3 *Changes downstream in river channel and valley.*

Discharge

Discharge is the volume or flow of water passing a river measuring station at a particular time. The amount of water in the river fluctuates greatly. The factors causing the day-to-day and seasonal variations are nearly all physical; only land use is under the control (at least partly) of humans.

Ideal conditions for high river discharge after a rainstorm are illustrated in **Figure 5**. All favour rapid surface movement of water into the river:

- High land means heavier rainfall and lower temperatures with less evaporation.
- Steep slopes cause fast surface run-off.
- Granite (page 30) is an impermeable rock (it does not allow water to pass through it).
- Bare rock outcrops and lack of trees.
- Built up area with many hard surfaces.

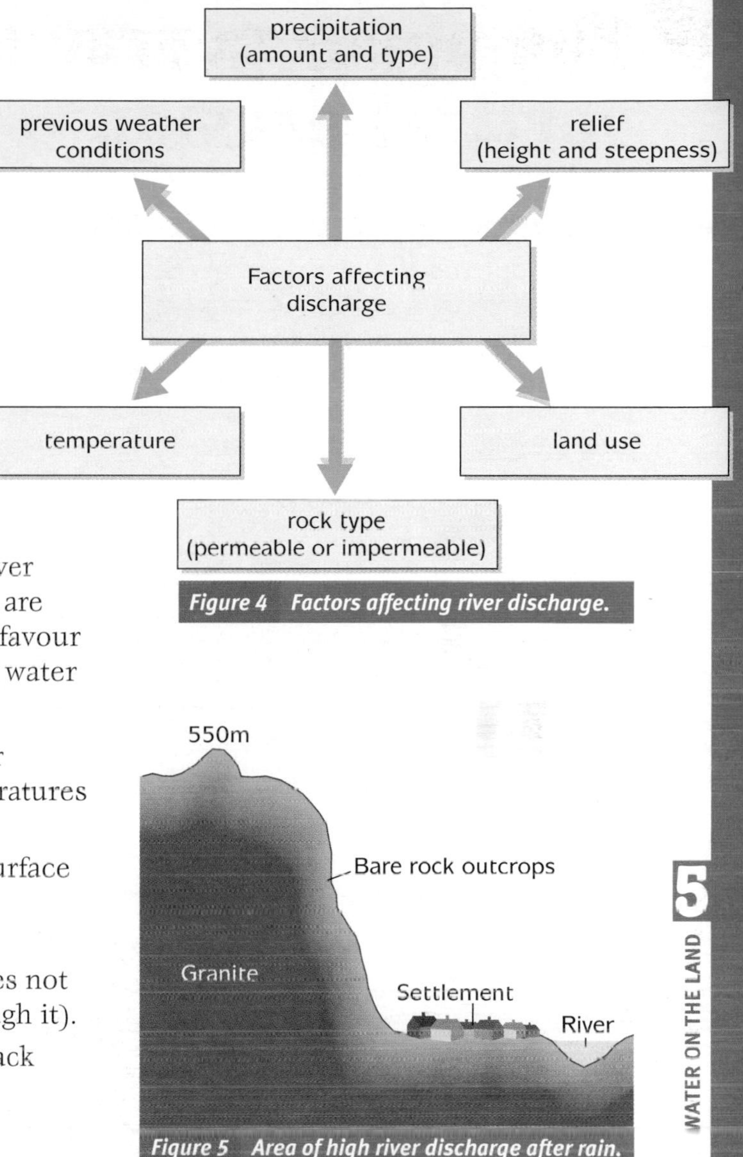

Figure 4 *Factors affecting river discharge.*

Figure 5 *Area of high river discharge after rain.*

ACTIVITIES

1. Draw a side view of a flowing river. Show the different ways in which boulders, pebbles, sand and silt are transported.
2. Fill in a copy of the table on the right using sketches and information from these two pages.
3. Draw a labelled diagram to show the opposite of **Figure 5**, an area of low river discharge after rain.

	Upper course	Middle course	Lower course
Long profile			
Valley cross-section			
Channel cross-section			

The upper course of a river – vertical erosion

How do rivers erode the land? Why do rivers form waterfalls and gorges in upland areas?

In the upper course of a river, in the mountains and hills, vertical erosion is the dominant process. As they are flowing high above sea level, rivers concentrate on cutting downwards. The four processes by which rivers erode and wear away the land surface are shown in **Figure 1**.

Hydraulic action

This is the force of the water on the bed and banks of the river. It is particularly powerful when the river is in flood. The force of the water removes material from the bed and banks of the river.

Abrasion

The river carries with it particles of sand and silt and moves pebbles and boulders at times of high flow. This material rubs against the bed and banks of the river and wears them away.

Solution

Some rock minerals, such as calcium carbonate in limestone and chalk, slowly dissolve in river water, which is sometimes slightly acid.

Attrition

The load being carried by the river collides and rubs against itself, breaking up into smaller and smaller pieces. The rough edges also become smooth, forming smaller, rounded material. Eventually the particles are reduced to sand and silt-sized particles.

Figure 1 Processes of erosion.

Landforms in the upper course

V-shaped valleys and interlocking spurs

The vertical erosion in the upper course creates a **V-shaped valley** (**Figure 2**), which is steep-sided and narrow. As the river erodes downwards, soil and loose rock on the valley sides are moved downhill by slopewash or soil creep. The river also winds its way around **interlocking spurs** of hard rock (**Figure 2**), which should not be confused with meanders! There is no flat valley floor and the valley gradient is steep.

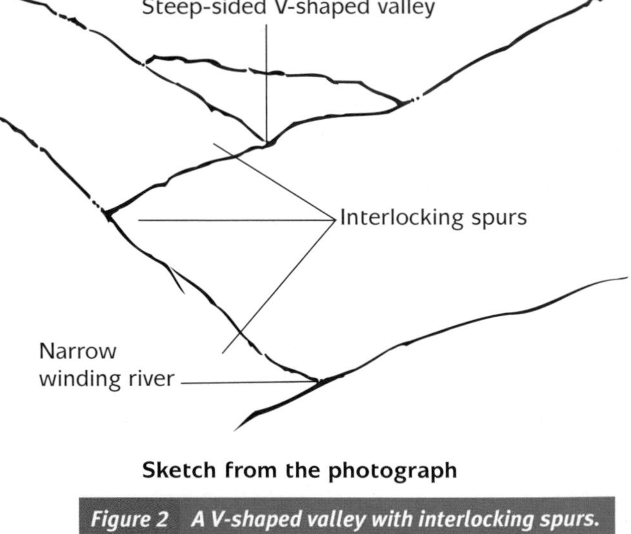

Sketch from the photograph

Figure 2 A V-shaped valley with interlocking spurs.

The river channel

The river channel is narrow and shallow; it is often lined with large angular boulders. The gradient of the river may be quite steep, and waterfalls and rapids may be found along the river. The velocity of the river is high at waterfalls and rapids but may be quite low in other stretches because so much energy is used in overcoming friction with the rocky bed and banks of the river. The water is often quite clear because the river is not carrying much load in suspension. The river has not had time to grind down the boulders into fine sand and silt-sized particles by abrasion and attrition.

Waterfalls and gorges

A waterfall is a steep drop in the course of a river. It has a high head of water and a characteristic plunge pool at the base. The rocks at the top of the waterfall are often hard and resistant, forming a cap rock, and softer rocks below are undercut (**Figure 3**). The waterfall may lie within a gorge.

Waterfalls often form when a band of resistant rock lies over softer, less resistant rocks. The softer rock is eroded more quickly, causing undercutting of the hard rock. The hard rock overhangs until it can no longer support its weight. The overhang then collapses, adding large blocks of rock to the base of the waterfall. The great power of the water falling to the base moves the material around, eroding the base into a deep plunge pool. The bed of the river below a waterfall contains boulders eroded by splashback from behind the waterfall, and some blocks of rock from the collapse of the hard cap rock.

Over a very long time the process of undercutting and collapse is repeated many times, causing the waterfall to retreat upstream. The retreat creates a steep-sided **gorge**. At the same time chips of the hard cap rock are eroded away, which reduces the height of the waterfall.

***Figure 3* The formation of a waterfall.**

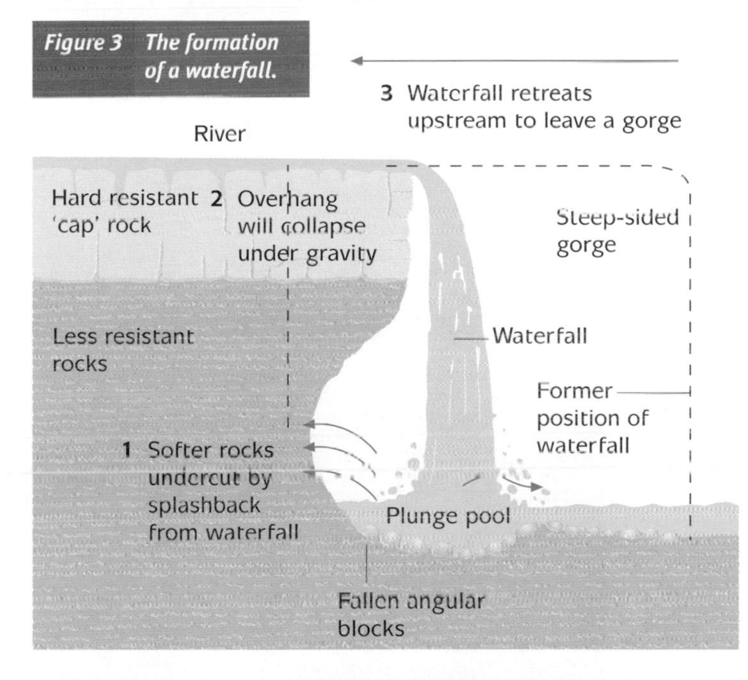

***Figure 4* The Iguaçu Falls.**

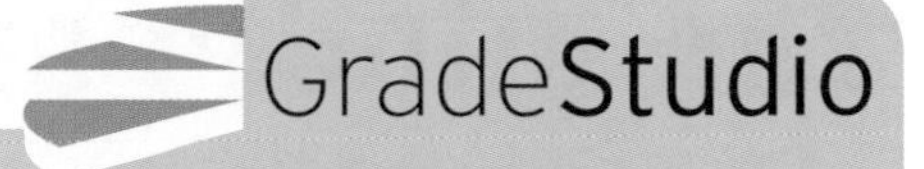

1 Draw a labelled sketch of **Figure 4** to show the features of the river, channel and valley. (4 marks)

2 a What is meant by vertical erosion? (1 mark)

b Describe the evidence that vertical erosion is taking place in the areas shown in **Figures 2** and **4**. (2 marks)

c Which two processes of erosion are likely to have the greatest effect in the Iguaçu River (**Figure 4** and page 79)? Explain your choices. (3 marks)

3 a Why are rocks of different hardness needed for the formation of waterfalls? (2 marks)

b Explain how the gorge below a waterfall is formed. (3 marks)

Exam tip

When asked to draw a labelled sketch (or diagram), remember that more marks are given for the labels than for the drawing. Add more labels than would seem to be needed for the number of marks.

***Figure 5* Upper course of the River Tees. Waterfalls form where whinstone rock (a hard, resistant igneous rock) outcrops. A: High Force, England's highest waterfall (21m high). B: River channel below High Force.**

The middle and lower courses of a river – lateral erosion and deposition

Down valley, why do lateral erosion and deposition replace vertical erosion? How big a role does flooding play in the formation of river landforms in the lowlands?

The middle course

As the river flows downstream, the gradient over which it flows becomes less steep and the river is not as high above its base level. Lateral erosion (sideways erosion, erosion on the sides of the channel and valley) increases in importance. When the river emerges from its upland area it begins to **meander** in order to use up surplus energy. The erosion on the outside of meanders removes the ends of the interlocking spurs and the valley becomes wider and has a more recognisable valley floor.

Meanders are bends in the river's course (**Figure 1**). On the outside of a meander the water is deeper and the current flows faster. The force of the water erodes and undercuts the outside bend by abrasion, forming a steep bank called a **river cliff**. On the inside bend there is slack water and the current is less strong, which encourages deposition. Sand and small pebbles are deposited, creating a gentle **slip-off slope**. An underwater current spirals down the river, carrying the eroded material from the river cliff to the slip-off slope.

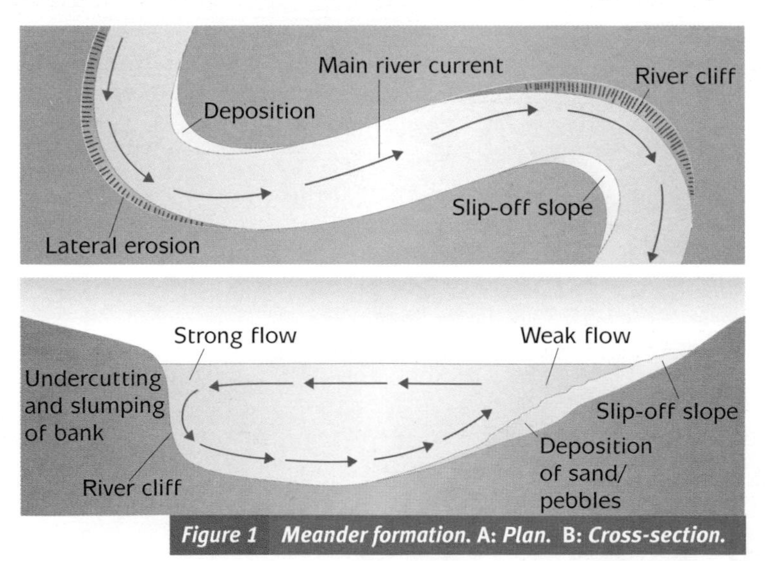

Figure 1 **Meander formation. A: Plan. B: Cross-section.**

The lower course

In the lower course the river channel becomes wider and deeper. The velocity is often greater than in the upper course because the channel is more efficient with less friction. The channel is almost semi-circular and much smoother because of deposits of sand and mud. In the lower course a river flows through a wide, flat valley called the flood plain (**Figure 2**).

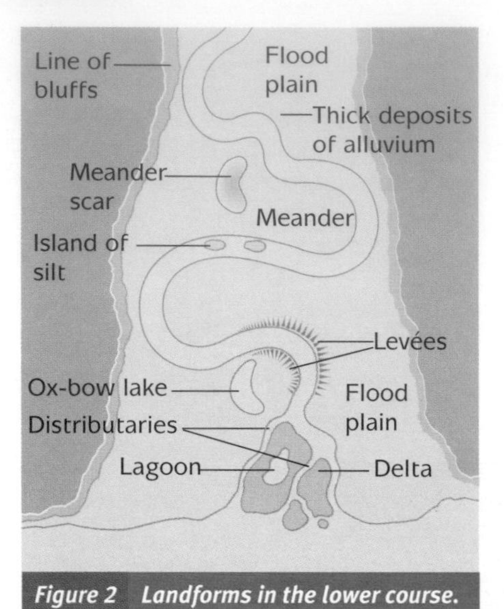

Figure 2 **Landforms in the lower course.**

Figure 3 **Meandering river in the Amazon Basin.**

The meander bends become even larger in the lower course as the river meanders more vigorously (**Figure 3**). Continued erosion on the outer bends and deposition on the inside of the bends may eventually lead to the formation of an **ox-bow lake** (**Figure 4**). The neck of the meander narrows as erosion continues on the outside bends. Eventually the neck is broken through, creating a straight channel. This often happens during a flood when the river is particularly powerful. As the flood

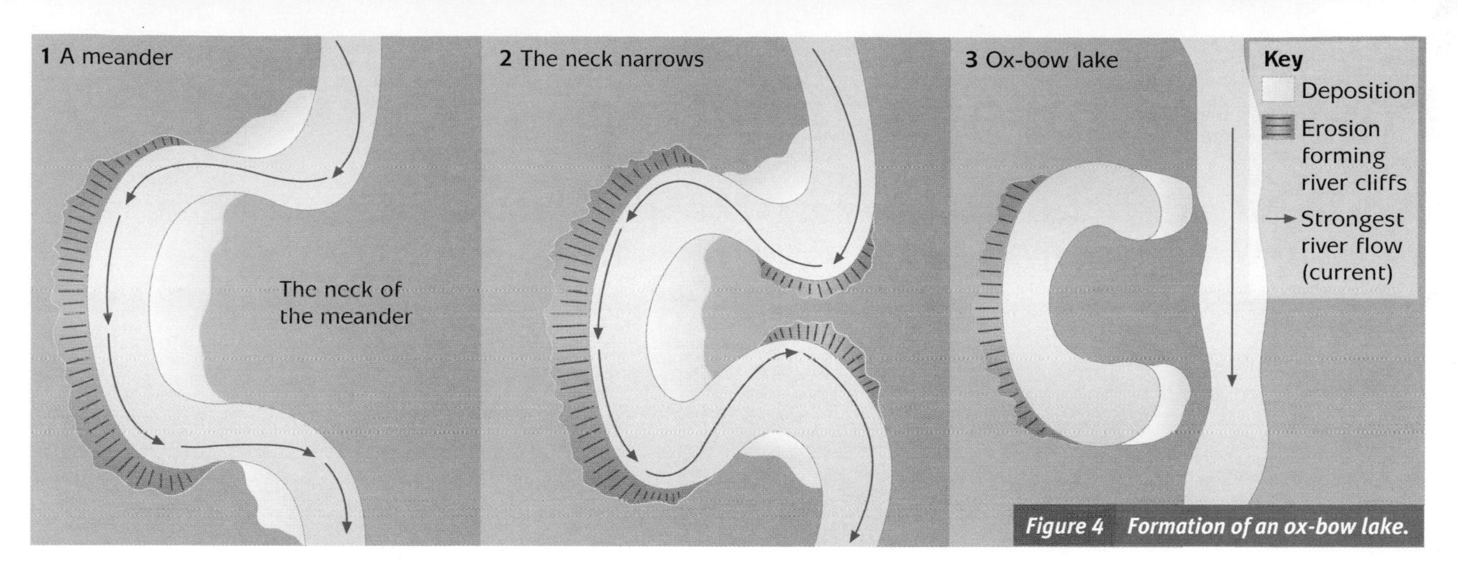

Figure 4 Formation of an ox-bow lake.

waters fall, and at times of low flow, alluvium is deposited, which seals off the old meander and forms an ox-bow lake. Gradually the ox-bow lake dries up, forming a **meander scar**.

What encourages river deposition?

Factors that encourage deposition include:

- a river carrying a large load of sediment, providing a great deal of material for deposition
- a reduction in velocity, as at the inside bend in a meander
- an obstruction, e.g. a river enters a lake and velocity falls, or it meets waves and currents, or bridge supports interrupt flow
- a fall in the volume of river water, e.g. at times of low flow during a period of drought.

Deposition of sand and silt, called **alluvium**, becomes the most important process in the lower course. Alluvium (often simply referred to as **silt**) is found in great thicknesses on the flood plain, especially where levées are formed.

Levées

Levées are natural embankments of silt along the banks of a river, often several metres higher than the flood plain. Levées are formed along rivers that flow slowly, carry a large load and periodically flood (**Figure 5**).

Every time the river leaves its channel, the greatest amount of sediment is deposited on the edge of the channel where the loss of flow due to increased friction is most pronounced. Big rivers like the Mississippi have built up huge natural levées of 10 metres and higher in places.

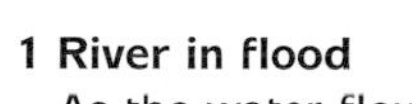

1 River in flood
As the water flow slows, energy is lost. Coarser, heavier material is deposited on the bank and finer material further away.
Coarser material
Fine material
Flood plain
Channel
2 River at low flow
During a dry spell the river's velocity slows down and the volume falls. This causes the deposition of material on the bed.
3 After repeated floods
After many floods the river banks form levées and the bed may be raised so much that the river rises above the flood plain. This can lead to more flooding.

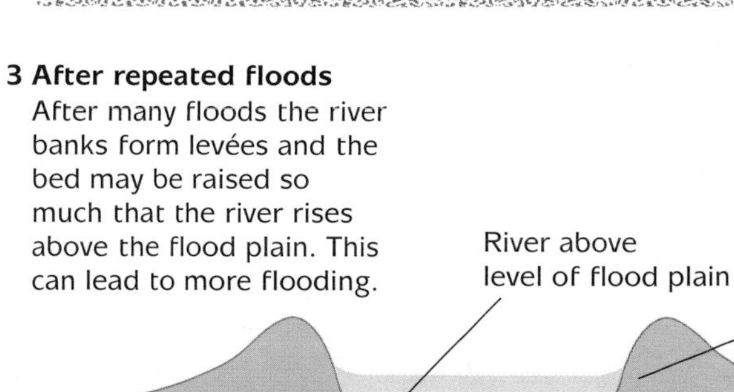

Figure 5 The formation of levées.

ACTIVITIES

1 (a) What is meant by lateral erosion?
(b) Where and when does it occur?
(c) Which landforms are formed by it?

2 (a) Draw a labelled diagram to show the differences between a river cliff and a slip-off slope.
(b) Explain why they are found on different sides of the channel.

3 Explain how river flooding leads to the formation of (a) levées and (b) ox-bow lakes

Flood plains and flooding

Why is the flat land next to a river known as the flood plain? How much are people to blame for river floods? Why do some rivers flood more quickly than others?

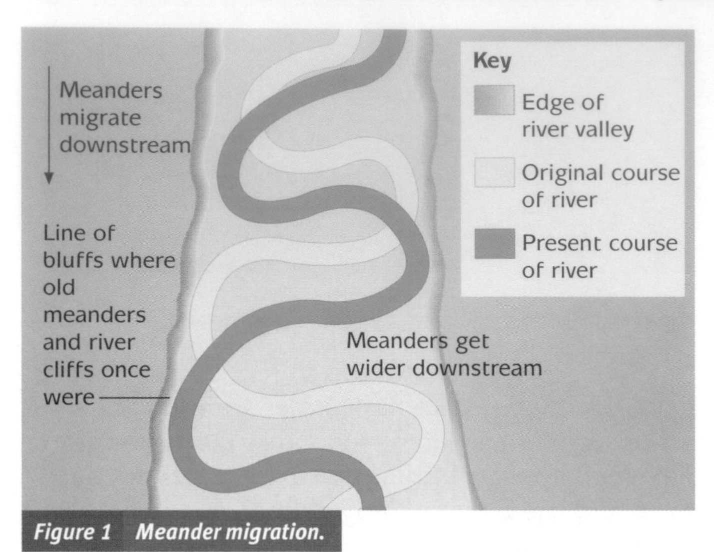

Figure 1 **Meander migration.**

The flood plain is the wide, flat area of land either side of the river in its lower course. The flood plain is formed by both erosion and deposition. Lateral erosion is caused by meanders and the slow migration downstream to widen the flood plain (**Figure 1**). The deposition on the slip-off slopes provides sediment to build up the valley floor. This is added to during a flood when the river spills over its banks onto the surrounding land. The river carries with it large quantities of suspended load. As the water floods onto the flood plain there is greater friction, the water is shallow and the river's velocity falls so its load is deposited onto the flood plain as alluvium. Over many thousands of years these deposits build up into great thicknesses of alluvium.

Most British rivers enter the sea through a wide river mouth known as an **estuary**. Near the sea, where the river is tidal, the flood plain is covered by salt marsh (**Figure 4**, page 129). At low tide the exposed sides of the river channel are full of mud. Big rivers in other parts of the world carry much larger loads of sediment; when this is deposited, many of them split up into separate channels before reaching the sea. A triangle- shaped area of land is built out to sea to form a **delta**. In the UK the most likely place to see small river deltas is where streams flow into lakes; meeting the still lake waters causes instant river deposition.

What to look for in Figure 2:

- The River Earn becomes tidal at 107194.
- Flood plain – where is the greatest width of flat land below 10 metres?
- River meanders – size of the largest, their frequency (how close together)?
- Ox-bow lakes – sizes and locations?
- River levées or embankments – marked in square 1119.

Flooding

Flooding is a normal, natural and regular occurrence in the lower course of a river. This is how flood plains are formed by rivers. A flood occurs when the discharge is so great that all the water can no longer be contained within the channel, so that the river overtops its banks. Most floods occur because of the weather. For example:

- *prolonged rain* – long, continuous periods of rainfall, leading to saturated ground, as happened in many parts of the UK in 2000 and again in 2007 and 2008 (pages 88–9)
- *heavy rain* – a cloudburst in a thunderstorm, which causes large amounts of run-off in a short time, as in Boscastle in 2004 (pages 88–9)

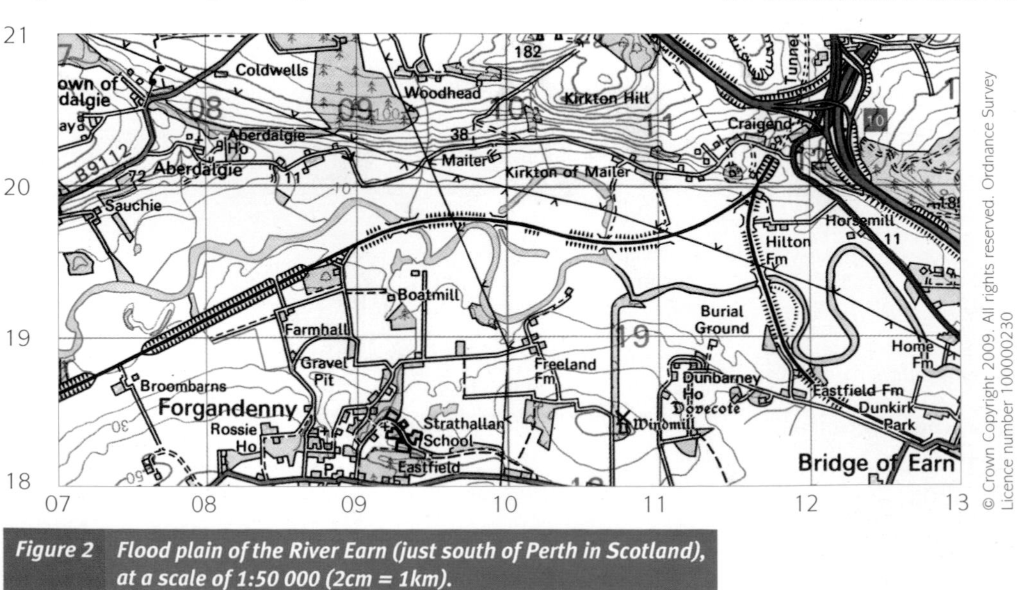

Figure 2 **Flood plain of the River Earn (just south of Perth in Scotland), at a scale of 1:50 000 (2cm = 1km).**

- *snowmelt* – a sudden increase in temperature that rapidly melts snow and ice. In winter the water cannot seep into the ground because it is still frozen.

Apart from occasional disasters, such as a dam burst, human activities are not a direct cause of flooding, but they can make it worse:

- Building on flood plains and increasing the size of urban areas create impermeable surfaces so that more water runs off the surface (and faster). Rainwater falling on hard surfaces (buildings and roads) is led rapidly into underground drains and straight into rivers, with little chance of evaporation into the atmosphere or infiltration into the ground.
- Deforestation reduces interception by the trees and rates of transpiration back into the atmosphere; more rainwater reaches the surface quicker, thereby increasing rates of run-off.

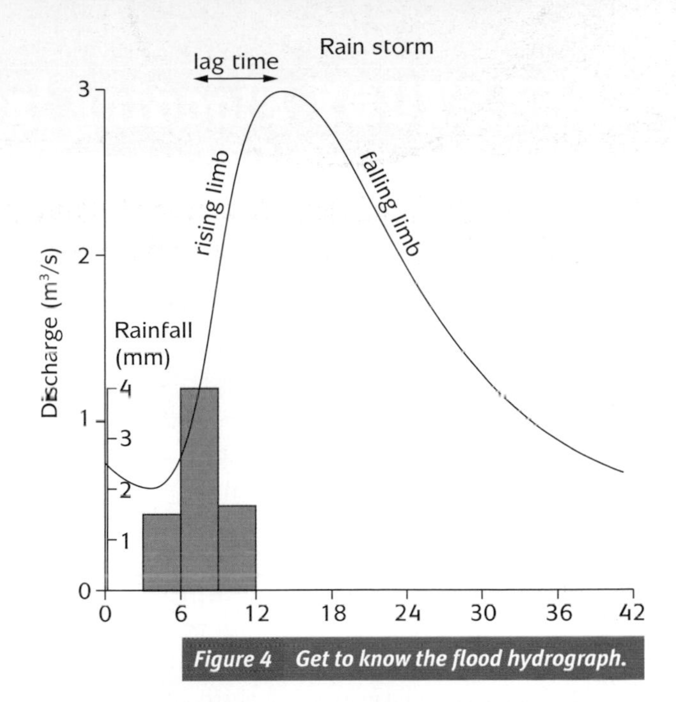

Figure 4 Get to know the flood hydrograph.

Flood hydrographs

After a rainstorm, the discharge of a river usually increases. Graphs called flood or storm hydrographs are used to show this (**Figure 4**). What is important for flooding is the length of the **lag time**, the difference in time between the peak of the rainstorm and the peak of the discharge. The shorter this is, and the steeper the rising limb line on the graph, the greater the risk of flooding.

Two drainage basins may react differently to a rainstorm producing identical amounts of rain. River basin A in **Figure 5** is an example of a '**flashy river**'. This means that it can rise very quickly after a rainfall event and fall equally quickly following a dry spell. It is likely to have many, if not all, of the drainage basin characteristics identified in **Figure 5** on page 81. The flood threat is high. In river basin B the rainwater is either being delayed from reaching the river by interception and gentle slopes; or a higher percentage is being lost to evapo-transpiration and to infiltration into the ground, most likely because of the presence of permeable rocks – rocks with spaces and gaps that allow water to pass through (page 35).

Figure 3 One possible combination for flooding – a summer thunderstorm in an urban area.

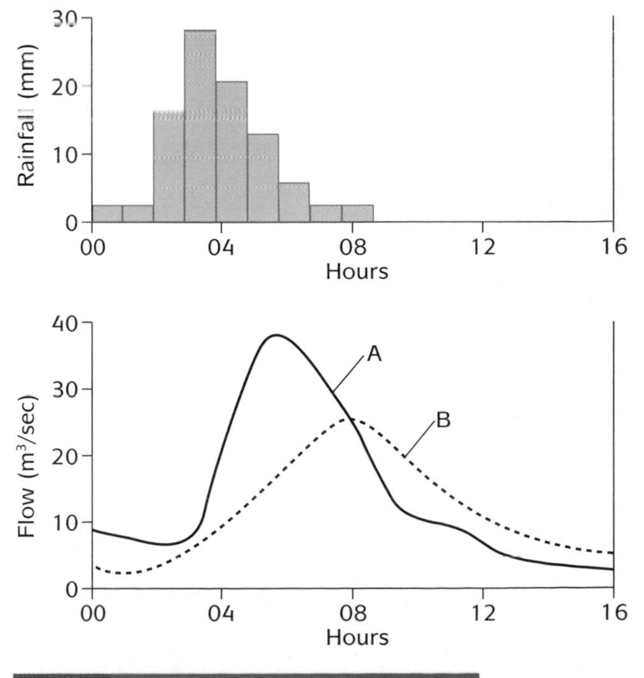

Figure 5 Discharge after the same rainstorm in two drainage basins.

ACTIVITIES

1 (a) Describe the main physical features of flood plains.
  (b) Why are they called flood plains?

2 Study **Figure 5**.
  (a) (i) Work out lag times from hydrographs A and B.
    (ii) Describe the other differences between them.
  (b) Explain how (i) relief and rock type (ii) land uses may have caused these differences.

Flood events in the UK since 2000

2000 – a very wet year in the UK

The year 2000 was the wettest for a century in the UK and one of the four wettest years since records began in 1766. April and December were particularly wet months (**Figure 1**); the annual average was 40 per cent above normal. This resulted in frequent and widespread river floods throughout the UK. These reached their peak in autumn, although parts of the north of England were flooded in summer as well; yet in terms of average for the UK, summer was the only consistently dry time of the year. For example, in north-east England up to 75 millimetres of rain fell during the first three days of June, leading to scenes like those shown in **Figures 2** and **3**.

Figure 2 **When the River Wear burst its banks in Durham City in early June 2000.**

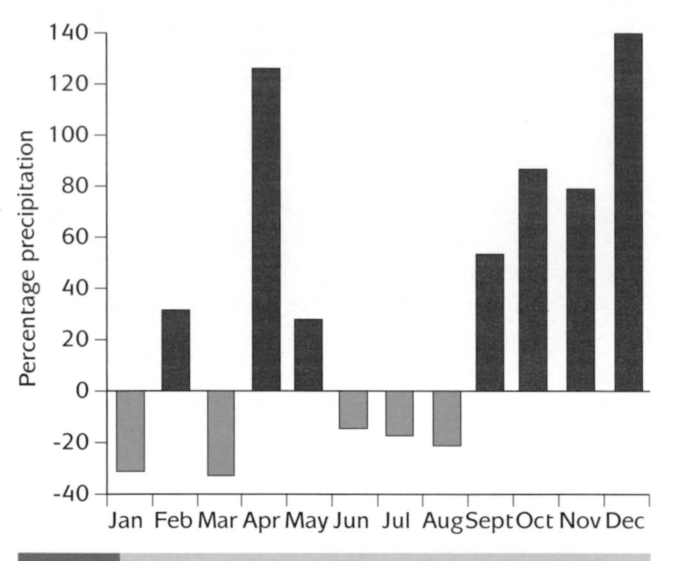

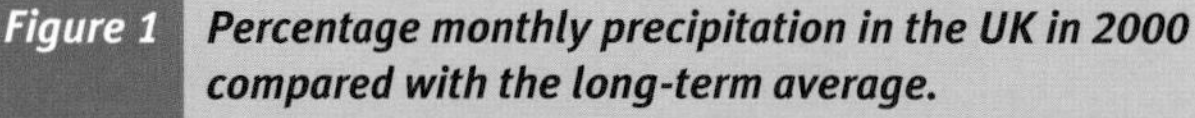

Figure 1 **Percentage monthly precipitation in the UK in 2000 compared with the long-term average.**

Figure 3 **After the River Wear flooded for a second time in October.**

Boscastle, Cornwall, August 2004

The flood event

- A summer storm dropped 200 millimetres of rain in four hours.
- In just one hour, 90 millimetres of this rain fell.
- It poured down the steep hillsides into the rivers Valency and Jordan.
- Meeting in Boscastle, these rivers caused a flash flood.
- A 3-metre-high wall of water swept through the village.
- Cars were washed into the sea, riverside shops were destroyed and houses badly damaged.

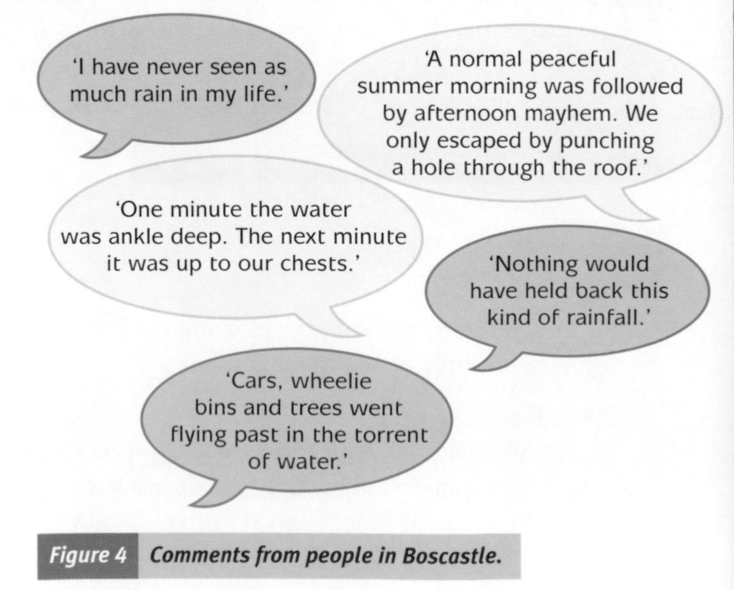

'I have never seen as much rain in my life.'
'A normal peaceful summer morning was followed by afternoon mayhem. We only escaped by punching a hole through the roof.'
'One minute the water was ankle deep. The next minute it was up to our chests.'
'Nothing would have held back this kind of rainfall.'
'Cars, wheelie bins and trees went flying past in the torrent of water.'

Figure 4 **Comments from people in Boscastle.**

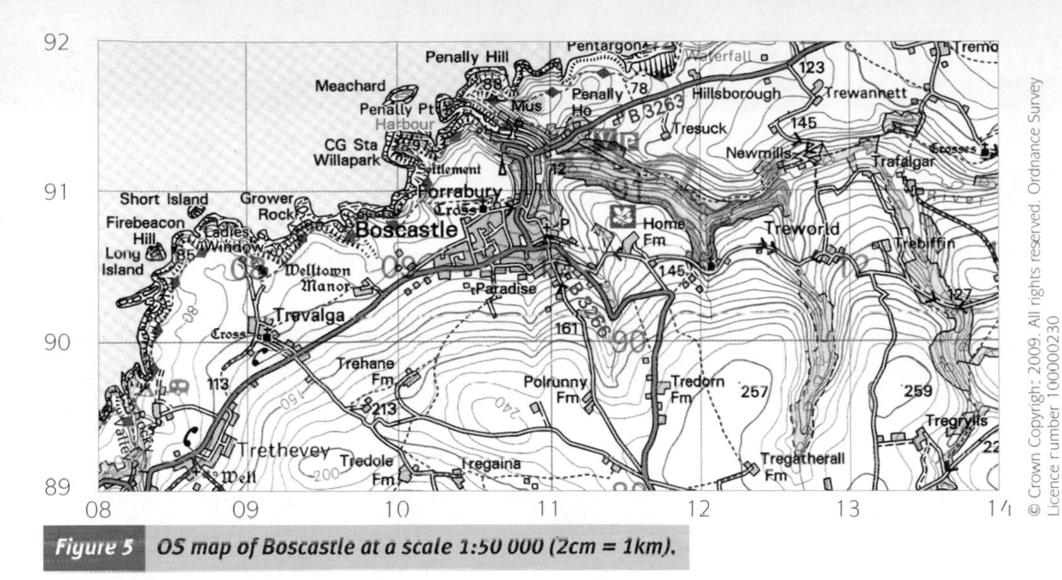

Figure 5 OS map of Boscastle at a scale 1:50 000 (2cm = 1km).

Note:

- The steepness of the valley sides.
- Only small areas of woodland.
- Water from all surface rivers meets in Boscastle.

FURTHER RESEARCH

Find out more about Boscastle on the weblink www.contentextra.com/aqagcsegeog.

Immediate responses

- Emergency services responded speedily and efficiently.
- Helicopters airlifted to safety about 80 people from rooftops.

Medium- and long-term responses

- £800 000 flood defence scheme completed in Boscastle by April 2005.
- Engineers investigating future flood control works on River Valency.

The summer floods of 2007

Refer back to page 53 (Chapter 3). Compared with 2000, the flooding was even more widespread and locally more severe (**Figure 3**, page 53). However, the causes were similar – prolonged periods of wet weather, much higher than average rainfall, and concentrated periods of heavy rain, all of which precipitated the very serious local flood events. A year later, many of the affected households in south and east Yorkshire had still not fully recovered from the floods.

It happened again in 2008

This time it was the Northumberland market town of Morpeth that was most badly hit. In early September the River Wansbeck, which flows through the town centre, could take no more after a wet summer, and a particularly wet August was followed by more than the whole of September's rain falling within 48 hours. The heavy, persistent rain came from an active area of low pressure, which spent the weekend over north-east England, moving only a little.

Another flood event seemed to confirm what many people have been thinking – that the frequency of damaging floods in the UK is increasing. Some blame global warming for more extreme weather events. Others blame planners for allowing houses to be built on flood plains. Weather experts point out that weather across the British Isles is highly variable; wet summers happen when frontal depressions hit the UK by tracking further south than normal.

GradeStudio

1 Study **Figure 1**.
  a In how many months was rainfall above average in the UK in 2000? (1 mark)
  b Which was the driest season in 2000? (1 mark)
  c Suggest reasons why flooding was more widespread in autumn than in spring. (3 marks)

2 Describe similarities and differences between the two scenes shown in **Figures 2** and **3**. (4 marks)

3 Read these two comments about flooding in the UK in 2000:
  A 'In our overcrowded country we are building too close to rivers. We are building too many new housing estates and shopping centres with hard surfaces; heavy rain just bounces off them.'
  B 'We must learn to accept natural disasters. Nothing can be done to protect us when one month's rain falls in a few hours.'
  a Explain each of the comments about the causes of the floods in 2000. (4 marks)
  b With which one do you agree most? Explain your choice. (2 marks)

Exam tip

Using the 2007 floods as the main case study:

- Be guided by the headings in activity 2 on page 53.
- Causes, effects and responses are the most likely exam question themes.

CASE STUDY Flooding in a poor part of the world

The 2004 floods in Bangladesh

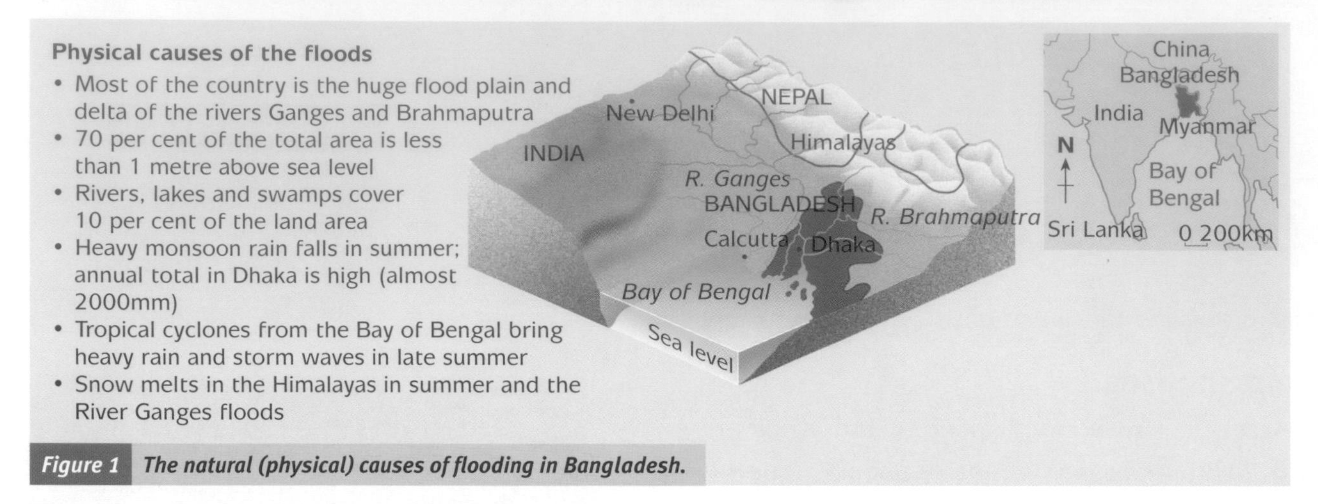

Physical causes of the floods
• Most of the country is the huge flood plain and delta of the rivers Ganges and Brahmaputra
• 70 per cent of the total area is less than 1 metre above sea level
• Rivers, lakes and swamps cover 10 per cent of the land area
• Heavy monsoon rain falls in summer; annual total in Dhaka is high (almost 2000mm)
• Tropical cyclones from the Bay of Bengal bring heavy rain and storm waves in late summer
• Snow melts in the Himalayas in summer and the River Ganges floods
New Delhi
NEPAL
INDIA
Himalayas
R. Ganges
BANGLADESH
R. Brahmaputra
Calcutta
Dhaka
Bay of Bengal
Sea level
China
Bangladesh
India
Myanmar
N
Bay of Bengal
Sri Lanka
0 200km

Figure 1 *The natural (physical) causes of flooding in Bangladesh.*

There are several **physical reasons** (relief, drainage and climate) why Bangladesh suffers from flooding almost every summer (**Figure 1**). Most of the country is delta and flood plain, landforms formed by regular river floods. The amount of surface water makes them challenging environments for people. At the same time, they are attractive environments because the deep, constantly renewed silt soils are extremely fertile. These support some of the highest agricultural densities of population in the world, especially in a country where padi rice is the main food crop.

Bangladesh is a populous country of 150 million people. When the flooding is more severe than usual, as it was in 2004, considerable loss of life and great suffering result (**Figure 2**). In September 2004, Dhaka had its worst rains for 50 years; on 13 September, 350 millimetres fell in 24 hours.

Immediate effects
- floods covered more than half of Bangladesh
- 760 people were killed
- 8.5 million people were left homeless
- more than 35 million people were affected
- rice growing and fish farming were disrupted
- roads and bridges were damaged and destroyed

Later effects
- more than 1 million children suffered from malnutrition and disease in the following months
- government rebuilding costs for roads and industry were estimated at US$ 2–3 billion
- emergency food aid was needed until the following year's harvest

Figure 2 *The effects of flooding in 2004.*

Possible human causes of the floods

The sources of the Ganges and Brahmaputra rivers are in Nepal and Tibet. In recent years the populations there have grown rapidly, causing the removal of vast areas of forest to provide fuel, timber and grazing land. In Nepal, 50 per cent of the forest cover that existed in the 1950s has been cut down. The forests absorb water from the ground, bind the soil particles and reduce the impact of rain droplets on the ground surface. The removal of the forest cover has increased soil erosion and overland flow. The soil is deposited in the river channels, causing the river beds to rise, and reducing the capacity of the rivers. It has been estimated that the river bed of the Brahmaputra is rising by 5 centimetres each year. The building of the Farakka dam in India in 1971 is blamed for raising the river bed of the River Hooghly, a tributary of the Ganges. This increases the risk of flooding.

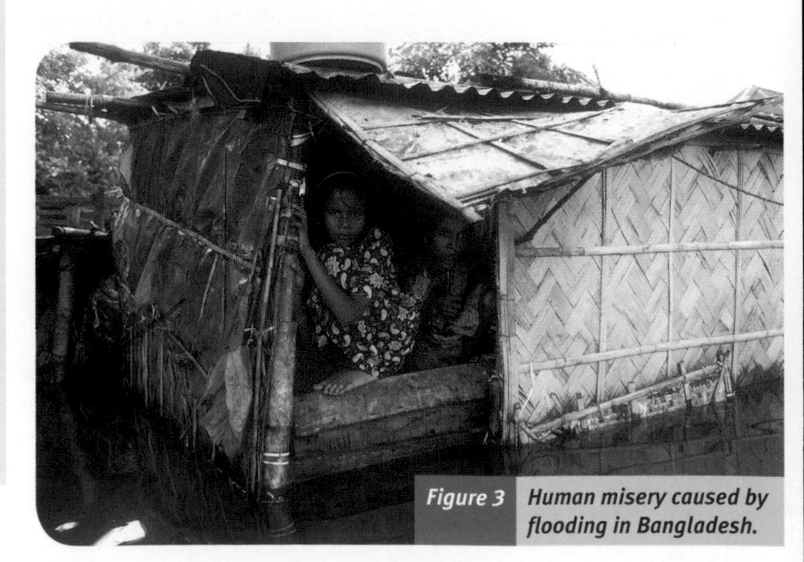

Figure 3 *Human misery caused by flooding in Bangladesh.*

It has been noticed that the interval between big floods is shortening – it used to be between 10 and 15 years, but the 2004 flood came only six years after the major disaster in 1998. This has led to the suggestion that human activities are a contributing factor of increasing importance. If they are, Bangladesh has little control over this. Low-lying, delta countries like Bangladesh are the first to feel the effects from rising sea levels due to global warming.

Responses

Bangladesh is one of the world's poorest countries, with a Gross National Product (GNP) of only US$ 380 per head. In the *short term*, the prime concern is always for the health, survival and suffering of the people affected. A heavy reliance is placed upon emergency aid (food, drinking water, medicines, plastic sheets, boats for rescuing people and animals) from international organisations such as the United Nations, governments in rich countries and charities. One big problem is distribution, because so much of the country is under water. As the flood waters recede, it becomes easier to set up medical treatment centres, distribute water purification tablets and provide help with repairing houses and restarting economic activities such as farming and fishing.

However, these actions can do nothing to manage the flood problem in the *long term*. In July 1987 the World Bank prepared an Action Plan for Flood Control (**Figure 4**). The plan involved the completion of 3500 kilometres of coastal and river embankments and included seven large dams – partly to stop water from reaching the land and partly to provide up to 15 floodwater storage basins.

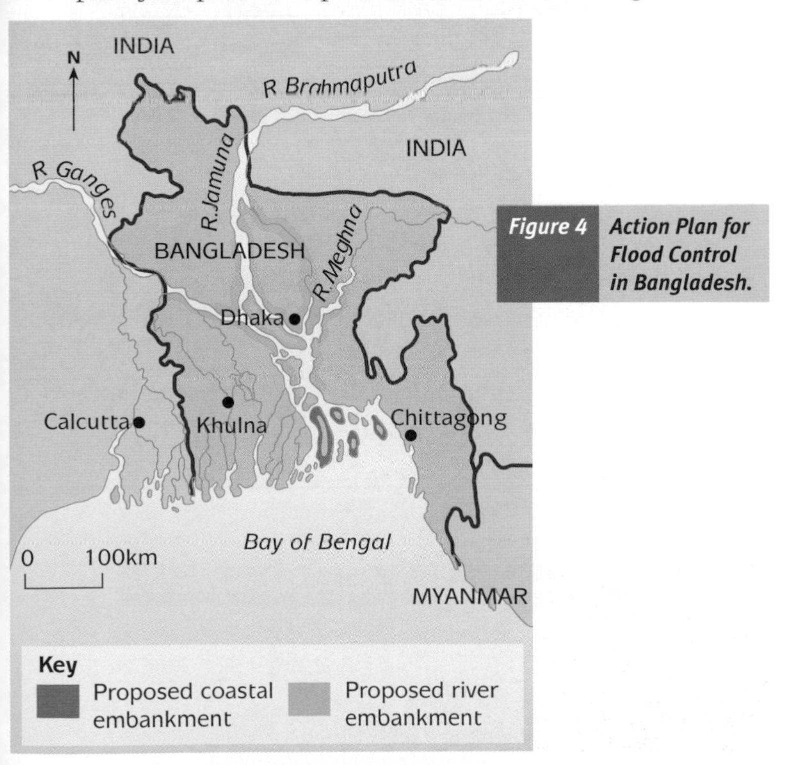

Figure 4 *Action Plan for Flood Control in Bangladesh.*

This is an example of the **'hard engineering'** approach to flood management (page 92). Millions of dollars of aid were poured into these engineering projects, but the scheme remains unfinished due to a mixture of corruption and inadequate funding.

From the beginning, some people suggested that what Bangladesh really needed was a mixture of strategies involving flood forecasting and early-warning schemes on the one hand and more flood shelters stocked with emergency supplies on the other. It was argued that these would be both cheaper and more appropriate for farming and fishing communities in rural areas. They would be more in keeping with local knowledge, skills and income levels. They would have the additional advantage of being less likely to damage delicate ecosystems, thereby making a contribution to sustainable development.

ACTIVITIES

1 Put together the information needed for a 'Case study of flooding in a poorer area of the world – Bangladesh 2004' using the headings:
- Causes of the floods – physical, human and the relative importance of each of these
- Effects of the floods – immediate, later and their overall scale
- Strategies for managing floods – short-term, long-term and relative chances of success.

2

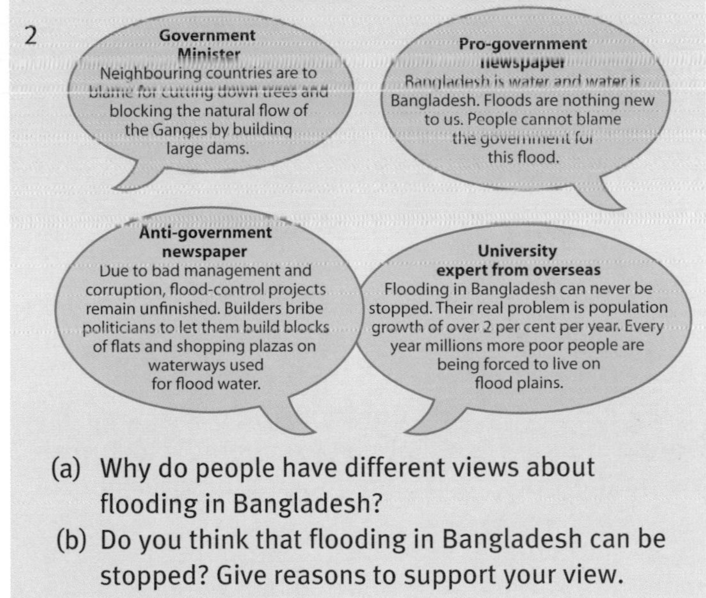

Government Minister
Neighbouring countries are to blame for cutting down trees and blocking the natural flow of the Ganges by building large dams.
Pro-government newspaper
Bangladesh is water and water is Bangladesh. Floods are nothing new to us. People cannot blame the government for this flood.
Anti-government newspaper
Due to bad management and corruption, flood-control projects remain unfinished. Builders bribe politicians to let them build blocks of flats and shopping plazas on waterways used for flood water.
University expert from overseas
Flooding in Bangladesh can never be stopped. Their real problem is population growth of over 2 per cent per year. Every year millions more poor people are being forced to live on flood plains.

(a) Why do people have different views about flooding in Bangladesh?

(b) Do you think that flooding in Bangladesh can be stopped? Give reasons to support your view.

FURTHER RESEARCH

Check for information about more recent floods on the weblink www.contentextra.com/aqagcsegeog.

Flood control – should hard or soft engineering be used?

What are the options for dealing with the problem of flooding, which seems to be on the increase? What are the costs and benefits of hard and soft engineering options?

This comes under the general heading of **river basin management**. Management means planning ahead and controlling development and change. Rivers have many valuable uses for people – how many can you name? In addition, hundreds of millions of people throughout the world live close to rivers. Management methods are needed to try to reduce the risks from natural flooding that occurs beside all rivers, especially in their lower courses.

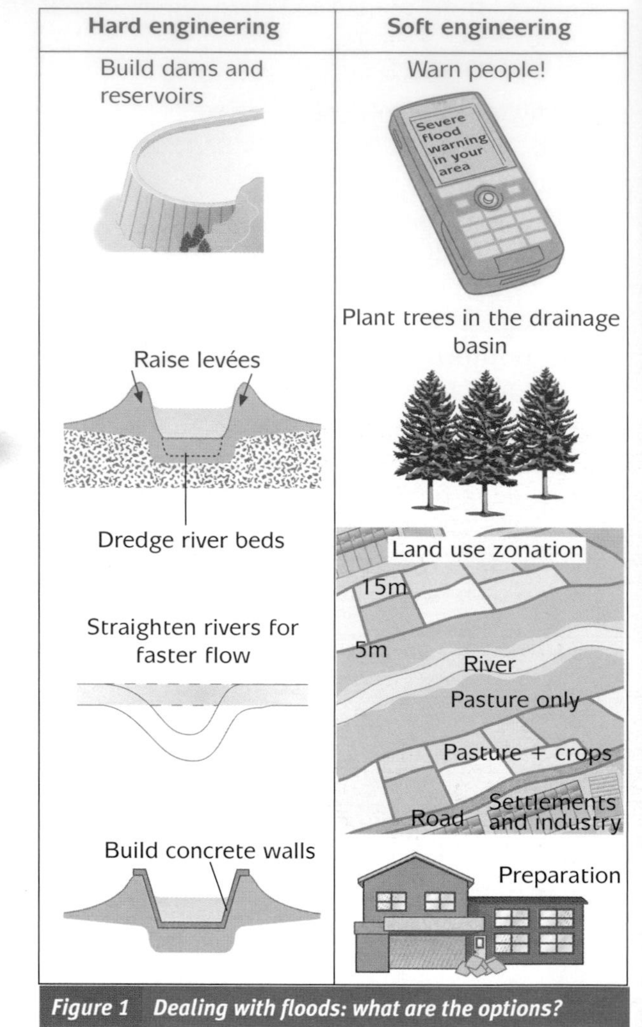

Figure 1 Dealing with floods: what are the options?

Strategies for dealing with river floods

Some of the options for dealing with floods are shown in **Figure 1**. Building retaining dams, increasing the height of levées with concrete embankments, and building walls to confine the channel course through urban areas are examples of hard engineering. In the short and medium term, these probably afford the greatest protection, but they are expensive to build and require costly maintenance. However, where the economic, social and political need is high, and the flood risk great, their use is not optional. The Thames Barrier is the clearest example of this in the UK (page 128). Here the potentially lethal combination of

- river flooding risk from inland
- tidal flooding dangers from the sea and
- the largest concentrations of people and commercial activities in the UK meant that London had to be protected, whatever the economic and environmental costs of barrier construction.

In the long term, **soft engineering** options, such as zoning land uses so that no permanent structures are built on the flood plain, offer a more sustainable way. By planting many more trees in the drainage basin, particularly in places where slopes are steep, more rainwater will be delayed before reaching the river. In other words, a river will become less 'flashy' with lower and later peak discharge (page 87). **Figure 2** shows how important the role of trees in flood prevention can be. The problem, of course, is that many homes and businesses have already been built on flood plains. The first demand of people in villages and towns affected by floods, like those described on page 53, is new and better flood defences, of the hard engineering type, and quickly.

	evapo-transpiration	groundwater	run-off
All coniferous trees	50%	40%	10%
Half coniferous trees, half buildings	25%	30%	45%

Figure 2 How clearing half the trees in an area affected run-off.

CASE STUDY Flood control

The River Tees

The River Tees rises in the Pennines where precipitation is high (over 2000 millimetres per year), rocks are impermeable and slopes are steep. It is a flashy river, with a long history of flooding in its lower course. The river is managed partly to control flooding, but also for its usefulness for water supply, recreation and tourism.

A *Upper course – Cow Green reservoir* (**Figure 1**, *page 97*)
Built in 1970, mainly to provide water for the growing industries on Teesside, it traps heavy rainfall and water from snowmelt and allows water flow to be regulated downstream.

B *Lower course – Yarm's flood defence scheme*
Yarm, a historic market town and once an inland port, has a long record of floods, going back centuries. The most recent serious flood was in January 1995. Since then a new flood defence scheme costing £2.1 million has been implemented, with:

- reinforced concrete walls with metal flood gates for access by people and vehicles (**Figure 3**)
- earth embankments
- gabions (baskets filled with stones) to protect walls and embankments from erosion
- fishing platforms, street lighting and replanting to improve the environment
- building materials approved by English Heritage to remain in keeping with existing architecture.

Figure 3 ***High walls and metal gates that can be closed in times of flood protect expensive housing on the banks of the River Tees in Yarm.***

C *Near the tidal mouth – river straightening, dredging and the Tees Barrage*
As early as 1810, one large meander near Stockton, the Mandale Loop, was cut off, shortening the river by 4 kilometres. That was to aid navigation. Since then, other stretches have been straightened to allow water to move faster along the channel, reducing the flood risk. The Tees estuary is periodically dredged to maintain the deep water channel for shipping and speed of water flow. In 1995 the Tees Barrage (**Figure 4**) was completed at a cost of £54 million, with multiple objectives – to maintain high water levels for shipping in the estuary, reduce the flood risk at very high tides or during a storm surge, and act as a catalyst for commercial, residential and leisure developments in the surrounding area.

ACTIVITIES

1 Make a copy of the table below and complete it for the hard and soft engineering options in **Figure 1**. The 'Do nothing' option has been done for you, but feel free to amend or add to it.

Option	Impact	Benefits	Costs
Do nothing	Little or none if not built-up Great in built-on areas Discourages new flood plain development	Cheap River naturally floods Fertile silt/ water supply for farming	Floods homes, fields, roads, services, etc. Costly to clean up and repair damage

2 (a) Identify and arrange the different methods used to manage the River Tees under the headings hard and soft engineering strategies.
(b) Which one dominates – hard or soft? Explain why.

3 Explain why the attitudes of the following groups of people to methods of flood control are often different:
(a) resident householders
(b) Local Authorities and government agencies paying for the works
(c) environmental groups based in large cities such as London.

Figure 4 ***The Tees Barrage.***

Rivers and water supply in the UK

The UK has a wet climate – so why is water supply an issue in some parts of the country? What are the issues arising from water transfers within the UK?

Three-quarters of the UK's water supply comes from mountain lakes, upland reservoirs and river intakes; the remaining 25 per cent is obtained from underground stores in layers of porous rock (**Figure 1A**). Demand for water continues to go up. The UK's population is now over 60 million and water consumption per head is high. Many formerly manual domestic tasks are now done by machines; automatic washing machines, dish washers and jet hose car washers all consume more water than doing the same jobs by hand. Electricity power stations (both thermal and nuclear) use water for cooling (**Figure 1B**). Demand for electricity in offices, industries and homes, and for transport, is increasing as well; think about your own family's dependence on electricity. Most people are unaware of the amount of water needed to produce electricity.

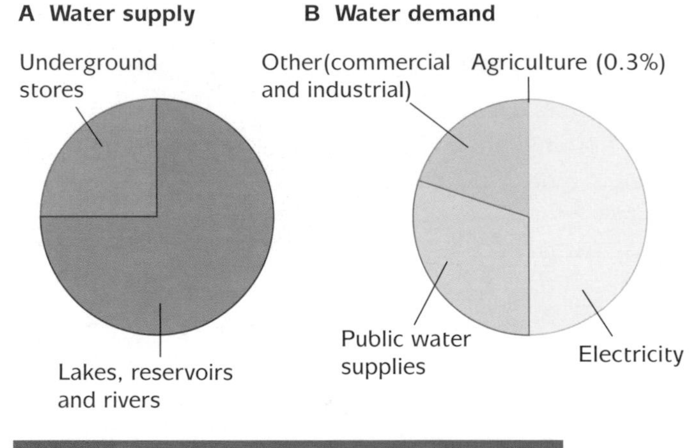

Figure 1 **A: Water supply B: Water demand in the UK.**

Why are river water transfers needed?

Scotland, Wales and Northern Ireland have a relative abundance of unpolluted water from upland sources; the water **surplus** is the result of a combination of upland relief, high precipitation, plentiful surface rivers and lakes, and low population densities. South-east England is the complete opposite – about 50 per cent of the UK's population live south and east of the Severn-Wash line, and it includes the single greatest concentration of people in the UK in and around London (**Figure 2**). At the same time, this area has the lowest annual precipitation in the UK, along with generally low and gentle relief. It is the main area of water deficit. This is the region with greatest dependence on groundwater sources, but these sources are already being used to their full capacity (page 39). Although the water companies are being coerced into reducing leaks and consumers are being led towards more efficient water use by metering and education, demand can only be satisfied by river transfers from wet west to dry east.

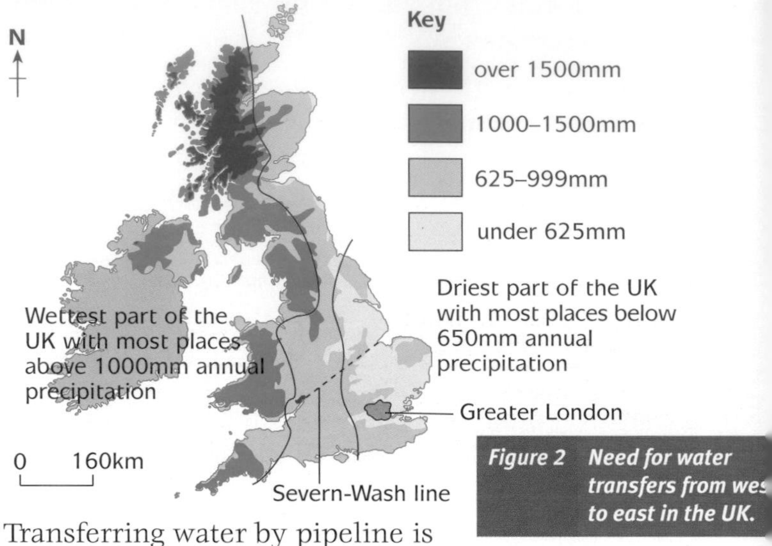

Figure 2 **Need for water transfers from west to east in the UK.**

Transferring water by pipeline is costly. Some aqueducts were built many years ago, such as the one which transfers water from Lake Vyrnwy in mid-Wales to Merseyside. Transfers by river are easier and cheaper, but they raise more environmental concerns. A proposed Severn-Thames transfer, which would increase London's water supplies by a very valuable 20 per cent, has highlighted several issues in connection with river transfers, human as well as environmental.

- Severn river water has different mineral, nutrient, acidity and temperature characteristics from Thames river water – what will the effects of 'foreign water' on insect, fish and plant life be in the Thames?
- The total bill for increasing dam sizes in Wales, and building tunnels between the Severn and Thames valleys, would be hundreds of million pounds – can the economic cost be justified?
- Land will be lost, and river habitats and salmon migration disturbed, without any benefits to Welsh people and the local economy – is it acceptable, in a time of political devolution, for a poor area of Wales to suffer disadvantages while the richest region of England benefits?

The basic problem of water supply in the UK is not going to go away – the highest demand is where rainfall is lowest (and vice-versa).

Kielder Water in Northumberland

Kielder Water, over 10 kilometres long and located close to the Scottish border, is the largest artificial lake in the UK. Planned in the late 1960s to satisfy an expected increase in demand for water from the then growing chemical and steel industries on Teesside, it was opened in 1982. The site offered both physical and human attractions for reservoir construction.

- The River North Tyne valley had a large, relatively wide floor, with steep sides.
- Annual precipitation was high (1370 millimetres).
- Few people lived in the valley; only a few families needed to be removed and re-housed.
- The land was mainly poor-quality farmland, remote from markets in the lower Tyne valley.
- A limited variety of lost wildlife habitats, in an area of rough grazing and coniferous woodlands.

Therefore, physical factors were favourable and perceived economic need was strong, while there were very few of the social, economic and environmental issues normally associated with big dam construction.

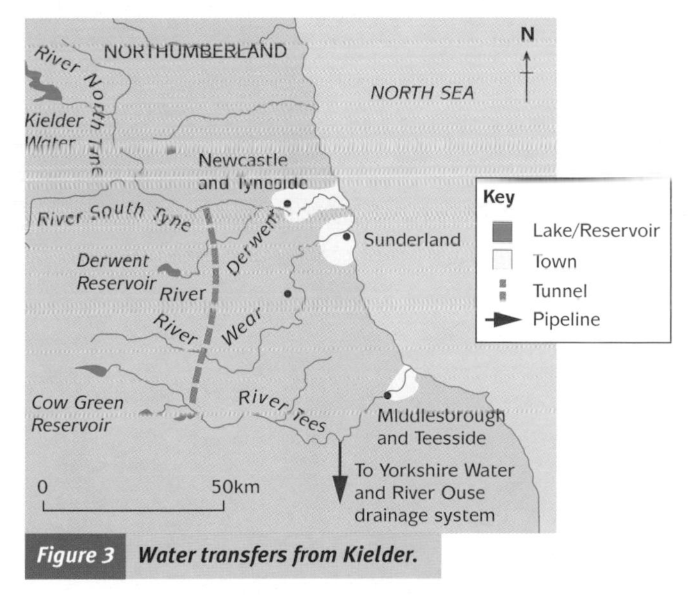

Figure 3 *Water transfers from Kielder.*

Water supplies from Kielder

Figure 3 shows how water is transferred to the populated parts of the north-east. Water is released directly into the River North Tyne; from the River Tyne at Riding Mill it is pumped through a tunnel under the high land between river valleys. Then it can be fed into other east-flowing rivers, especially the River Tees. While the Derwent reservoir remains the primary source of water for Tyne and Wear, water from Kielder is used to supplement the River Derwent flow when the reservoir is low and fed into the pipe distribution system. Biological studies of river life in the other rivers receiving Tyne water have discovered no adverse effects, probably because all have sources in the northern Pennines with many similar drainage basin characteristics.

The north-east enjoys the most reliable and sustainable water supply in England. Restrictions on water use, such as hosepipe bans, are unknown in this region, even in the driest of summers, such as 1995 when even Lake District reservoirs dried up (**Figure 4**, page 51). Drought in Yorkshire in 1995 led to the building of a 13-kilometre pipeline link south from the lower Tees to the Ouse river system for use in York. Not only this, but the economy of remote northern Northumberland has been revitalised. Water- and land-based leisure activities attract over a quarter of a million visitors each year. Around it is Kielder Forest, the largest woodland in England. Job opportunities in the water industry, tourism and forestry exist for people in surrounding villages, while previously there was only farming and little else.

ACTIVITIES

1 Average daily water use per family (litres):
UK 180; Bangladesh 45.
Average income per head (US$):
UK 35 760; Bangladesh 380.
   (a) How many times greater is family daily water use in the UK than in Bangladesh?
   (b) Suggest reasons why people in rich countries use more water.

2 In the UK, where rainfall is highest, demand for water is lowest.
   (a) Describe how **Figure 2** supports this statement.
   (b) What are the (i) environmental (ii) economic and (iii) social issues associated with river water transfers in the UK?

3 (a) Draw a sketch map to show how rivers are used for water transfers from Kielder Water.
   (b) Does Kielder provide sustainable water supplies? Explain your answer.
   (c) Undertake a cost-benefit analysis for Kielder Water by (i) listing costs and benefits in a table (ii) explaining which is the greater.

Practice GCSE question

See a Foundation Tier Practice GCSE Question on the weblink www.contentextra.com/aqagcsegeog.

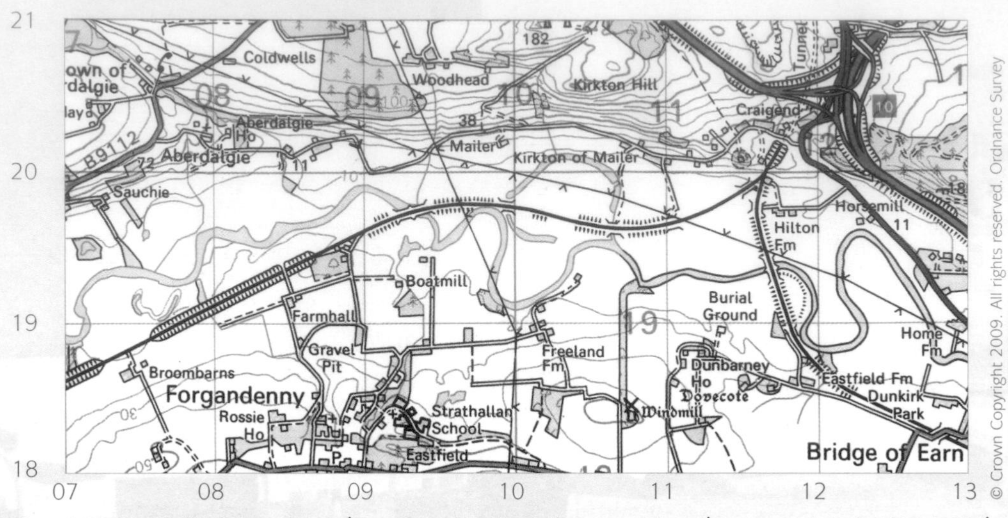

Flood plain of the River Earn (just south of Perth in Scotland) at a scale of 1:50 000 (2cm = 1km).

5 (a) Study the 1:50 000 OS map extract of the River Earn.

(i) State the direction of the windmill in square 1018 from the school in Fogandenny. **(1 mark)**

(ii) Why is it a good site for a windmill? **(1 mark)**

(iii) State two characteristics of the River Earn. **(2 marks)**

(iv) Describe the physical features of the valley of the River Earn. **(3 marks)**

Figure 1 Upper course of a river in the Pennines.

Figure 2 River Cuckmere in Sussex.

(b) Study **Figure 1**. Describe and explain the formation of the river landforms shown. **(6 marks)**

(c) Study **Figure 2**. Explain what could happen in the future at the point marked X. **(4 marks)**

(d) With reference to case studies, describe and explain similarities and differences in the effects of flooding between rich and poor parts of the world. **(8 marks)**

Total: 25 marks

Exam tip

The question for this topic will be Question 5 in Section B in Paper 1.

Improve your GCSE answers

How to recognise river and river valley landforms on OS maps

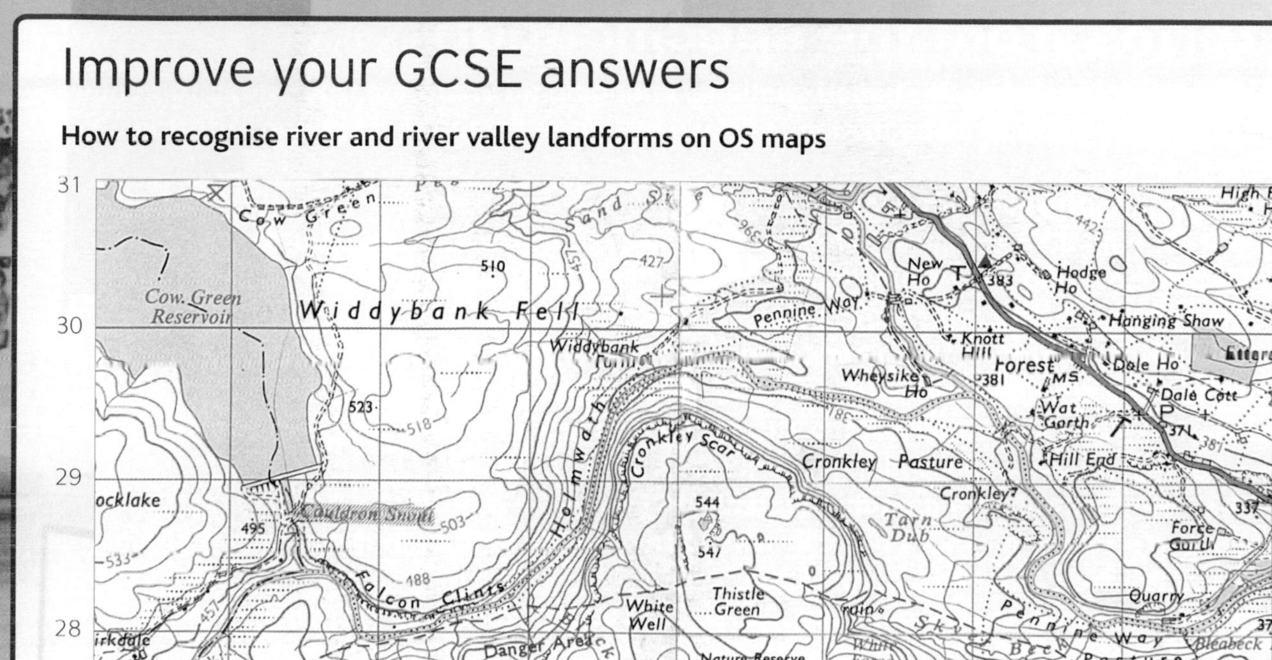

Figure 1 OS map of part of the upper course of the River Tees at a scale of 1:50 000 (2cm = 1 km)

OS maps showing upper courses of rivers, as in **Figure 1**, are not easy to interpret because of the number and closeness of the contours – but look carefully, and know what you are looking for.

What is the evidence for these landforms in **Figure 1**?

1. Steep-sided valley – shown by many contour lines, close together, in some places running very close to the river sides.
2. Gorge – shown below Falcon Clints and between Holmwath and Cronkley Scar by the signs for rock outcrops and the tightly packed contour lines, hugging the river sides.
3. Steep river gradient and waterfalls – look for contour lines crossing the river, especially where they cross close together. Perhaps the best place to look in **Figure 1** is just below the label Cauldron Snout where at least two contours cross the river. Cauldron Snout is a waterfall, as also is High Force (square 8828 and **Figure 5A**, page 83) with its downstream gorge. They are not easy to recognise from maps; you often need another clue, such as a tourist sign or a name highlighting them.

Notice that the River Tees does not flow straight – no river does unless it has been 'managed' by people. However, the bends in **Figure 1** are not real meanders. Compare the course of the Tees here with that of the Earn (OS map on page 86) and you will see the differences. Also, look back to **Figure 5** on page 89 to see good examples of steep-sided V-shaped river valleys. Remember that the centre of the V in a contour always points upstream.

What is the evidence in **Figure 1** that the River Tees is a managed river? 'Reservoir' in the name and the dam marked on the southern side show that Cow Green is not a natural lake. What physical and human advantages to building a dam and reservoir here are suggested by the map evidence?

ExamCafé

REVISION

Key terms from the specification

Exam Café

Cross profiles of river valleys – V-shaped sections, changing downstream from steep to gentle
Erosion processes – wearing away the land surface by hydraulic action, abrasion, attrition and solution
Discharge – amount of water in a river at any one time
Flood plain – flat land built of silt on the sides of a river, usually in its lower course
Flooding – water covering land that is normally dry after a river bursts its banks
Gorge – steep narrow valley, with rocky sides
Hard engineering strategies – strong construction methods to hold floodwater back or keep it out
Levée – raised bank along the sides of a river, made of silt from river floods
Long profile of a river – a summary of the shape and gradient of a river bed from source to mouth
Meander – bend in a river, usually along its middle or lower course
Ox-bow lake – semi-circular lake on the flood plain of a river, a cut-off meander
Soft engineering strategies – more natural ways to reduce the impact of flooding on humans, with less intervention and more preparation
Transportation processes – movement of sediment by traction, saltation, suspension and solution

Checklist

	Yes	If no – refer to
Can you name three different ways in which a river and its valley change from source to mouth?		pages 80–81
Can you explain how waterfalls and gorges are formed by river erosion?		pages 82–3
Do you understand why rivers form meanders and ox-bow lakes in their lower courses?		page 85
Can you name two landforms formed by river flooding and explain their formations?		pages 85–6
Do you understand why some rivers are described as flashy, while others are slow to flood?		page 87
Can you give some of the costs and benefits of different strategies used for dealing with river floods?		pages 92–3
Can you explain why long-distance water transfers from west to east are needed in the UK?		page 94

Case study summaries

Flooding in a rich part of the world (e.g. UK)	Flooding in a poor part of the world (e.g. Bangladesh)	Dam/reservoir in the UK
Location	Location	Location and advantages
Causes	Causes	Issues about construction
Effects	Effects	Benefits
Responses	Responses	Dealing with issues for sustainable supplies

Chapter 6
Ice on the Land

The Andes in southern Chile, where the effects of frost shattering and glacial erosion can be seen in the mountain peaks. The undulating lowland (grazed by guanacos) shows the results of glacial deposition.

QUESTIONS

- **How do we know that most of the UK was once covered by snow and ice?**
- **What are the differences between glacial landforms in mountainous areas and those in lowland areas?**
- **Why do glaciated mountain areas like the Alps attract visitors both in summer and winter?**
- **What is the evidence that most of the world's glaciers are retreating today?**

Ice sheets and glaciers

What are glaciers? When was most of the UK covered by ice and snow? Why are glaciers powerful agents of erosion?

Freshly fallen snow is composed of ice crystals and many air spaces. When you make a snowball, you compress the snow and remove the air spaces. The same happens naturally when snow accumulates; the weight of snow above compresses the air out of the snow below and converts snow into ice. As the ice becomes thicker it will move down the slope, propelled by its own weight. When ice moves it is called a **glacier**. Glaciers are of two main types.

1 **Valley glacier** – a moving mass of ice in which the movement is confined within a valley. It begins in an upland area and follows the route of a pre-existing river valley (**Figure 1**). Today most valley glaciers are found near the tops of young fold mountain ranges, such as the Alps, Andes, Rockies and Himalayas. Examples include the Mer de Glace near the ski resort of Chamonix in the Mont Blanc region of south-east France and the Rhône glacier in south-east Switzerland, the source of one of Europe's largest rivers.

2 **Ice sheet** – a moving mass of ice that covers the whole of the land surface over a wide area. Today these exist only in Antarctica and Greenland, which account for 96 per cent of the Earth's ice-covered land. In Antarctica, where only the peaks of some high mountains stick through the ice, just a tiny strip of bare rock is exposed along a few parts of the coast in summer (**Figure 2**).

EXAM PREPARATION

Pleistocene Ice Age

- It began 2 million years ago.
- At its peak 30 per cent of the Earth's surface was covered by ice.
- Ice reached as far south as New York in North America and Berlin in Europe.
- It ended just 10 000 years ago.

Figure 1 ***A valley glacier reaching the sea in southern Chile.***

Figure 2 ***Antarctica covered by its ice sheet with just enough bare rock for the location of a base for Chilean scientists.***

Distribution of glacial landforms in the British Isles

If you are sitting north of the line from London to Bristol and take a look out of the window, it must be difficult for you to imagine that just 40 000 years ago all the land you can see would have been part of a snow- and ice-covered white wilderness.

The British Isles was invaded by ice sheets from Scandinavia during the **Pleistocene Ice Age**, which covered everywhere except for the extreme south of England. In the higher areas, such as the Cairngorms, the Lake District and Snowdonia, heavy snowfall led to the accumulation of snow and ice in hollows on the rocky mountainsides. These were the sources for valley glaciers, which flowed down valleys previously eroded by rivers. Over the past 10 000 years the world has warmed up. Today no part of the British Isles lies above the **snow line** (the line above which snow and ice remain all year). However, the present-day landscapes of the British Isles show plenty of signs that ice sheets and glaciers once covered the land (**Figure 3**).

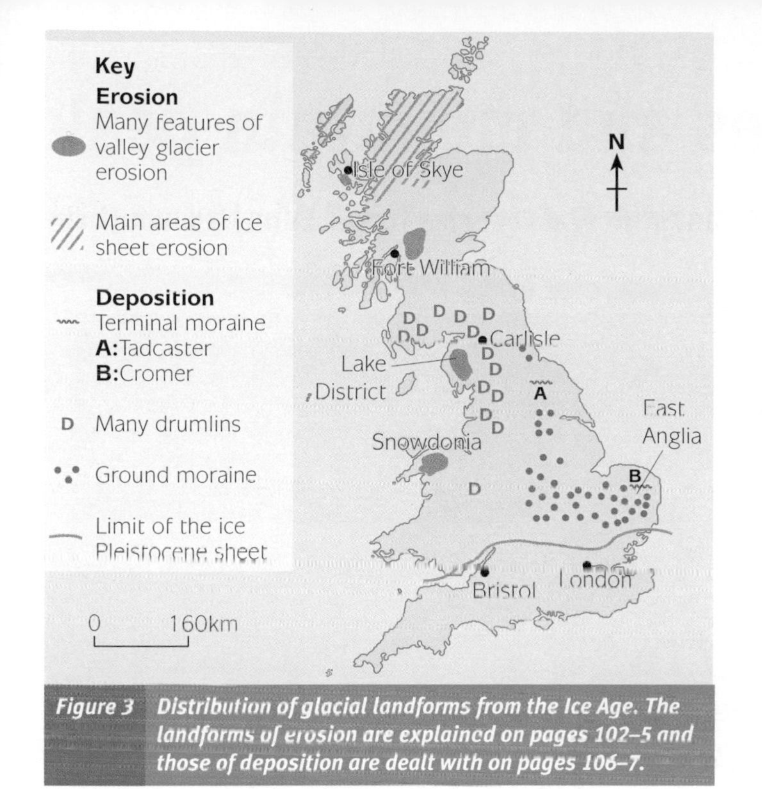

Figure 3 ***Distribution of glacial landforms from the Ice Age. The landforms of erosion are explained on pages 102–5 and those of deposition are dealt with on pages 106–7.***

Processes of glacial erosion

There are two main processes (or ways) of glacial erosion.

1. **Abrasion** rocks and rock particles embedded in the bottom of the glacier wear away the rocks over which the glacier passes. These sharp-edged pieces of rock of all sizes are held rigid by the ice above and are used as the tools for abrasion. Smaller rock particles have a sandpaper effect on the rocks over which the ice passes, while the sharp edges of the large rocks make deep grooves, called **striations** (**Figure 4**).
2. **Plucking** – this is the tearing away of blocks of rock from the bedrock as the glacier moves. These blocks of rock had been frozen to the bottom of the glacier where water had entered joints in the rock and become frozen. The blocks of rock between the joints are pulled away or plucked.

Glaciers would be less effective at eroding the landscape without the help of **freeze–thaw** weathering (page 28). Before the ice advanced, freeze–thaw left many frost-shattered rocks that

Figure 4 ***Striations on the hard rocks that outcrop in Central Park in New York. They are useful to geologists for working out the direction of ice movement.***

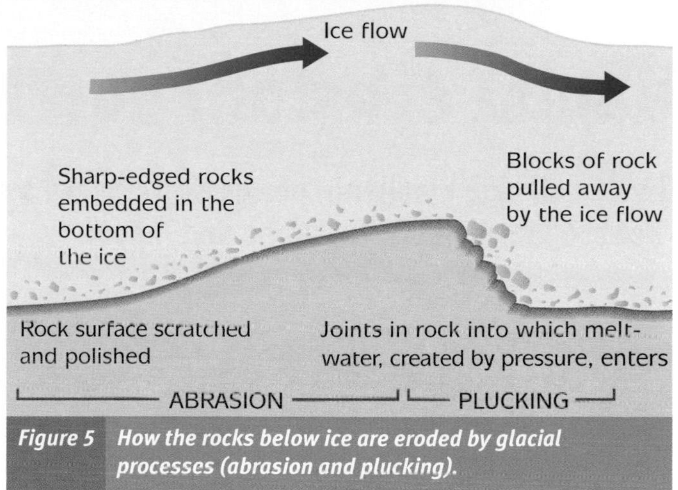

Figure 5 ***How the rocks below ice are eroded by glacial processes (abrasion and plucking).***

were easily removed by the glacier and then used as tools for abrasion. Even when the ice is present, freeze–thaw action affects rocks that outcrop above the surface of the ice because, in a cold climate, there are likely to be many changes of temperature above and below freezing point.

Of the two types of glacier, valley glaciers are considered to be more effective agents of erosion than ice sheets. Confined in a valley, the ice touches both the floor and the sides so that there is more contact between the ice and the rock and therefore more erosion. Also, valley glaciers flow more quickly, partly because of steeper gradients and partly because more meltwater is present to lubricate their flow. There is a plentiful supply of rock fragments from the frost-shattered peaks above, so these glaciers are well supplied with tools for abrasion. However, ice sheets cover and therefore erode a much greater area, so although they erode more slowly, they can still remove a large total amount of rock. As with the other agents of erosion, rocks that are soft or have weaknesses, such as many joints, are eroded more quickly, both by valley glaciers and by ice sheets.

ACTIVITIES

1. State the similarities and differences between glaciers and ice sheets.
2. (a) Draw labelled diagrams to show how each of the following processes operates:
   (i) freeze–thaw weathering
   (ii) plucking
   (iii) abrasion.
   (b) Explain why the breakdown and removal of rock is quicker when:
   (i) all three processes operate in the same area
   (ii) rocks have many lines of weakness.

Glacial erosion: corries and mountain peaks

Why are all high mountain peaks knife-edged and sharp? How do corries form? What allows them to be recognised in photographs and maps?

Figure 1 Panoramic view of part of the Swiss Alps taken from a tourist leaflet for the Jungfrau region. You should be able to identify the corries, arêtes and pyramidal peaks after studying these two pages. How many visitors to the region can do this?

Glaciers modify and enlarge landscape features that existed before the Ice Age. The effects of glacial erosion upon the landscape are greatest in upland areas where glaciers have been present for the longest time and the ice was deeper. In general, slopes are steeper and peaks narrower, especially in areas where processes of glacial erosion are still occurring (**Figure 1**).

Corrie (cirque)

The first landform formed by a glacier is the **corrie**. This is a circular rock hollow (hence the alternative name of **cirque** used by some geographers), usually located high on the mountainside, with a steep and rocky backwall up to 200 metres high in the UK, but much higher in the Alps.

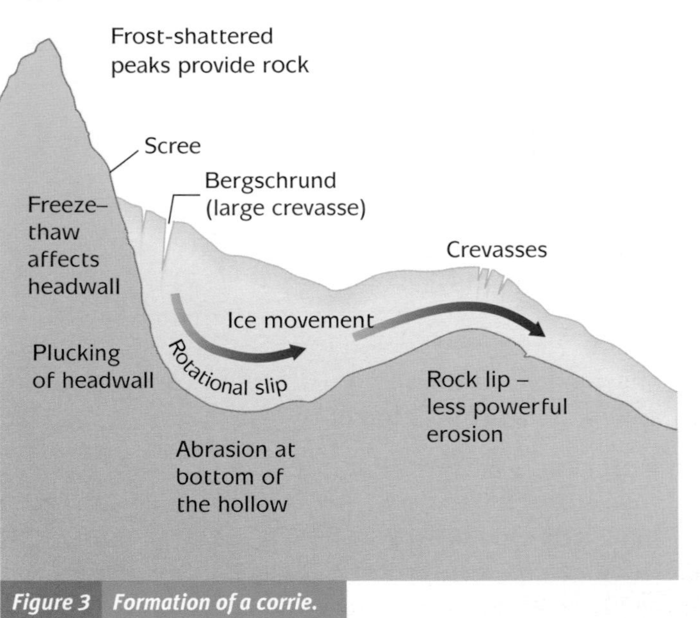

Figure 3 Formation of a corrie.

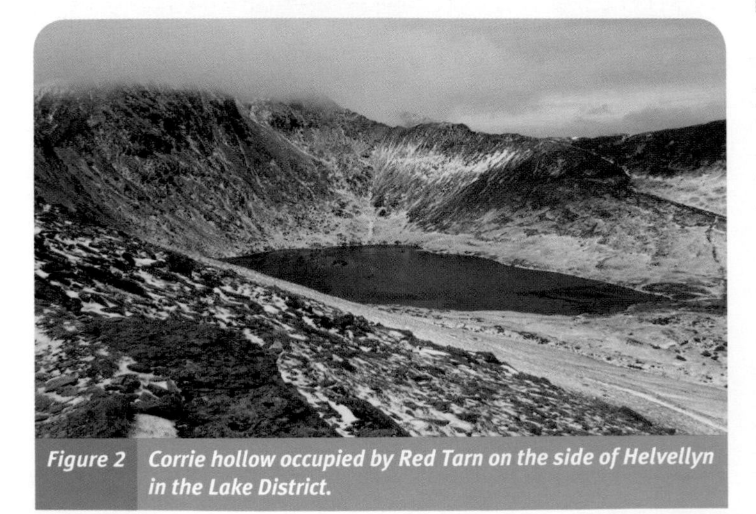

Figure 2 Corrie hollow occupied by Red Tarn on the side of Helvellyn in the Lake District.

Although most of the corrie is ringed by steep rocks leading to sharp rocky ridges, the front is open, with nothing more than a small rock lip on the surface. The hollow is typically filled with a small round lake, called a **tarn**, after the ice has melted. The information above *describes* the corrie; **Figure 2** shows a corrie that matches this general description.

Corries begin where snowfields (called **névés**), which accumulate below the mountaintops, form ice and grow. As with many landforms, it is necessary to refer to several different processes in order to explain the *formation* of the corrie (**Figure 3**).

- *Freeze–thaw weathering* plays a part in its formation. Frost action on the mountaintops and slopes above supplies loose rocks (scree). Water seeps down the bergschrund crevasse onto the headwall, increasing the amount of freeze–thaw activity, cutting back the headwall and making it steeper.
- Ice sticking to the headwall pulls away blocks of rock by *plucking* as the glacier moves.
- Loose rocks obtained from freeze–thaw and plucking are embedded in the ice and act as tools for scraping out the bottom of the hollow by *abrasion*.
- As a result of the *rotational slip movement* of the ice, there is greater pressure from the ice at the bottom of the headwall and in the base of the hollow than near the front, where the glacier leaves the corrie hollow to flow down the valley; the rock lip forms near the exit as a result of less powerful erosion.

When all the ice has melted, the corrie provides an ideal place for a tarn to form. There is a natural ice-carved hollow in which the water can accumulate. The rock lip acts as a natural dam on the one side that is not surrounded by steep slopes. Location in upland areas means that precipitation is likely to be high and there will be a large amount of run-off down the steep sides of the corrie, because the corrie forms a natural catchment area. A tarn fills the floor of most well-developed corries.

Figure 4 The Matterhorn near Zermatt in Switzerland.

FURTHER RESEARCH

Look at maps of Snowdonia on the weblink www.contentextra.com/aqagcsegeog to identify landforms of glacial erosion.

Arête and pyramidal peak

Look at a photograph showing the peaks of any of the world's high mountain ranges, not just the Alps, and you will find pointed mountain peaks and long and narrow knife-edged ridges. An **arête** is a two-sided sharp-edged ridge, whereas the **pyramidal peak**, as its name suggests, is a three-sided slab of rock, of which the most famous example is the Matterhorn (**Figure 4**) with its three near-vertical rock faces. Both landforms are created by the cutting back of the headwalls of corries on the slopes below the peaks by the processes of freeze–thaw weathering and plucking. For an arête, two corries, one on each side of the ridge, cut back until only a narrow piece of rock is left as the ridge top. For a pyramidal peak, three corries cut back. All the peaks continue to be sharpened by frost action.

ACTIVITIES

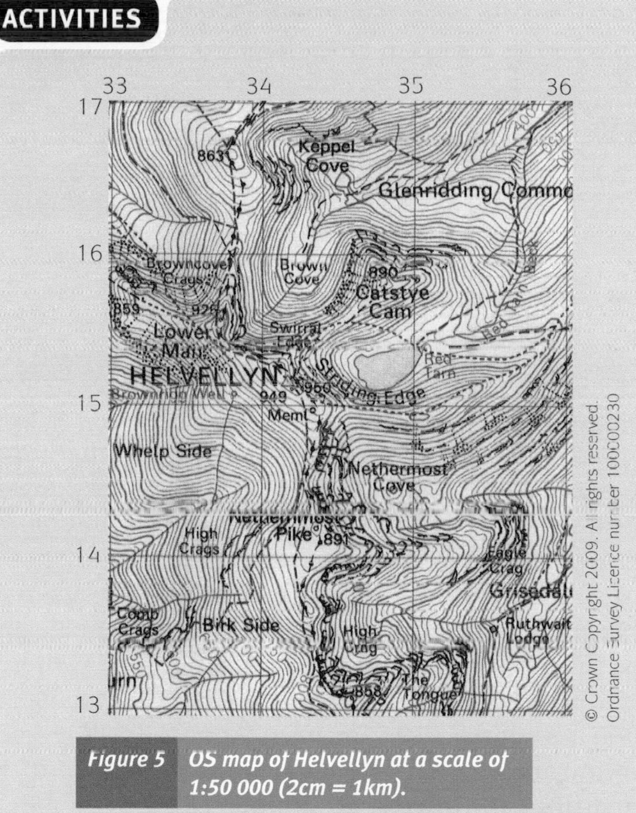

Figure 5 OS map of Helvellyn at a scale of 1:50 000 (2cm = 1km).

1 (a) Using **Figures 2** and **5**, describe fully the features of the corrie occupied by Red Tarn.
  (b) Explain how it may have been formed.

2 From **Figure 5**, draw a labelled sketch map to show how corries and arêtes can be recognised on OS maps.

3 (a) What evidence can you find on the map that visitors come to the area?
  (b) Why is the area uninhabited?

ICE ON THE LAND 6

Glacial erosion: valley landforms

How are glaciated valleys different from river valleys? Why are they usually wider, deeper, steeper and straighter?

Figure 1 *The Lauterbrunnen valley in Switzerland, carved out by glacial erosion. The valley is too large to have been eroded by the small stream that flows in it today.*

The **glacial trough**, often more simply called a U-shaped valley, is an impressive landscape feature (**Figure 1**). These glaciated valleys can be hundreds of metres deep with vertical rock walls, down which waterfalls cascade from **hanging valleys**. On the top of the valley sides the land often flattens out to form a **high-level bench**, known as an 'alp' in the Alps of Switzerland (the area in the foreground in **Figure 1**). The width and flatness of the valley floor are in marked contrast to the steepness of the sides. These valleys are drained by **misfit streams**, which are dwarfed by the size and scale of the new glaciated valley. In some glacial troughs, lakes fill parts or the whole of the valley floor; these lakes are **ribbon lakes**, so-called because of their shape, which is long and thin. In the lower parts of the valley, examples of landforms of glacial deposition, such as **terminal moraines**, are found. The valley's *long profile* is characterised by its irregular shape (**Figure 2**), providing many hollows for lake formation.

Formation of valley landforms

Everything about a glacial trough shows the power of ice to erode. The former V-shaped river valley is widened, deepened and straightened by the valley glacier into a U-shaped valley. Before the ice, river erosion was confined to the small part of the valley where the river flowed; the glacier, however, fills the whole valley. This means that ice is in contact with the entire floor and with both valley sides, so erosion is no longer confined to the centre of the valley. The V-shaped river valley is changed into the U-shaped glacial valley because glacial erosion by abrasion and plucking occurs everywhere in the valley where the ice is in contact with rock. The river moved around obstacles in its path, and its winding course created interlocking spurs. The more powerful glacier cannot flow so freely around corners, and it pushes straight forward, cutting off the edges of interlocking spurs to form **truncated spurs** and straight valley sides. The ice is thicker in the main valley because it is fed by all the glaciers from tributary valleys. In each tributary valley there was a smaller, less powerful glacier than the main glacier. When only the rivers remained after the ice melted, those in tributary valleys were left hanging well above the level of the main valley floor. The streams from these hanging valleys fall as **waterfalls** into the main valley (**Figure 3**).

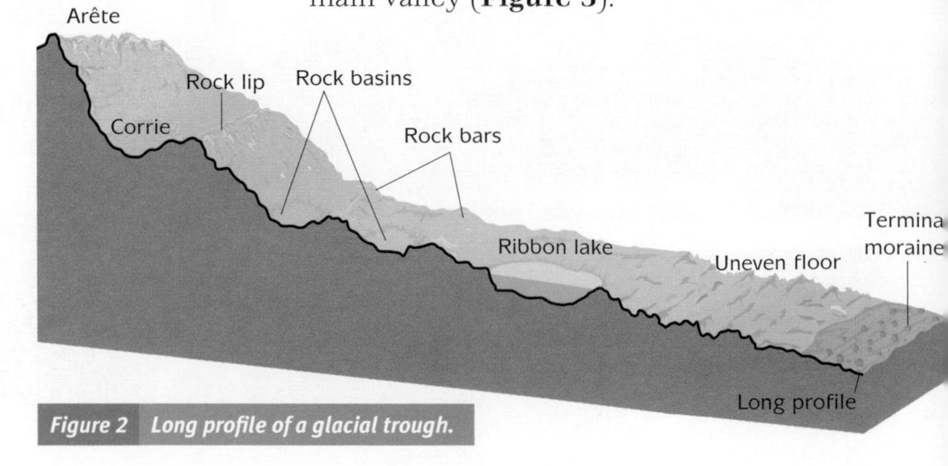

Figure 2 *Long profile of a glacial trough.*

Figure 3 Waterfall from a hanging valley along the side of the Lauterbrunnen valley.

As a glacier flows down a valley it picks out weaknesses in rocks, eroding those rocks that are soft and well-jointed more rapidly than those that are hard and resistant. In those places where outcrops of hard and soft rocks alternate, the glacier erodes the soft rock more quickly and more deeply, by abrasion and plucking, forming a **rock basin**. The hard rock is left as a **rock bar**. After the ice melts, the rock basin is left as a hollow on the valley floor between two rock bars, and it is soon filled up by water to form a ribbon lake.

The map (**Figure 4**) shows that the centre of the rock basin that contains Wast Water is over 70 metres deep. Note how the submarine contours indicating water depth show the U-shaped cross-profile of the glacial trough. The steep valley sides continue below the water until the flat valley floor in the centre of the lake is reached.

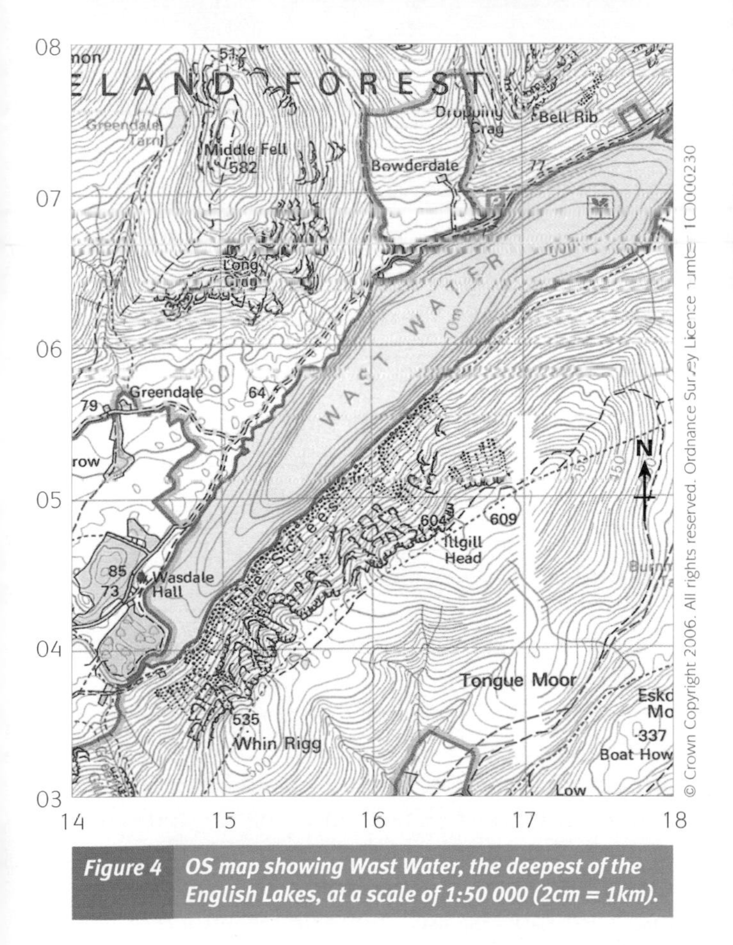

Figure 4 OS map showing Wast Water, the deepest of the English Lakes, at a scale of 1:50 000 (2cm = 1km).

Figure 5 Wast Water and its scree slopes.

ACTIVITIES

1 Make a sketch of **Figure 1.** Name and label the features of glacial erosion shown.

2 Explain how the screes shown in **Figure 5** have formed (page 28).

3 (a) From **Figure 4,** draw a sketch cross-section across Wast Water valley.
(b) Explain how the lake was formed.

Landforms of glacial transportation and deposition

When, where and why do glaciers deposit their loads? What are the similarities and differences between the resulting landforms?

Ice behaves in the same way as all the other agents of erosion:

- it wears away the land surface – *erosion*
- it carries away the materials eroded – *transportation*
- it dumps elsewhere the materials it is carrying – *deposition*.

Valley glaciers erode with so much power, and ice sheets erode such great expanses of land, that large amounts of loose rock are available for transport. Glaciers can transport enormous loads. Look at **Figure 1**, which shows an **erratic**, the name given to a boulder dropped by ice in an area where it does not belong. This big grit boulder has been dumped on top of the local white limestone rock. This boulder is just one of hundreds that an ice sheet deposited in the same area. Can you imagine a river having the power to transport one of these boulders, never mind hundreds of them? In a river channel a boulder of this size would need to be broken down by corrasion and attrition into small pieces before it could be transported; the ice simply carries it.

All materials transported by glaciers are called **moraine**. Although most are carried in the glacier's base, some are carried on the surface. These show up as dark lines of moraine (**Figure 2**) on the top of the glacier. Piles of material along the sides are called **lateral moraines**; those somewhere in the middle of the glacier, formed after valley glaciers join together, are called **medial moraines**. Two separate lateral moraines unite to form one medial moraine. The material for these moraines is broken off from the rocky peaks above by freeze–thaw weathering and it falls down the valley sides on to the top of the ice.

Figure 1 Erratic block of grit perched on limestone in the Yorkshire Dales National Park.

Figure 2 The Grosser Aletsch glacier in Switzerland with lateral and medial moraines on the ice surface.

Figure 3 Boulder clay.

However, not even a glacier can keep on growing for ever. It reaches a point where the ice loss is greater than the amount of new ice supplied. For example, most valley glaciers begin to melt when they reach lower ground where temperatures are higher. Only a few reach the sea before they have completely melted (see **Figure 1** on page 100). As the ice melts and thins, its carrying capacity is reduced. When the glacier reaches the point of overload (load greater than carrying capacity), it must deposit some or all of its load. Any obstacle along its course encourages deposition.

The general name given to all materials deposited by ice is **boulder clay**. As its name suggests, this is usually clay that contains numerous boulders of many different sizes (**Figure 3**). It is an *unsorted* deposit. This means that large and soft rocks, as well as finer particles, are all mixed together. The boulders it contains are described as *angular*. They have sharp edges, not yet rounded off, as they would have been if they had been transported by rivers. The 'ingredients' of the boulder clay vary greatly according to what the glacier eroded before it reached the area. Sometimes the deposits are more sandy than clayey, which is why some physical geographers prefer to use the term **glacial till** instead of boulder clay to describe all ice-deposited materials. As the glacier continues to push forward, melting more and more all the time, it leaves a trail of boulder clay behind it, which forms a hummocky surface of **ground moraine**.

Drumlins

In many of the lowland areas of south-west Scotland and north-west England, glacial deposition has produced a distinctive landscape of many low hills, each one typically about 30–40 metres high and 300–400 metres long. These hills all lie in the same direction and have similar shapes – blunt at one end and tapered at the other; in fact, each hill looks like an egg. Each of these hills is a **drumlin**. Drumlins occur in swarms and are said to form 'basket of eggs' topography, so-called because of the appearance of the landscape (**Figure 4**).

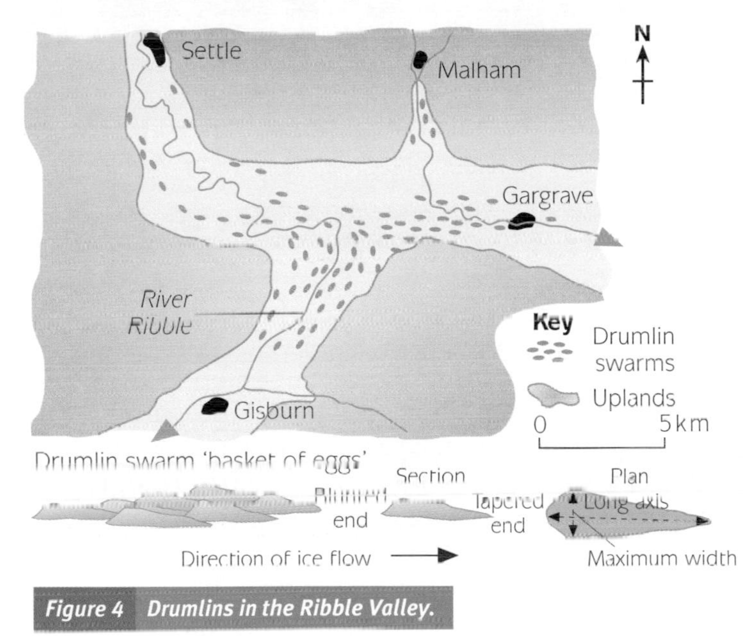

Figure 4 *Drumlins in the Ribble Valley.*

Drumlins form when the ice is pushing forward across a lowland area, but it is overloaded and melting. It does not need much to encourage more deposition; any small obstacle, such as a rock outcrop or mound, is sufficient. Most deposition occurs around the upstream end of the obstacle, which forms the drumlin's blunt end. The rest of the boulder clay that is deposited is then moulded into shape around the obstacle by the moving ice to form the tapered end downstream. The drumlin is another landform from which it is possible to detect the direction of ice movement.

Terminal moraine

All the remaining load is dropped and dumped at the glacier's **snout** – the furthest point reached by the ice. This point is marked by a ridge of boulder clay across the valley or lowlands, running parallel to the ice front, and is called a terminal moraine (page 104). Where ice sheets remained stationary for a long time, as in central Europe during the main ice advances in the Ice Age, sufficient boulder clay was deposited to form ridges more than 200 metres high. More typically, terminal moraines formed by valley glaciers are 20–40 metres high. Terminal moraines that cross valleys form natural dams behind which river water can collect and form lakes. These lakes are also long and thin and are called ribbon lakes (page 104). This tells you that landforms with the same appearance can have different methods of formation.

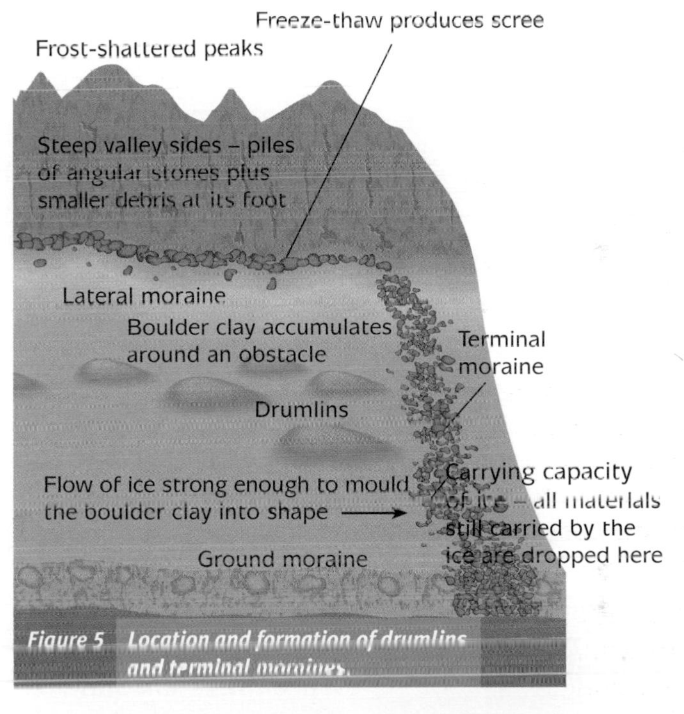

Figure 5 *Location and formation of drumlins and terminal moraines.*

GradeStudio

1 a State two features of boulder clay. (2 marks)
  b Explain why these features show deposition by ice instead of rivers. (2 marks)

2 a Name one example of a terminal moraine. (1 mark)
  b State one difference and one similarity between terminal and lateral moraines. (2 marks)
  c Explain why terminal moraines are usually larger than lateral moraines. (2 marks)

3 a From **Figure 3** on page 101, describe the distribution of drumlins in the UK. (2 marks)
  b Explain why drumlins are 'egg-shaped' and are found in 'swarms'. (4 marks)

Exam tip

Be prepared – make sure that you can give named examples for the major landforms, such as corrie, arête, pyramidal peak, glacial trough and terminal moraine.

Glaciers – advance and retreat

Why is it normal for glaciers to keep on advancing and retreating over time? How much evidence of glacier retreat is there today?

As with climate (page 54), the majority of people seem to think that physical things just stay the same. They don't, and this includes glaciers. The size and length of glaciers are always fluctuating; they advance and retreat. Before 1850 glaciers were advancing rapidly in Europe, during what was called the 'Little Ice Age'; today they are retreating, and fast. It all depends upon the **glacial budget**.

Glacial budget

Glaciers form where the amount of snow that falls during the year is greater than the amount that melts. This is the **zone of accumulation** (**A** on **Figure 1**). As snowfalls continue to accumulate, ice forms as the air spaces are sealed. The ice is carried downslope by the glacier's movement. At a certain point downslope, as summer melting of the ice (on the top, at the sides and in the bottom of the glacier) increases, water loss begins to exceed the winter gains from snowfall. This is the **zone of ablation** (**B** on **Figure 1**), where there is a net loss of ice. Looking at **Figure 1**, where would you draw the line between the zone of accumulation and the zone of ablation?

Figure 1 Fox Glacier, New Zealand.

Remember that the position of the dividing line between the zones of accumulation and ablation changes from year to year, and over longer periods of time. A glacier is a system of inputs and outputs. Changes to one affect the other. Therefore, if precipitation that falls as snow increases, the greater volume of ice pushes the ice front forwards. At present, in this time of global warming (**Figure 1**, page 54), winter snowfall is decreasing and summer melting is increasing, which means that we are living in a period of glacial retreat (**Figure 2**). Each summer it appears that the surface ice cover in the Arctic Ocean is melting more and becomes thinner, while even bigger icebergs are calving (breaking off) from the Antarctic ice sheet. Some of these icebergs are the size of an island such as Malta.

Figure 2 Many of the rock outcrops seen below this glacier in South Georgia, on the edge of Antarctica, have only been exposed in the last 30 years. Notice also the terminal moraine (beyond the king penguins) deposited by the melting glacier.

Glacier watch

Glaciers and ice sheets are being watched, monitored and measured as never before. Use of satellites for taking photographs and measurements now makes this much easier. However, in some places there is a long history of study and measurement. Swiss geographers began measuring the country's 121 glaciers in 1881 and they continue to do so. They record not only length, but also mass balance – the relationship between winter accumulation and summer melting. The general trend is clear (**Figure 3**). The recent trend post-1980 is even clearer.

Figure 3 Cumulative changes in ice mass for a glacier in south-east Switzerland (1900–2007).

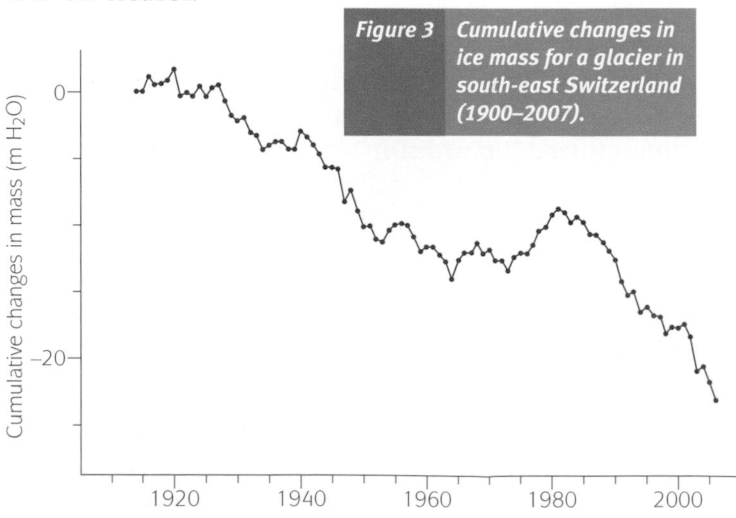

Grosser Aletsch glacier

Europe's longest glacier (23 kilometres long today), this is a typical valley glacier (**Figure 2**, page 104). It originates in snowfields on the southern sides of the Jungfrau and Monch, on the other side of the peaks shown in **Figure 1** (page 102), flowing towards the Rhône valley. The maximum depth of ice is just below the peaks, about 900 metres. Several side glaciers join it, which is why there are parallel lines of medial moraines on the ice surface, about 12–15 metres high. The glacier is moving continuously, up to 200 metres a year, or about half a metre a day.

Figure 4 *The 1860 ice level of the Grosser Aletsch glacier is marked X.*

In 1860 ice levels in the lower part of the glacier shown in **Figure 2** (page 104) were 200 metres higher and 3 kilometres longer than today. When viewed more closely in **Figure 4**, this ice level can be seen to have left its mark on the valley sides. It is shown by the distinctive broad band of lighter-coloured rock, against which the ice previously flowed. This marks the end of a colder climatic interlude in Europe, which lasted from 1350 to 1850. This 500-year period reversed the general warming up of the Earth that has continued since the end of the Pleistocene 10 000 years ago.

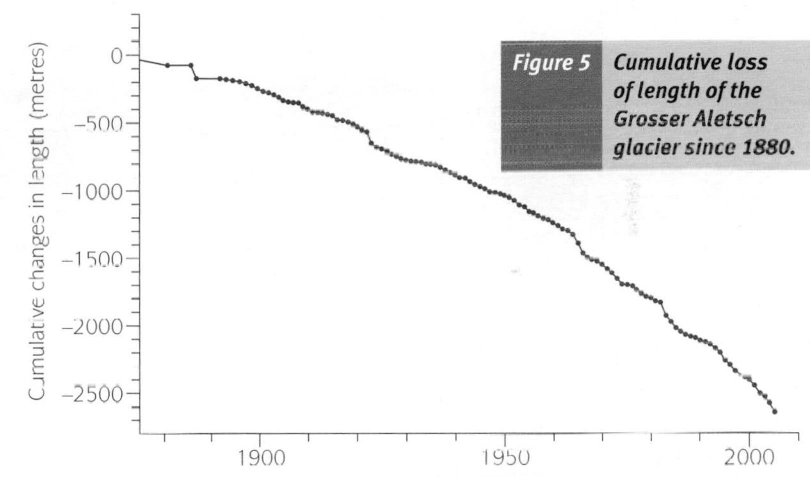

Figure 5 *Cumulative loss of length of the Grosser Aletsch glacier since 1880.*

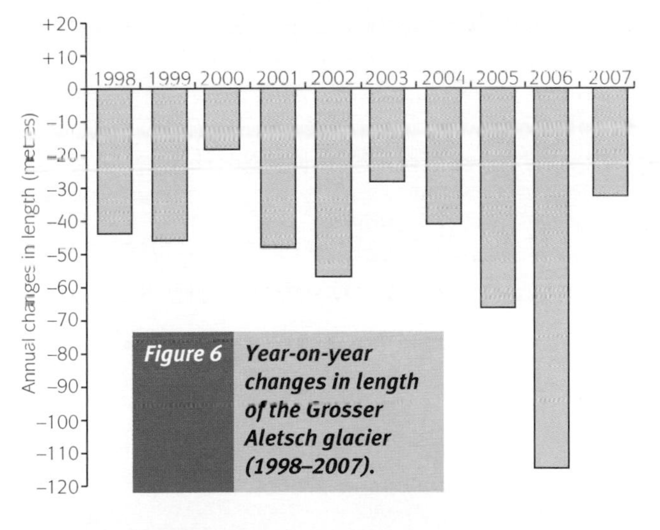

Figure 6 *Year-on-year changes in length of the Grosser Aletsch glacier (1998–2007).*

Since 1860 the Grosser Aletsch glacier has been in almost continuous retreat (**Figure 5**). In recent years thinning at its sides has become more noticeable. **Figure 6** is a record of the year-on-year changes for the ten years up to 2007. The big loss in 2006 was caused by the combination of a hotter-than-average summer and lower winter snowfall. One source summarised the glacier's current state as 'retreating by up to 50 metres per year'. How fair a summary is this?

GradeStudio

1 Draw a labelled sketch to show three types of moraine. (3 marks)

2 Study **Figure 1**.
   a Name one type of moraine that can be seen and state where it is located. (2 marks)
   b Describe the differences between the top part of the glacier (around A) and the lower part of the glacier (around B). (4 marks)
   c Give reasons for these differences. (4 marks)

3 a Most of the world's glaciers are retreating. Using examples, describe the different types of evidence for this. (5 marks)
   b How reliable is the evidence? Explain your view about this. (2 marks)

Exam tip

Always look at the number of marks for the question. Make your answer match the marks – and a bit more, for safety.

Jungfrau region of Switzerland

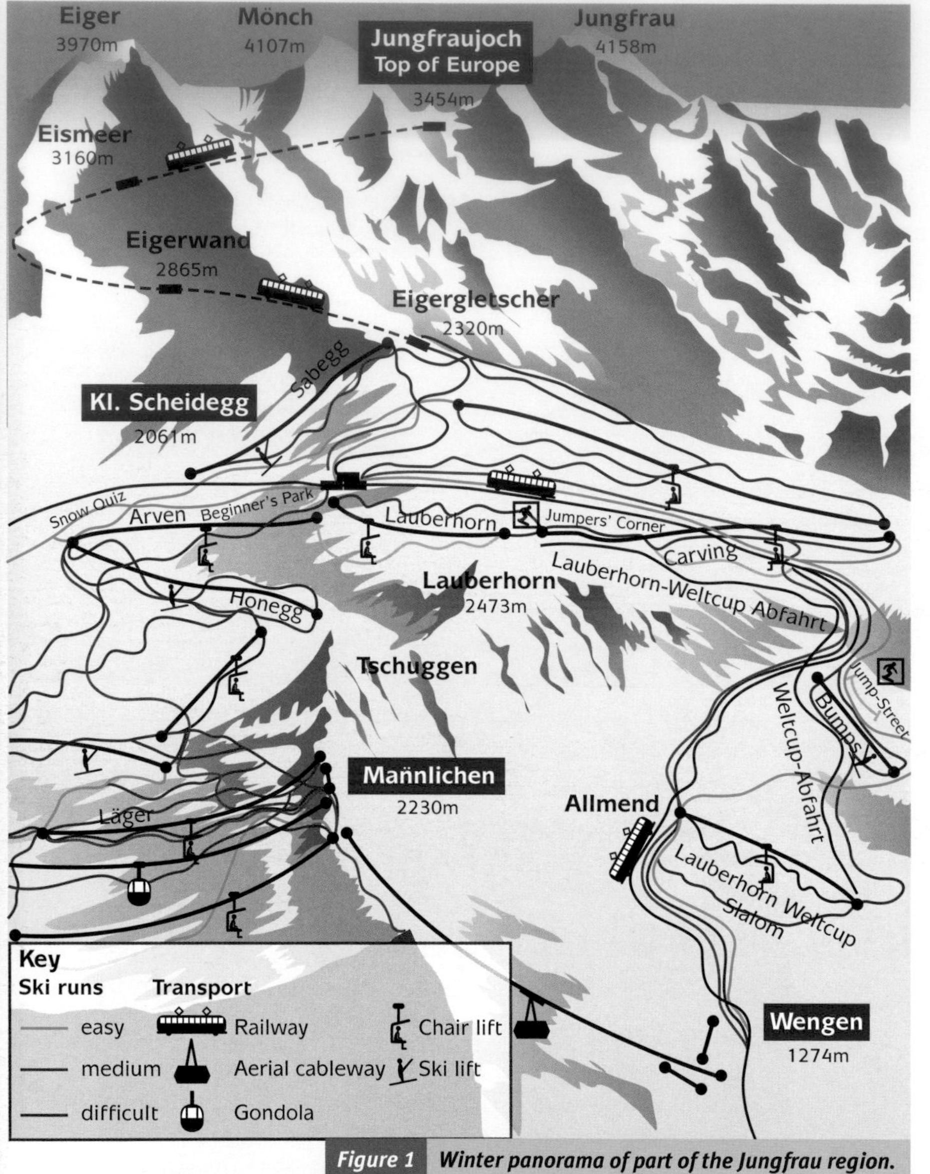

Figure 1 *Winter panorama of part of the Jungfrau region.*

Figure 1 shows some of the many natural and human attractions of the region for visitors:

- winter snow, and all-year glaciers and snowfields at high levels
- some of the best high mountain scenery in the Alps
- easy access using the dense transport network of mountain railways, cable cars and gondolas
- the highest railway station in Europe (Jungfraujoch – Top of Europe)
- many facilities for skiers such as chair- and ski-lifts, accommodation and food in resorts such as Wengen.

As income from traditional Alpine activities such as mountain agriculture and forestry (pages 10–11) began to dwindle, tourism revived and expanded the economy of the region. Winter skiing in particular is an activity of high economic value; skiing holidays are not cheap, a lot of local services are used and much winter employment is generated. Mountain farmers use winter tourism to supplement their low incomes. It has stopped the outward migration to the cities. Over 80 per cent of the jobs in the area are tourist-dependent. Facilities built for winter sports can be used by summer tourists, who enjoy the mountain views and walks.

Negative impacts of tourism

The switch to a tourist-based economy in the region has hastened the decline of farming, which cannot compete economically with tourism. Traditional ways of life are under pressure. The social make-up of the region is changing; tourism pulls many outsiders into the region, many of them just as temporary workers, and a lot of them from other countries.

EXAM PREPARATION

Avalanches

- Sudden downhill movement of snow and ice (with or without rocks).
- Descend at great speeds; average 40–60 kilometres per hour.
- Sweep away and cover everything in their paths.
- Highly unpredictable.

The environment is suffering. Alpine forests are cleared to make way for ski-runs and chair-lifts. Alpine habitats for wild flowers and wildlife are being lost. Remaining trees are damaged by skiers knocking off branches and killing off saplings. Increased vehicle emissions from visitors are contributing to acid rain; half the trees are dying from pollution. Building work for new ski facilities and to re-shape ski slopes increases slope instability. Without the protection of vegetation, the **avalanche hazard** is now higher than ever (see Exam Preparation Box).

Economic pressures for development are strong, but this is a **fragile environment**. In cold environments like this, with severe winters, vegetation struggles to survive even without human interference. At the end of the winter season, the grass on many ski slopes is badly worn; patches of bare ground show through where people have skied on snow that was too thin. After the snow melts, litter is left strewn all over the ski slopes. How great are the chances of trees, grasses and alpine plants recovering before the next winter visitor invasion?

Figure 2 *Alpine slopes showing wear and tear after winter skiing.*

Figure 3 *A snow-blower in operation.*

How is tourism managed in this region?

Car access is restricted to towns around the edges of the area shown in **Figure 1**. All the transport services shown run on clean electricity from hydroelectric power stations in the Alps. Small electric vehicles are used in Wengen for essential local services such as waste collection and deliveries. At the end of each ski season, a survey of environmental and other damage is undertaken. Areas of bare slopes are fenced off and re-seeded to make them green again; the course of the winter ski pistes may be altered to allow more time for recovery. People are employed to clear the slopes of rubbish and litter as soon as the snow melts, and before cattle return to the summer pastures.

The Swiss government, well aware of environmental pressures, has signed up to the Convention on the Protection of the Alps, which prohibits further development of new ski stations and promotes sustainable development. Since the Jungfrau region is well established, the demand for new building works is less intense than in many other parts of the Alps; however, all new proposals are carefully scrutinised and monitored for their long-term sustainability.

The effects of climate warming

Low-level resorts below 1500 metres, like Wengen, are already feeling the effects of a shortening of the ski season, caused by reduced winter snowfalls. What happened in the winter of 2006–7 is almost becoming the norm – little snow fell before Christmas, while mild weather from February onwards shortened the benefits from the good New Year snowfalls. Snow-making machines provide only a partial solution (**Figure 3**); they are expensive and noisy to operate, and consume millions of litres of water.

FURTHER RESEARCH

Read 'Melting glaciers will destroy Alpine resorts within 45 years, says report' at the weblink www.contentextra.com/aqagcsegeog.

Predictions for the future make gloomy reading for anyone with economic interests in Alpine skiing.

- Up to 75 per cent of Alpine glaciers will disappear within the next 50 years.
- Risks of avalanches and floods will greatly increase.
- More rain and less snow will close all ski resorts below 2000 metres.
- Surface erosion will increase as slopes are graded to allow skiing on thinner snow cover.

For resorts to be profitable, the tourist season needs to be busy and prolonged. More ski resorts will need to diversify into other leisure areas, such as spas, and to promote more summer tourism. Given its wonderful scenery and efficient transport systems, the Jungfrau region seems likely to enjoy economic prosperity long after Alpine resorts elsewhere have gone out of business.

ACTIVITIES

1 Explain, with examples, the meaning of these key syllabus terms: (a) avalanche hazard (b) fragile environment (c) environmental impact (d) management strategy.

2 Make exam notes for a case study of an Alpine area for tourism using these headings:
(a) Attractions for visitors (**Figure 1**)
(b) Impacts from visitors (economic, social, environmental) (c) Management strategies for visitors, and comment on their level of success (d) Planning and management for the future.

Practice GCSE question

See a Foundation Tier Practice GCSE Question on the weblink www.contentextra.com/aqagcsegeog.

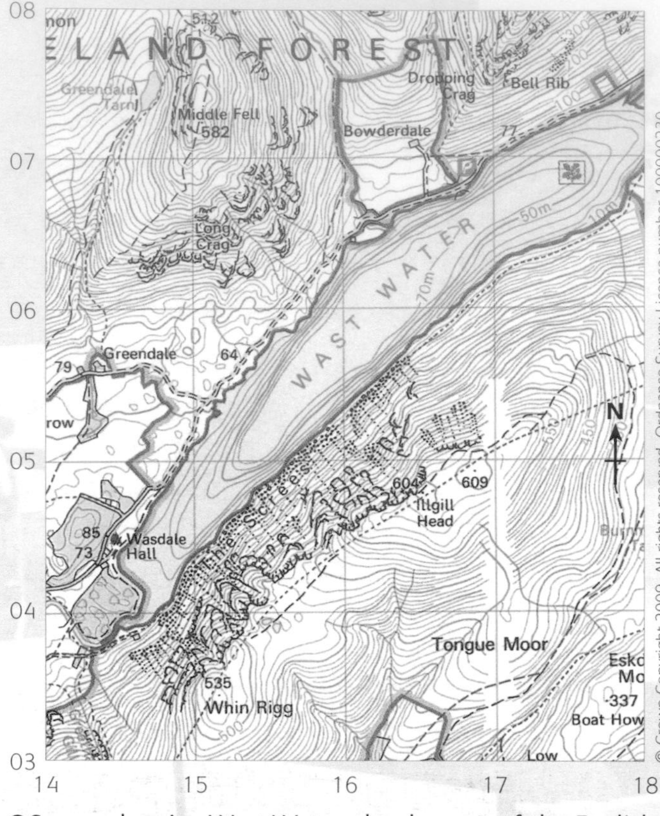

OS map showing Wast Water, the deepest of the English Lakes, at a scale of 1:50 000 (2cm = 1km).

Figure 1 Aerial view of part of the Alps.

6 (a) Study the Ordnance Survey map extract of Wast Water, in the Lake District.

(i) Give the six-figure grid reference for the highest point on the map extract (609m). **(1 mark)**

(ii) Measure the greatest width and length of Wast Water lake shown on the map extract. **(2 marks)**

(iii) Describe the physical features of the Wast Water valley that show it is a glacial trough. **(4 marks)**

(b) Study **Figure 1**, which is an aerial view of part of the Alps.

(i) Name the type of lake shown and describe its shape. **(2 marks)**

(ii) Explain how the lake may have been formed. **(4 marks)**

(iii) Describe how the appearance and shape of the peaks show the effects of freeze–thaw weathering. **(2 marks)**

(iv) Explain how freeze–thaw weathering works. **(4 marks)**

(c) Using examples of landforms of deposition, explain why they are of many different shapes and sizes. **(6 marks)**

Total: 25 marks

Exam tip

The question for this topic will be Question 6 in Section B in Paper 1.

Improve your GCSE answers

How to recognise glacial landforms from contour patterns and OS signs

A Landforms of erosion

Look at **Figure 1**. Can you pick out the following glacial landforms? They are arranged in alphabetical order of name, not letter order on **Figure 1**.

arête **corrie** **glacial trough**
hanging valley **pyramidal peak**
ribbon lake **tarn**

It should be easier for you to pick out these landforms from **Figure 1** than on the OS maps used in **Figure 5**, page 103 and **Figure 4**, page 105. This is because the contour interval is much larger (75 metres instead of 10 metres). Remember, however, that you need to focus on *contour shapes*. With all their contours, the OS maps have the advantage of showing the steepness and ruggedness of glacial uplands more clearly.

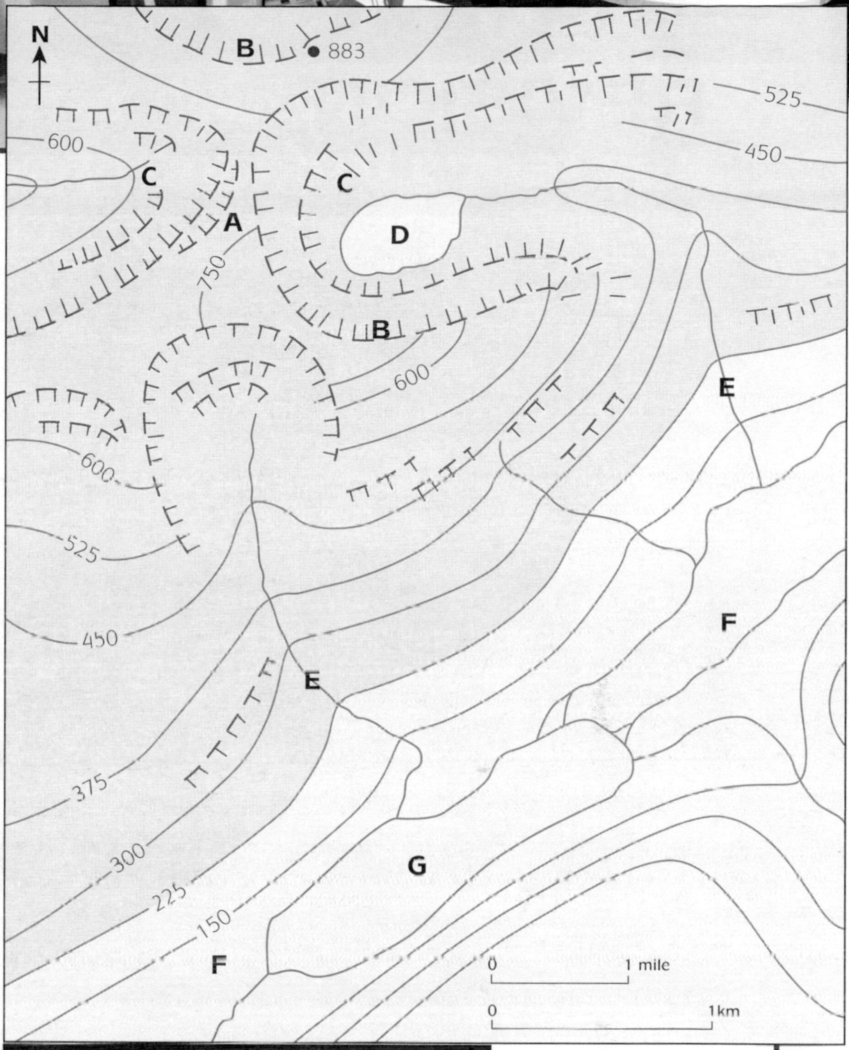

Figure 1 Contour map of glaciated upland area.

B Landforms of deposition

Landforms of glacial deposition are mainly low hills, rarely more than 40–50 metres high. The only ones that can be recognised with any certainty on OS maps are drumlins. Drumlins are distinctive because many are found together in one area, all aligned in a similar direction.

- In which direction are the drumlin hills aligned in **Figure 2**?
- What does this tell you about the direction of ice flow?
- Where are areas of marshy ground found? Why?

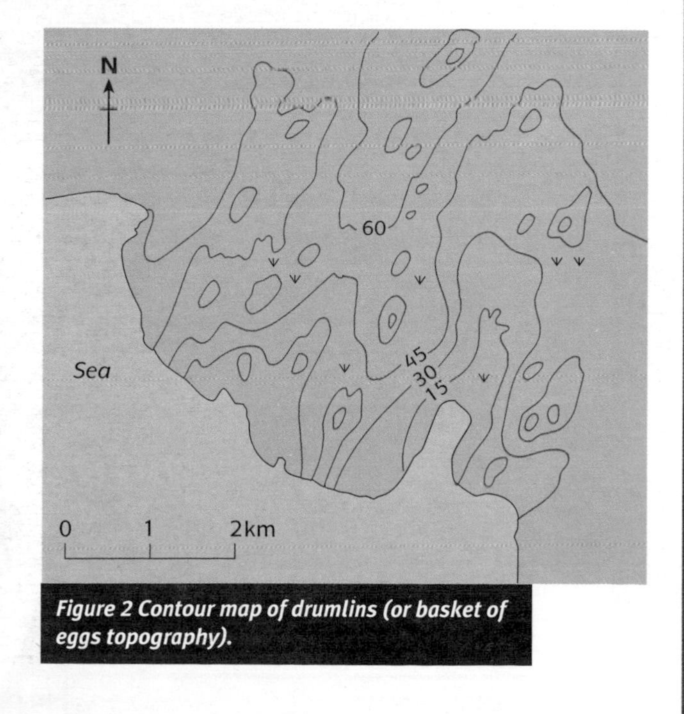

Figure 2 Contour map of drumlins (or basket of eggs topography).

REVISION

Key terms from the specification

Arête – sharp-edged two-sided ridge on the top of a mountain
Boulder clay/till – all materials deposited by ice, usually clay containing sharp-edged boulders of many sizes
Corrie – circular hollow, high on a mountainside, surrounded by steep rocky walls except for a rock lip on the open side
Glacial trough – U-shaped valley, with flat floor and steep sides, formed by a valley glacier
Hanging valley – tributary valley, high above the main valley floor, with a waterfall
Ice sheet – moving mass of ice that covers all the land over a wide area
Moraine – materials deposited by ice, with different names according to where they were deposited
Ribbon lake – long and narrow lake in the floor of a glaciated valley
Scree – pieces of rock with sharp edges, lying towards the foot of a slope
Striations – deep grooves in surface rocks, made by the sharp edges of stones carried in the bottom of moving ice
Tarn – circular lake in a corrie hollow, where water is trapped by the steep sides and rock lip
Truncated spurs – higher areas on the straight rocky sides of a glaciated valley
Valley glacier – a moving mass of ice confined in a valley

Checklist

	Yes	If no – refer to
Do you know the difference between a valley glacier and an ice sheet?		page 100
Can you name the two processes of glacial erosion and explain how they erode the land?		page 101
Do you understand how a corrie is formed?		pages 102–3
Can you recognise corries, arêtes and tarns on OS maps?		page 103
Do you know when, where and why glaciers deposit their loads?		pages 106–7
Can you name four different landforms of glacial deposition and describe their different characteristics in appearance and location?		page 107
Do you understand why glaciers keep advancing and retreating?		page 108
Can you list the attractions of the Alps for winter sports and the impacts of these sports (good and bad)?		pages 110–11

Case study summaries

A glacier	Alpine area for winter sports
Location	Location
Evidence for advance	Attractions for tourists
Evidence for retreat	Impacts on the area
Reasons for these shifts	Management and future issues

Chapter 7

The Coastal Zone

The Norfolk village of Happisburgh (pronounced *'Hayes-borough'*) under attack from the sea. What was lost in the ten years from 1998 to 2007? What else will have gone by now?

1998

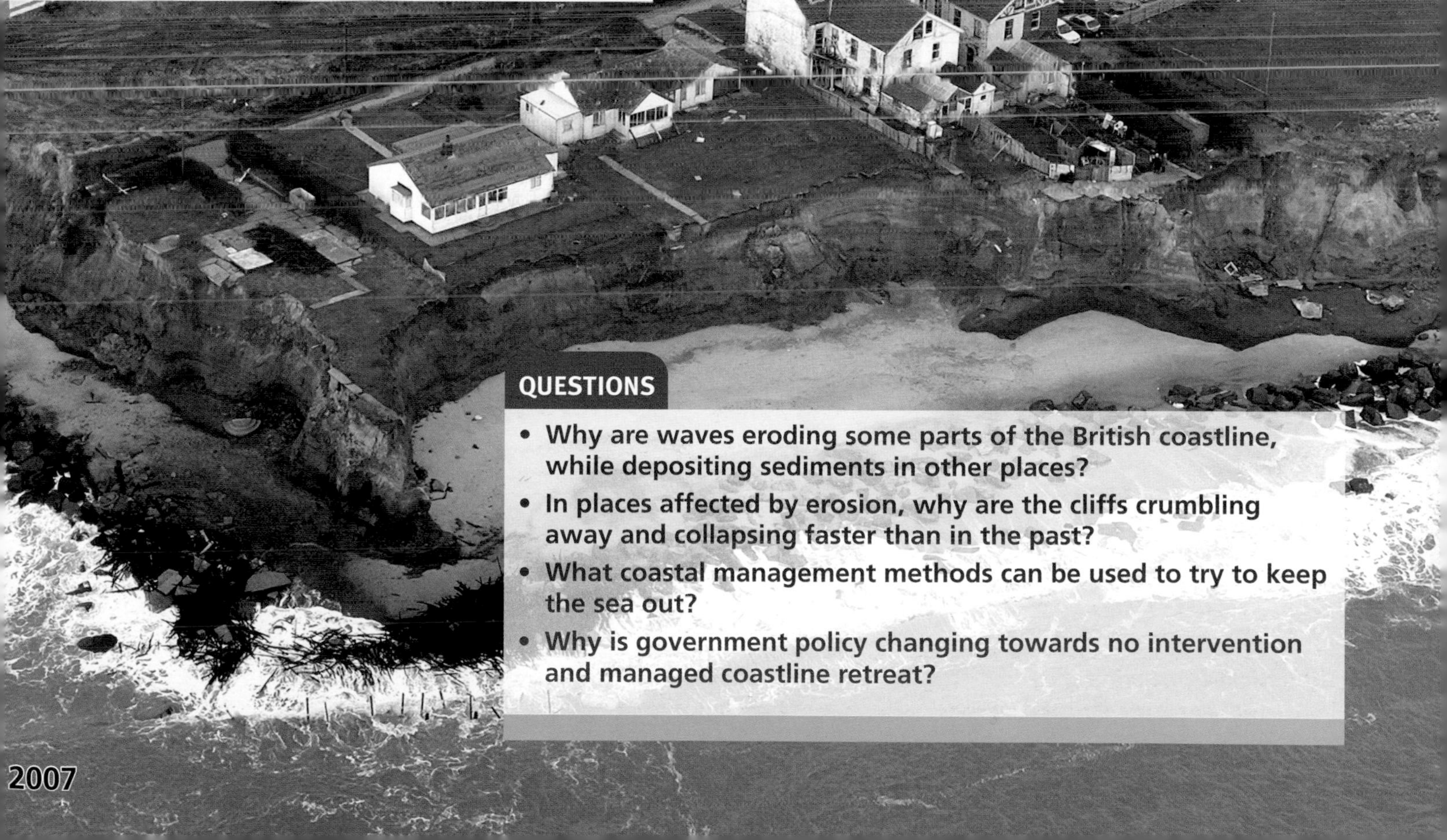

2007

QUESTIONS

- **Why are waves eroding some parts of the British coastline, while depositing sediments in other places?**
- **In places affected by erosion, why are the cliffs crumbling away and collapsing faster than in the past?**
- **What coastal management methods can be used to try to keep the sea out?**
- **Why is government policy changing towards no intervention and managed coastline retreat?**

How is the coast eroded?

Why is the sea a powerful agent of erosion? How do waves erode the coast? What else helps to speed up rates of coastal erosion?

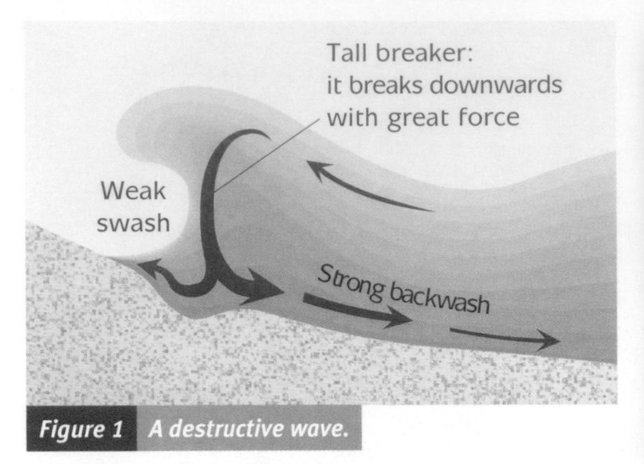

Figure 1 *A destructive wave.*

Figure 2 *Chalk cliffs at Beachy Head, created by destructive waves.*

Figure 3 *'Waves lash the seafront at Lyme Regis' was one newspaper headline: other newspaper headlines about storms are given below.*

Waves are responsible for most of the erosion along coasts. Wind blowing over a smooth sea surface causes small ripples, which grow into waves. When a wave approaches the coast its lower part is slowed by friction with the sea bed, but the upper part continues to move forward. As it is left unsupported, it topples over and breaks forward against the cliff face or surges up the beach. The waves that erode most are called **destructive waves** (**Figure 1**).

The power of destructive waves

Destructive waves have three main features:

1 They are high in proportion to their length.
2 The backwash is much stronger than the swash so that rocks, pebbles and sand are carried back out to sea.
3 They are frequent waves, breaking at an average rate of between eleven and fifteen per minute.

The height and destructiveness of these waves depend upon the distance over which the waves have travelled and the wind speed. If the waves driven by the wind have crossed over a large area of ocean, they have had time to build up and grow to their full height. A lot of energy is released when they break against the coastline. The length of water over which the wind has blown is called the **fetch**. The greater the fetch and the stronger the wind, the more powerful the wave, especially when driven onshore by storm-force winds. From time to time ideal conditions occur for the formation of huge destructive waves, as they did in Dorset on 3 November 2005 (**Figure 3**). Onshore south-westerly winds were strong, with gusts over 120 kilometres per hour after a long journey over the Atlantic Ocean. Imagine the weight and force of the water crashing against the coastline under these storm conditions.

Insurance companies
face £500 million payout
for storm and flood damage
We don't like to be
beside the quayside
Two flee for their lives as
4x4 ends up in the sea

The newspapers were mainly interested in telling their readers about the damage caused. Can you suggest a caption for **Figure 3** that a geographer would be more likely to write?

Processes of coastal erosion

There are many similarities in the ways in which rivers and waves erode, which is why the names used for the processes of erosion are the same or similar.

What else speeds up rates of coastal erosion?

One factor is the **weather**. The highest rates of erosion are recorded under storm conditions, when waves are at their most destructive, and hydraulic power and abrasion are strongest (**Figure 3**). Another is what the **rock** is like. Rock faces that are riddled with joints, bedding planes or faults (**Figure 3**, page 26) are eroded more quickly than those that are in massive blocks, because there are lines of weakness that waves can exploit. Cliffs made of a hard rock, such as granite in Cornwall and Devon (page 26), resist erosion longer than those made of soft sediments, such as sands and boulder clay in Norfolk (page 115) and Yorkshire (pages 119 and 126–7).

The ways the rocks are arranged can also increase rates of erosion. Look at **Figure 4**. The layer of clay at the bottom of the cliffs is too weak to withstand strong destructive waves in the English Channel with their long Atlantic fetch. However, rainwater seeps down through the sand that lies above the clay. This saturates the base of the sand layer along the junction with the clay, causing landslides and slumping, leading to cliff collapse. These are examples of **mass movement** – where a mobile mass of vegetation, soil and rocks removes the surface covering further down the slope. All coastal rock outcrops, even those out of reach of the waves, are subject to **weathering** (mechanical, chemical and biological, pages 28–9). Weathering loosens and breaks off fragments of rock and thereby speeds up the processes of wave erosion.

Figure 4 *Cliff erosion at Barton on Sea, east of Bournemouth.*

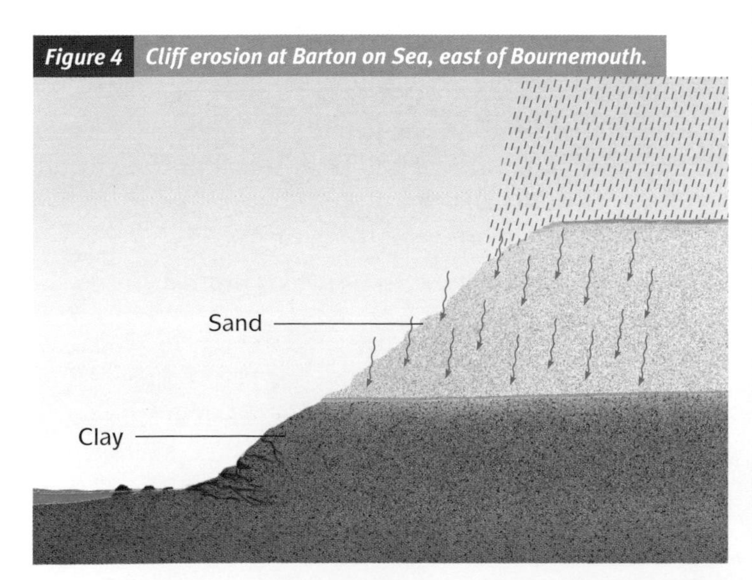

EXAM PREPARATION

Processes of coastal erosion

A Hydraulic power: This is the sheer weight and impact of the water against the coastline. It is greatest under storm conditions when hundreds of tonnes of water may hit the rock face. Also, air trapped in cracks and caves is suddenly compressed by the breaking waves, which increases the pressure on the rock.

B Abrasion: The breaking waves throw sand and pebbles against the rock face. These break off pieces of rock and cause undercutting. In large storms boulders will also be flung against the cliff face, causing even greater damage.

C Attrition: Particles carried by the waves are reduced in size as they collide with the rock face and one another. Boulders and pebbles are broken down into sand-sized particles, which are easier for the waves to carry away.

D Solution: This is the chemical action on rocks by seawater and is most effective on limestone rocks, which dissolve and are carried away in solution.

Exam tip

Refer to one or more of these processes when you are explaining the formation of any landform of coastal erosion.

ACTIVITIES

1 Show that you understand the differences between
   (a) mechanical and chemical weathering (pages 28–9)
   (b) hydraulic power and abrasion
   (c) processes operating at the base of the cliffs and those on the tops.

2 (a) Draw a labelled sketch of the cliffs at Beachy Head from **Figure 2**.
   (b) What shows that these cliffs are being actively eroded?
   (c) State two differences between the cliffs at Beachy Head (**Figure 2**) and Barton on Sea (**Figure 4**).

3 (a) Explain why so much coastal erosion takes place during storms.
   (b) Why will a storm cause more erosion along some parts of the coastline than others?

FURTHER RESEARCH

For information about the geology and coastal erosion of the Norfolk coast visit the weblink www.contentextra.com/aqagcsegeog.

What else shapes the coast?

How do waves transport sand and shingle along the coast? Where are they most likely to be deposited? How do headlands and bays show the effects of both erosion and deposition?

Transport

Loose, eroded materials of all sizes are transported by waves and deposited further along the coast. Rivers also carry sediment into the sea, which is picked up and carried away by the waves. The methods of transport are the same as those in the river channel (**Figure 2**, page 80): large boulders are rolled along the sea bed by waves (**traction**), smaller boulders are bounced (**saltation**), sand grains are carried in **suspension**, and lime from chalk and limestone rocks dissolves and is carried in **solution**. The transport of sand and pebbles along the coast by waves is called **longshore drift** (**Figure 1**). Waves often approach a coastline at an angle, but sand grains and pebbles roll back down the slope at right angles to the coastline because this is the steepest gradient. As **Figure 1** shows, a pebble will keep on being pushed up the beach by the waves at an angle, but every time it rolls back down the beach at right angles to the coastline. In this way the pebble is transported along the coastline.

The general direction of longshore drift around the coasts of the British Isles is controlled by the direction of the dominant wind (**Figure 1**). Prevailing south-westerly winds cause the drift from west to east along the Channel coast and from south to north along the west coast. The east coast is protected by land from the prevailing south-westerly winds. However, winds from the north cause longshore drift movement from north to south on the east coast. Northerly winds (winds from the north) have crossed a long stretch of open sea so that, although they do not blow as frequently as the westerly winds, they have the greatest influence overall. Longshore drift is important in the formation of all landforms of coastal deposition. Why do local authorities need to take the direction of longshore drift into account when planning to protect tourist beaches and construct sea defences?

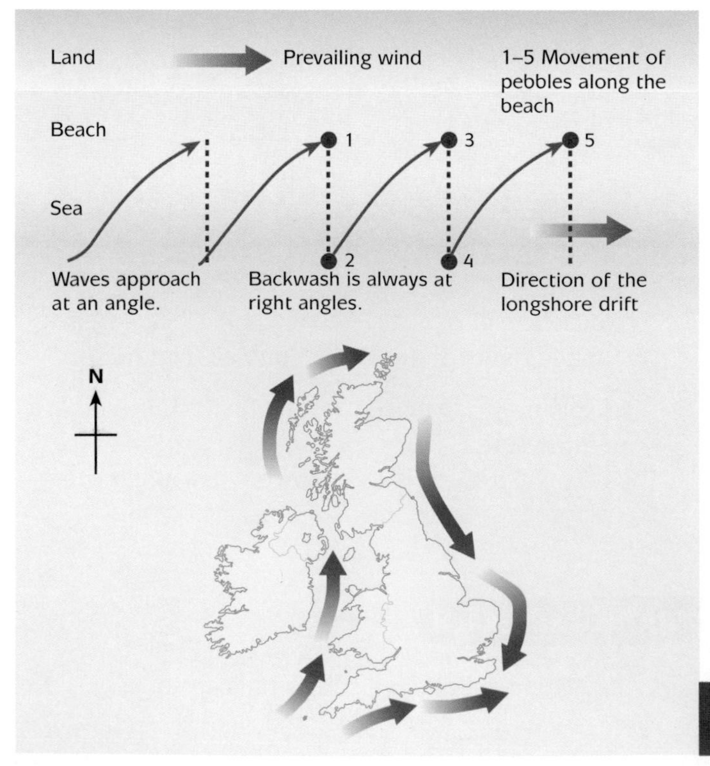

Figure 1 Direction of longshore drift around the British Isles.

Deposition

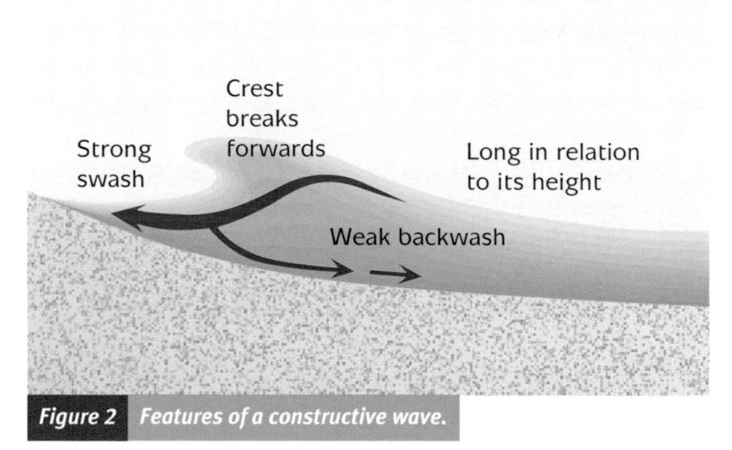

Figure 2 Features of a constructive wave.

The load of the waves – sand, shingle and pebbles – is deposited by **constructive waves** (**Figure 2**). Such waves add more material than they remove from the coastline. Constructive waves have three main features:

1 They are long in relation to their height.

2 They break gently on the beach so that the **swash** carrying materials up the beach is stronger than the **backwash** carrying them away.

3 They break gently, with between only six and nine waves per minute.

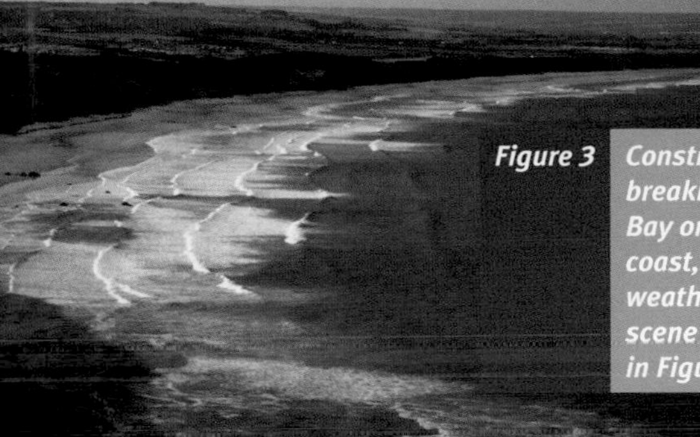

Figure 3 Constructive waves breaking gently in Filey Bay on the Yorkshire coast, during quiet weather. Compare this scene with the one shown in Figure 3, page 116.

These waves are associated with calm sea conditions when winds are light and are not blowing directly onshore. Therefore they occur more often in summer than in winter. Constructive waves operate most effectively in sheltered coastal locations such as in a bay sheltered by rocky headlands on both sides (**Figure 3**).Good places for deposition are in sheltered spots such as at the back of a bay or on a bend in the coastline, where sand and pebbles transported by the longshore drift are trapped.

Headlands and bays

These coastal landforms show the effects of both erosion and deposition. Where rocks of different hardness and resistance to erosion meet the sea, the weaker rocks are eroded back more quickly to form bays. In the shelter of the bay, constructive waves deposit and build up beaches. The stronger rocks resist wave attacks longer; they stand out as prominent rocky headlands, although a close look at them reveals how much they are being attacked by the waves (**Figure 4**). Flamborough Head is such a prominent headland because the chalk, although not a hard rock, is much more resistant than the weak boulder clay rocks that outcrop on the coast north and south of it (**Figure 5**).

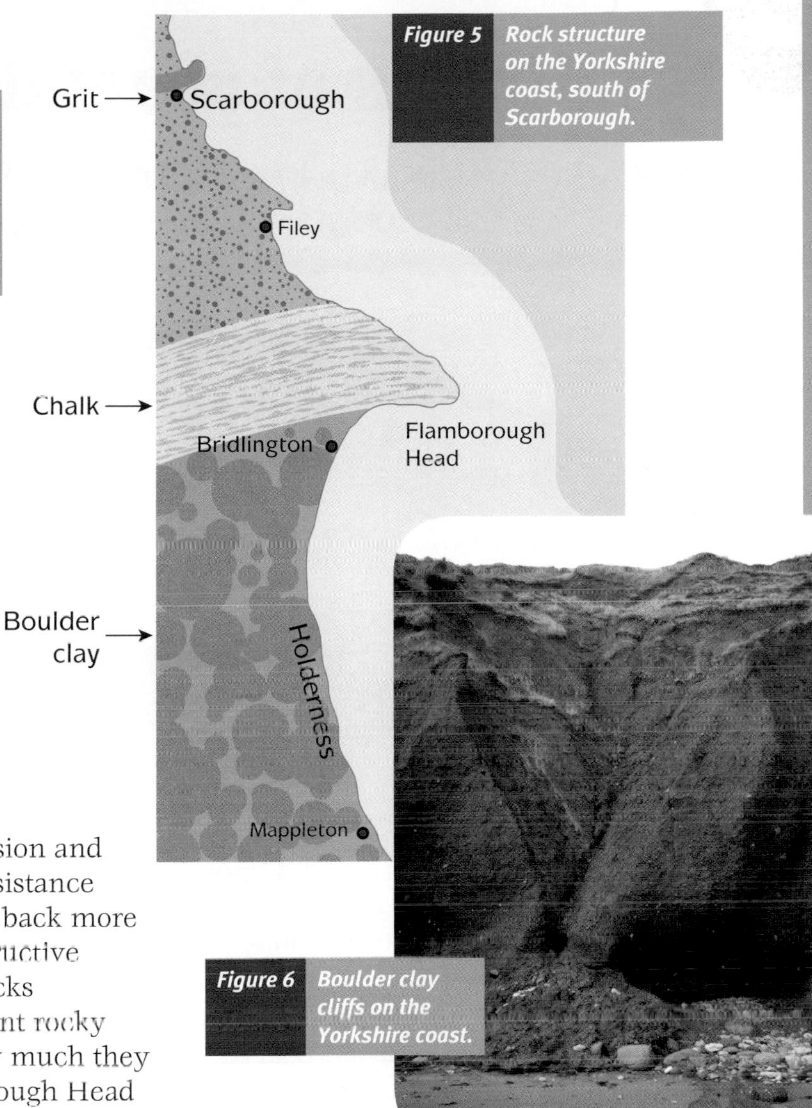

Figure 5 Rock structure on the Yorkshire coast, south of Scarborough.

Figure 6 Boulder clay cliffs on the Yorkshire coast.

Figure 4 Flamborough Head.

Figure 6 shows what boulder clay cliffs in the bays are like. Not only are they undercut by waves from below, but they are also washed away from above. What are the signs of rapid erosion? Have you noticed the scars from slumping and mass movement on the cliff top?

GradeStudio

1 a State two ways in which constructive waves are different from destructive waves. (4 marks)
  b Why are constructive waves more likely to deposit than destructive waves? (2 marks)

2 a Explain how sediment is transported along the coast. (4 marks)
  b Why is the direction of transport along the east coast different from that along the west and south coasts? (2 marks)

3 a Draw a sketch of the cliffs in **Figure 6** and label the evidence for rapid erosion. (4 marks)
  b Explain why chalk outcrops form headlands and boulder clay forms bays. (4 marks)

Exam tip

State *differences* so that they are two-sided (something about one and something about the other), especially when there are two marks per difference.

Landforms of coastal erosion

How are cliffs formed and why do they retreat? How can small weaknesses in rocks be enlarged to form caves, arches and stacks?

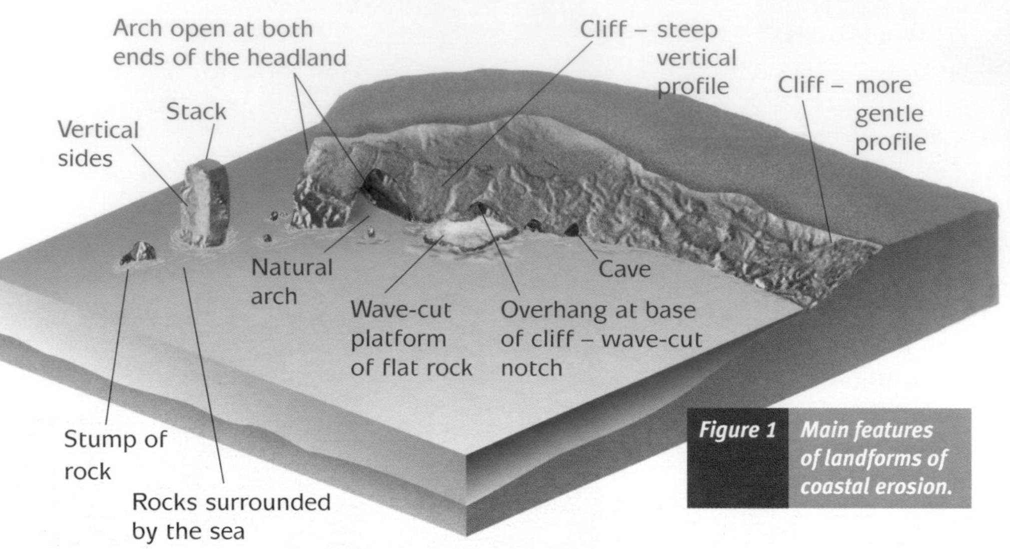

Figure 1 *Main features of landforms of coastal erosion.*

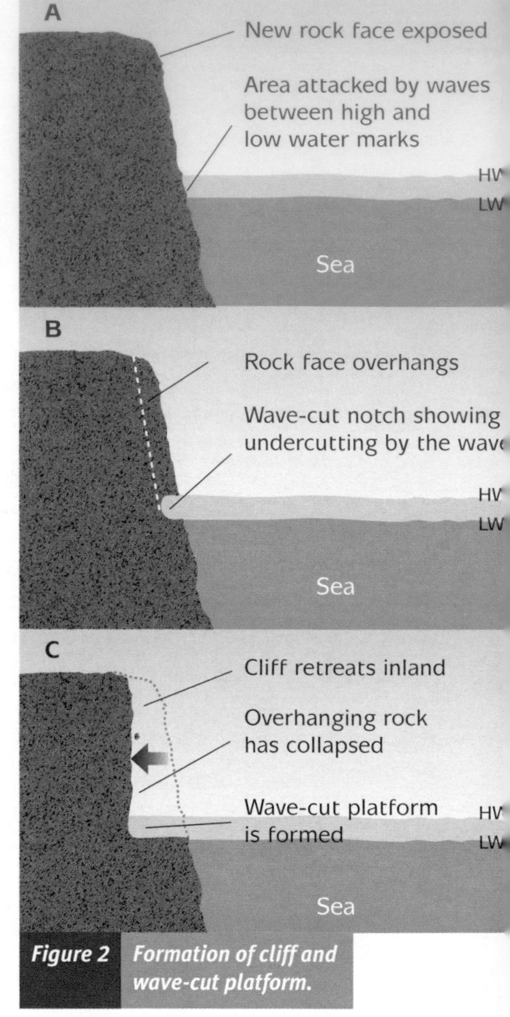

Figure 2 *Formation of cliff and wave-cut platform.*

The main landforms of erosion, which can be seen in many places around the British coast, are described in **Figure 1**. Pay particular attention to the labels that are used to *describe* the landforms. No attempt is made on **Figure 1** to *explain* their formation.

Cliffs and wave-cut platforms

The sea **cliff** is the most widespread landform of coastal erosion. Cliffs begin to form when destructive waves attack the bottom of the rock face between the high and low water marks. By the wave processes described on page 117, such as hydraulic power and abrasion, the waves undercut the face, forming a **wave-cut notch**. The rock above hangs over the notch. With continued wave attack, the notch increases in size until the weight of the overhanging rock is so great that it collapses. Once the waves have removed all the loose rocks and stones from the collapsed cliff, they begin to undercut the new rock face, which is now exposed to wave attack. Wave erosion, followed by cliff collapse, happens time and time again so that the cliff face and coastline retreat inland. Impressive cliffs are found where rocks resist erosion, such as the 'White Cliffs of Dover', which are built of chalk.

As the cliff retreats, a new landform, the **wave-cut platform**, is created at the bottom of the cliff face. This is the gently sloping rocky area between the high and low water marks (HW and LW in **Figure 2**). It is covered at high tide but exposed as the tide goes out. It is not a smooth platform of rock; rather its surface is broken by ridges and grooves. This is the area of flat rocks that holidaymakers often venture onto when the tide has gone out, looking for crabs, and where they are liable to get trapped as the tide races in! The wave-cut platform is formed where the rock above has been cut away by the waves to form the cliff. Because wave erosion is concentrated where the waves break between the high and low water marks, the rock below is less affected and is left as an area of flat rocks.

Figure 3 *Marsden Rocks – stacks near Sunderland. Notice in particular the amount of undercutting around the base of the rocks. The smaller stack has since collapsed without warning.*

Caves, arches and stacks

Waves are particularly good at exploiting any weakness in a rock, such as a joint. By the same processes of erosion, and particularly by hydraulic power and abrasion, any vertical line of weakness may be increased in size into a **cave**. However, the rock needs to be relatively hard or resistant, otherwise it will collapse before the cave is formed. Once a cave has formed, when a wave breaks it blocks off the face of the cave and traps the air within it. This compresses the air trapped inside the cave, which increases the pressure on the roof, back and sides. If the cave forms part of a narrow headland, the pressure from the waves may result in the back of the cave being pushed through to the other side so that it is open at both sides. The cave then becomes a natural **arch**. The base of the arch is attacked by waves, putting more and more pressure on the top of the arch. After continued erosion, and especially if there is a weak point at the top of the arch, the arch collapses and becomes a **stack**. The stack is a piece of rock isolated from the main coastline (**Figure 3**).

You can see that there is a sequence of features formed by wave erosion – cave, arch and stack (**Figure 4**). However, the stack itself is attacked by waves from all sides. It is gradually reduced in size and eventually it collapses so that all the signs of where the coastline used to lie disappear. When you look at a line of cliffs that mark the present-day coastline, as it is shown on maps, you must remember that it may be many kilometres further back than it used to be, as a result of the unceasing energy of destructive waves.

A
Waves erode weaknesses in the rock.
Largest cave eroded along greatest line of weakness.

B
Size of cave is increased by further erosion (corrasion, etc.) until the headland is opened out at both sides.
Other caves increase in size.
C
Stack separated off from rest of the land. Wave erosion at the base of the arch led to collapse of the roof.
The next cave is eroded and becomes an arch.

Figure 4 *Formation of caves, arches and stacks.*

ACTIVITIES

1. How many of the landforms in **Figure 1** can you find in **Figure 4**, page 119? Name them, and state the location for each one in **Figure 4**.
2. (a) On a sketch, show and label the coastal features on **Figure 5**.
   (b) Explain how they were formed.
3. Why are these cliffs being eroded less quickly than those at Barton on Sea a few kilometres east of Bournemouth (page 117)?

Figure 5 *Part of the coastline a few kilometres west of Bournemouth.*

Landforms of coastal deposition

Geographically, what is a beach and how does it form? How are spits and bars different from other beaches?

Wave deposition forms **beaches**. Everyone knows what a beach is, but can you describe it in geographical terms? The beach is the gently sloping area of land between the high and low water marks. Most of it is covered by the sea at high tide. However, beaches have many different characteristics.

- Some are straight and may extend for several kilometres – others, located in bays, are more likely to be curved.
- Some are made of sand – others are built of shingle and pebbles (**Figure 1A** and **B**).
- Sand beaches slope gently down to the sea, and are often long with sand dunes behind them (**Figure 1A**); shingle beaches have steep gradients with ridges running along them (**Figure 1B**).

Figure 1 *Two contrasting beaches: A is made of sand; B is made of shingle.*

What all beaches have in common is that they are constantly changing; the sand, pebbles and shingle on the beach keep being moved on and replaced by new materials. The materials forming the beach are carried by the longshore drift. If a coastline of weak rock, which is being greatly eroded, lies on the up-drift side, the waves will be heavily laden with material. Where there is a bend in the coastline, deposition by constructive waves is always more likely to take place because a more sheltered area has been created. Material accumulates over time and builds up the beach.

Characteristics and formation of spits

A **spit** is a long and narrow ridge of sand or shingle (**Figures 2**, **3** and **4**). One end is attached to the land while the other end lies in the open sea. It is really a beach that, instead of hugging the coastline, extends out into the sea. If the spit is formed of sand, sand dunes are usually found at the back of it. Behind the spit there is an area of standing water, some of which may have been colonised by marsh plants. Some spits, particularly those found along the coast of the English Channel such as Hurst Castle spit near Christchurch, have a hooked end (**Figure 2**). Others, particularly those found on the east coast, run parallel to the coast, perhaps for several kilometres. Some of these, such as Spurn Point (**Figure 3**), extend across estuaries, while others stretch across river mouths, diverting river flow southwards for a time behind the spit.

The formation of a spit begins in the same way as that of a beach. Eroded materials are carried along the coast by longshore drift (**Figure 4**). Deposition begins at a bend in the coastline. For a spit, however, the deposited materials accumulate away from the coast into the open sea until a long ridge of sand or shingle is built up. Fresh water and seawater are trapped behind this ridge as it forms. As the ridge extends into deeper and more open water, the end of the spit is affected by strong winds. These winds and sea currents help to curve the end of the spit. Do you understand how the direction of the longshore drift along a coastline can be worked out from the form of a spit?

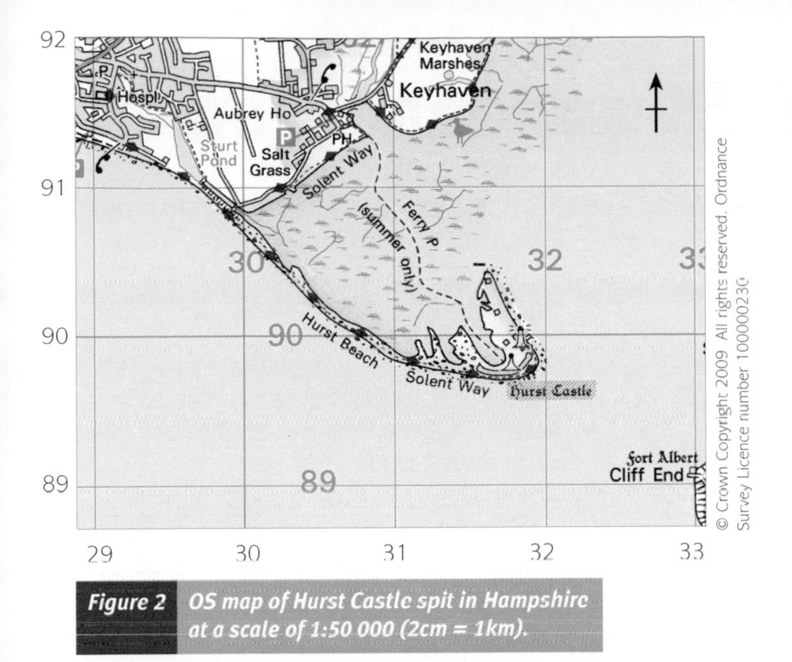

Figure 2 *OS map of Hurst Castle spit in Hampshire at a scale of 1:50 000 (2cm = 1km).*

Figure 3 *The spit at Spurn Point, Britain's longest spit (8 kilometres). Deposition began at a bend in the coastline formed by the Humber estuary (in the far distance). The spit is built of sand and clay from the rapidly eroding boulder clay cliffs in Holderness just to the north.*

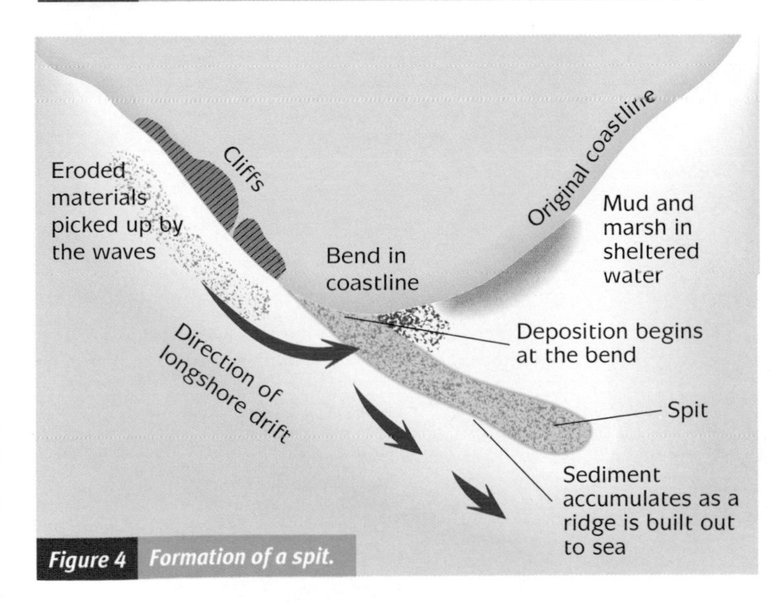

Figure 4 *Formation of a spit.*

How are bars different?

A **bar** is a ridge of sand or shingle across the entrance to a bay or river mouth. Some bars grow all the way across an entrance so that a stretch of water is cut off and dammed. A lagoon, such as the pool behind The Loe in Cornwall, is formed (**Figure 5**). The formation requirements (longshore drift with a large load, bend in the coastline, etc.) are the same as for spits, but the deposition continues in a line across an entrance until water is trapped behind in a lagoon. Refer to an atlas map of the Baltic Sea and look at the coasts of Germany and Poland. Sea currents are less strong in the sheltered Baltic than in the open seas around the UK so there is less to hinder sand accumulation, once it begins.

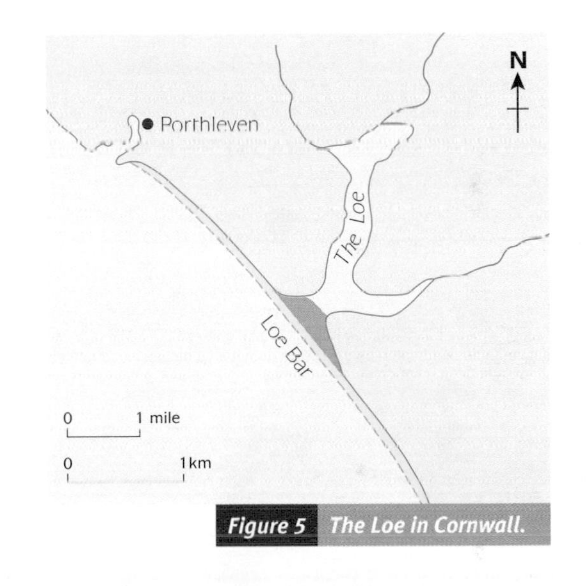

Figure 5 *The Loe in Cornwall.*

GradeStudio

1 a Name an example of a coastal bar. (1 mark)
b State one similarity and one difference between bars and spits. (2 marks)

2 a Study **Figure 1**. Describe two differences between the beaches in **A** and **B**. (4 marks)
b Explain the formation of beaches. (4 marks)

3 a Draw a labelled sketch of **Figure 3** to show the characteristics of the spit at Spurn Point. (4 marks)
b Explain why the spit at Spurn Point
(i) is longer than the Hurst Castle spit (2 marks)
(ii) has a different shape. (2 marks)

Exam tip

Know a *named example* of each of the major landforms, such as cliff, cave, stack and spit.
Showing characteristics is the same as *describing* the features, not *explaining* the formation.

Coastal issues in the UK and management

More than 15 million people in the UK live close to the coast – why are they getting more and more worried? Is the government willing or able to help them?

For people living close to the coast, two issues dominate – **coastal erosion** and the **threat of flooding**.

Overall, the UK is probably not losing any land, since erosion in one place is balanced by deposition in another. However, knowing this does not help the residents of Happisburgh (page 115), nor anyone else living next to the sea where cliffs are crumbling. **Figure 1** shows places at greatest risk from flooding in England and Wales. What do you notice about the distribution of these places?

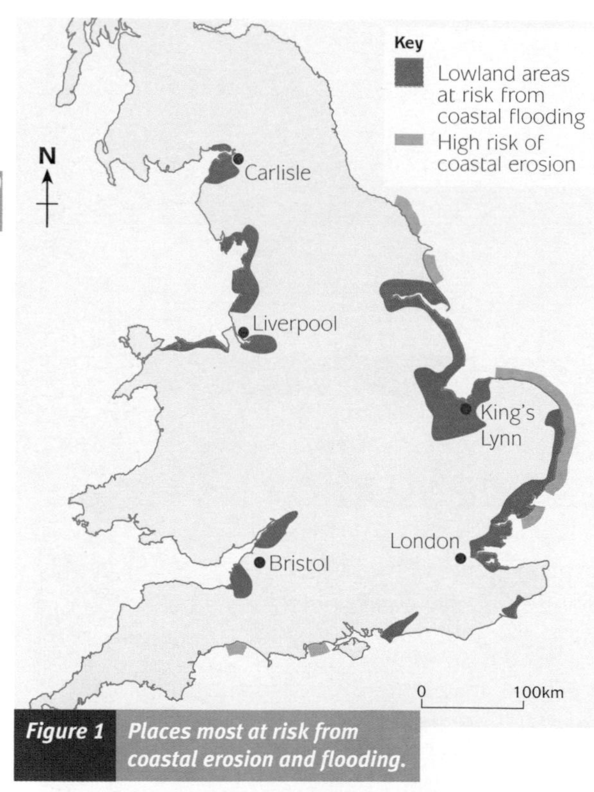

Figure 1 *Places most at risk from coastal erosion and flooding.*

What is worse, for a variety of reasons, rates of coastal erosion and frequency of flooding events are expected to increase in future years.

- **The Earth is warming up and sea levels are rising**
  In fact, both of these have been happening for 10 000 years, since the end of the Ice Age. At that time sea levels were 30 metres lower. Although people debate whether present-day global warming is natural or caused by people, it is a fact that glaciers and ice sheets are still melting into the seas and oceans.
- **Sea levels are rising faster in the east of the UK, where many lowland areas are located**
  On average present sea levels are increasing by 1–2 millimetres per year, but the increase is double the average in the east and south of England (**Figure 2**).
- **Strong wind and storm events are increasing**
  As the Earth warms up, there is more energy to drive Atlantic depressions. These can be associated with gale- and storm-force winds, leading to the formation of huge destructive waves, capable of causing more erosion in one night than in many years of normal wave activity.

Figure 2 *North-west Scotland goes up by 3 millimetres a year, because it is still recovering from the weight of ice sheets on top of it during the Ice Age. South-east England sinks.*

What can be done?

Option 1 Hard engineering – for total protection of the present coastline	*Option 2* Soft engineering – working with nature to help maintain the present coastline	*Option 3* Managed retreat – do nothing and allow present coastline to change (managed realignment)
• Maintain and extend existing sea defences • Build hard structures such as sea walls and flood barriers • Use **rock armour** to stop waves from reaching the coastline • Use groynes on beaches to trap the sand and keep the beach as wide as possible	• Beach nourishment – bring sand and shingle from elsewhere to maintain beach size and the natural protection it gives • Sand dune regeneration – plant grasses, bushes and trees to stabilise dunes and maintain their natural protection • Marsh creation – to provide a natural way of storing excess water during flooding	• Carry out no more repairs to old sea defences • Abandon the coast to the forces of nature • Allow the tide to invade low-lying land, forming salt marshes • Leave marshes as wildlife habitats for birds

Figure 3 *Three options for coastal protection in the UK.*

Figure 4 *Coastal management works for a new sea wall at Scarborough, completed in 2005 at a cost of over £50 million, double the original estimate. Blocks of granite (each weighing 6–10 tonnes) were placed in front as rock armour for protection. Concrete blocks that link together (called accropodes) were placed on top. Over 4000 were needed for the 2.1 kilometre wall.*

Coastal management involves controlling development and change in the coastal zone, as well as planning ahead. Of the three options given in **Figure 3**, hard engineering is the one most widely used, especially in coastal resorts. Most have a **sea wall**. The wall is often built with a curved lip at the top on the seaward side, to deflect the force of the waves away from the promenade behind. Such walls are expensive to build and costly to maintain (**Figure 4**). What was the cost per kilometre in Scarborough? Their construction can only be justified economically where there are many people and much property to defend.

The beach is the most important natural asset of any seaside resort. Many local authorities construct lines of wooden (or metal) **groynes**, built out into the sea at right angles to the coastline (**Figure 5**). They work by trapping sediment as it is transported along the coast by the longshore drift. They also reduce wave energy, making erosion of beach and cliffs less likely.

Groynes are a simple and effective way of protecting a beach, but local authorities rarely consider the consequences for other places. Once deprived of their load of sediment, waves remove sediment from beaches and cliffs further along the coast with renewed vigour. The same groynes that do a good job of protecting the beach and soft cliffs at Bournemouth are speeding up the already rapid erosion of the cliffs at Barton on Sea (page 117). Despite local authorities having spent more than £1 million protecting the cliffs here, they are still retreating, threatening holiday villages and housing estates. In other words, any interruption in sediment supply caused by groynes, breakwaters and harbour walls threatens the existence of cliffs and beaches further along the coast.

Figure 5 *Groyne at Bournemouth, aligned north–south. What is the direction of the longshore drift, and the evidence?*

What is needed is a full cost-benefit analysis before any money is spent on coastal protection (**Figure 6**). The balance of opinion within government and local authority circles has moved towards Option 3 in **Figure 3**; the case study on page 129 is an example. This option is more environmentally friendly, costs less and can be considered more realistic in the light of what is happening to sea levels.

Coastal management	
Costs	**Benefits**
• What will it cost to complete? • How much will it cost to maintain? • Who will be badly affected by it? • Which areas will be badly affected by it? • Will there be environmental damage?	• What are the advantages which justify the cost? • For how long will the benefits last? • Who will gain from it? • How large an area will gain from it? • Will it improve the environment?

Figure 6 *Costs and benefits of coastal management.*

ACTIVITIES

1 (a) From **Figure 1**, describe the distribution of places most at risk from coastal flooding and erosion.
(b) Explain why the erosion risk is higher in these places than in other coastal areas.

2 Choose two methods of hard engineering. For each one,
(a) describe how it works
(b) name and locate a place where it is used
(c) explain its advantages and disadvantages.

3 Some of the people likely to have views about coastal protection are:
A Bird watchers
B Holiday camp owners
C Londoners
D Council treasurers in resorts
E Farmers with land on cliff tops
F Economic advisers to the government
G People with coastal retirement homes
H Families who take summer holidays by the sea

(a) Draw a line like the one below. Mark the letters A–H along the line to show the likely feelings of the people above.

Option 1 ———————————— **Option 2**
Do nothing ———————————— Total protection

(b) Briefly explain your choice of position on the line for *four* of the people.

The case of Holderness, East Yorkshire

Figure 1 *1996 and a farmer on the brink in Holderness. What will remain now?*

Being number one in Europe for coastal erosion does not bring much pleasure to the people who live along the coast of Holderness. In a stormy year, the waves remove between 7 and 10 metres of land along unprotected stretches of coast. The coastline is today some 3–4 kilometres further west than it was in Roman times. Twenty-nine villages have been lost to the sea in the past thousand years. All along the Holderness coast farmers keep losing some of their land. Farmhouses are threatened (**Figure 1**); caravan sites and holiday homes have already been lost to the sea. The rate of erosion is unlikely to be reduced, because of a combination of land sinking on the eastern side of the British Isles and a possible sea level rise as a result of global warming. Sea levels in Holderness are estimated to be rising by 4 millimetres a year.

Why is the coastline being eroded so quickly? After all, the east coast does not receive as regular a battering from destructive waves as the west coast and exposed parts of the Channel coast. The main factor is rock type. The boulder clay is made up of soft clay and sands, which are not consolidated (cemented together).

The waves can wash away the clay and sands from between the boulders to leave them unsupported. Also, when it rains, water enters cracks and spaces in the rock; after heavy rains this makes the cliff top unstable and liable to slumping. Most erosion occurs when winds blow from the north or north-east along this coast because the waves cross a long stretch of open sea (a long fetch), which increases wave energy for erosion. There are problems in protecting a coastline that stretches for more than 50 kilometres.

Coastal defences at Mappleton

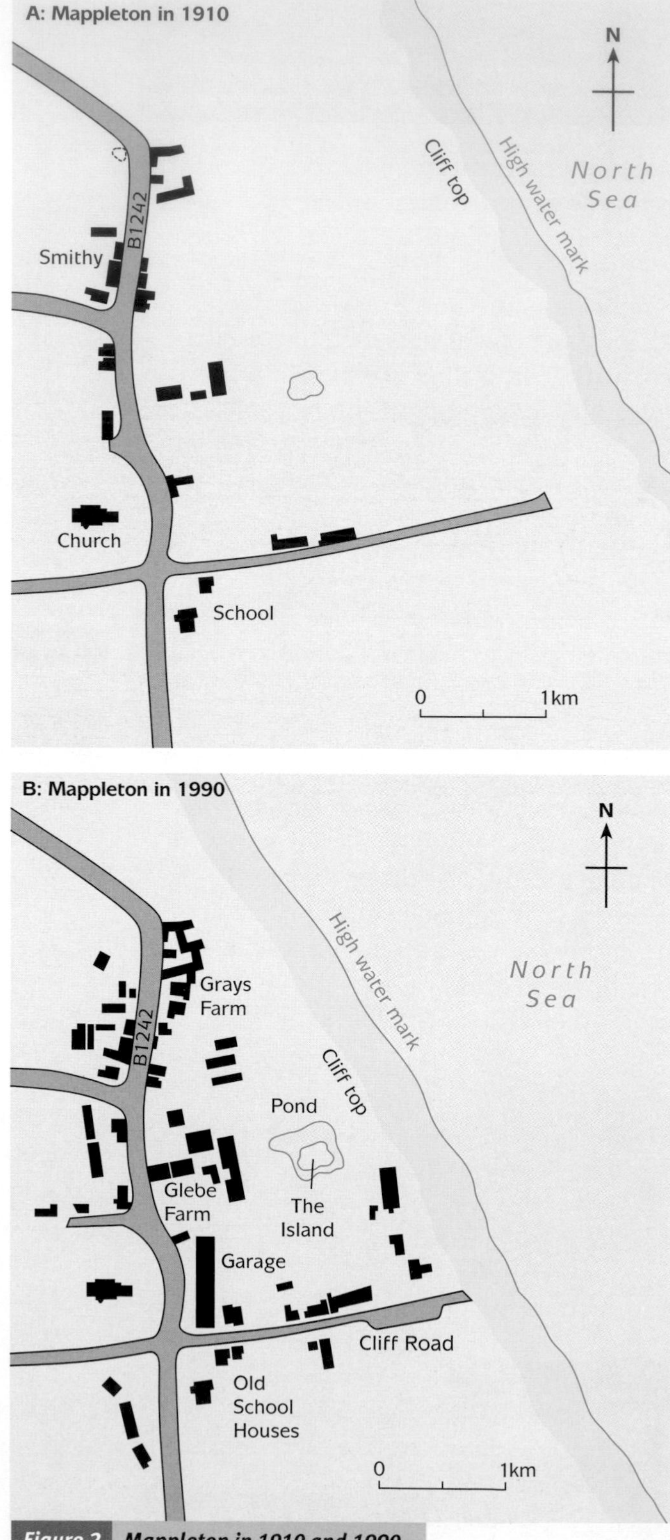

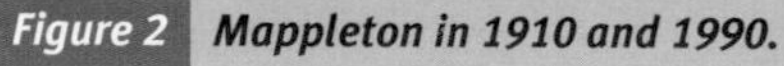

Figure 2 *Mappleton in 1910 and 1990.*

Coastal defences at Mappleton

Mappleton is a small village, which by 1990 was under real threat of becoming lost village number 30 along the coast of Holderness. The B1242 is the vital road link along this coast and it would have been expensive to find a new route for it. This helped to justify spending almost £2 million on a coastal protection scheme in 1991 for a village of about 100 people. Blocks of granite were imported from Norway for sea defences at the bottom of the cliff and for the two rock groynes. The purpose of the two rock groynes is to trap beach material, which will protect the rock wall from direct wave attack.

Figure 3 ***Mappleton and its sea defences. Notice the closeness of the village to the cliff top, the gentle angle of the soft sand and boulder clay cliffs, the rock wall at the bottom of the cliffs, and the rock groynes.***

Some views on coastal protection in Holderness

Holderness Council

We are a small authority with a total annual budget of only £4 million. Spending a large amount of money to protect a village is hard to justify. Many people agree that the village should be allowed to disappear ... but it is terribly difficult to say this.

Ministry of Agriculture

We are moving towards a policy of 'managed retreat'. Although towns, villages and roads would be protected, farmland and even isolated houses would be regarded as dispensable and allowed to disappear. There are food surpluses in the EU so we no longer need every bit of farmland.

Dr John Pethick – a top scientist at the University of Hull

Low-lying farmland should be abandoned and cliffs allowed to collapse (because they are the main sources of sand and silt that build up) to protect other parts of the coast, including towns and cities.

Farmer living just south of Mappleton

My farm is at greater risk from the sea than ever because of the coastal protection works at Mappleton.

ACTIVITIES

1. Describe the physical features of the coastline of Holderness as shown in **Figures 1, 2** and **3**.
2. (a) State three pieces of evidence from this page for rapid erosion along the coast of Holderness.
   (b) Give the physical reasons for this rapid erosion.
   (c) Why is the natural rate of erosion not expected to decrease in the future?
3. (a) From **Figure 2** state:
   (i) by how many metres the sea has invaded inland between 1910 and 1990
   (ii) the distance between the B1242 road and the top of the cliffs in 1990.
   (b) With the help of **Figure 3**, describe the methods used to protect Mappleton.
   (c) Explain why the farmer who owns the land south of the cliff road, which can be seen in **Figure 3**, is complaining about the coastal protection works at Mappleton.
4. (a) Make two columns and list the arguments for and against protecting the coastline of Holderness.
   (b) What do *you* think would be the best policy? Explain and justify your view.

London and the Thames estuary

The lower Thames estuary including central London is vulnerable to flooding. This was made clear by the terrible flooding in late January 1953, from a storm surge that was funnelled down the North Sea, which narrows towards the Thames estuary. When storm surges coincide with high spring tides, dangerously high water levels can occur in the Thames estuary. The old, poorly maintained sea defences that existed in the Thames estuary in 1953 were no match for the power of storm surges and they collapsed. Worst affected were places in the low-lying parts of the estuary, such as Canvey Island, which was simply submerged by the flood waters, resulting in loss of life, evacuation of homes and damage to businesses.

The Thames remains the focus of economic life in London. Historically it was the port of London with its docks and warehouses; today it is the 'City' of London with its banks, insurance companies and companies' headquarters. The river banks are lined with a mixture of modern office buildings, recently built and converted residential apartments, and remnants of London's long history such as the Tower of London and St Paul's Cathedral. Even the Houses of Parliament, the seat of political power in the UK, has a riverside location (**Figure 1**).

Figure 1 ***Houses of Parliament and the River Thames – what would be the political impact of London flooding?***

If London flooded, it would create nothing less than a 'financial disaster zone'. It would lead to social distress on a scale not seen in other UK towns and cities flooded in recent years (page 53). Political paralysis and panic would be likely results.

1. 1.5 million commuters work in Thames-side locations
2. property assets are estimated at around £100 billion
3. 1.25 million people live in the Thames flood zone
4. 69 tube stations, 400 schools and 16 hospitals are also in the zone
5. infrastructure value is too high to calculate.

This is why the Thames Barrier (or Barrage) was constructed between 1974 and 1984 (**Figure 2**). Built at a cost of £534 million (£1.6 billion at today's prices), the barrier is raised on average four to six times a year, but with increasing frequency. It was closed twice in one day when a storm surge similar to the one in 1953 occurred. Up to 2007 it was closed more than 100 times. Has it been value for money?

The Thames Barrier is expected to cope with current revised estimates for sea level rises until 2030–50, not the distant future as was expected when it was built. The idea has already been proposed that a much larger and more ambitious 16-kilometre-wide barrier across the river from Sheerness in Kent and Southend in Essex might soon be needed.

Figure 2 ***Thames Barrier. Gates across the openings are raised from the river bed during high tides when flooding threatens. Flood defences extending 18 kilometres downstream from the barrier were also raised and strengthened.***

The Essex marshes

The coastal zone of Essex shown in **Figure 3** is a mixed landscape of mudflats, sand and shingle banks, salt marsh, coastal grassland and sea walls between broad river estuaries and large sea inlets. The Essex marshes is one of the top five coastal wetlands in Britain in terms of its value to wildlife. It provides valuable habitats for marine life (shore crabs and herring), birds feeding on the marshes (redshank and oystercatcher), and salt-loving marsh-plant communities (marsh samphire).

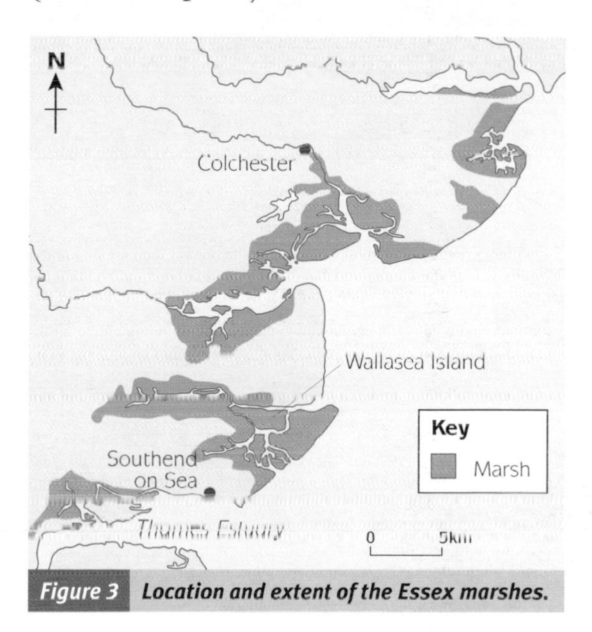

Figure 3 *Location and extent of the Essex marshes.*

Marshes under threat

Despite its flat, uninviting landscape (**Figure 4**), farmers have gradually reclaimed land from the sea during the last 400 years. Sea walls were built first to allow grazing, and later arable farming. Only about 2500 hectares of natural coastal marsh remain, compared to 30 000 hectares in 1600. The need for conservation becomes ever more urgent as threats to the area increase – human from urban development spreading north from the Thames estuary, physical from accelerating coastal erosion.

Strategies for conservation

Figure 4 *View over the Essex marshes.*

The main strategy involves reversing 400 years of reclamation history by deliberately allowing rising sea levels to flood large areas. This is to be done by making gaps in existing sea defences. In a trial scheme in 2006, the government Department for Environment, Farming and Rural Affairs (DEFRA) destroyed 300 metres of sea wall and allowed the tide to wash in. Already saltwater plants and wildlife have moved in. Much larger is the Royal Society for Protection of Birds' (RSPB's) proposed project at Wallasea Island near Southend. The £12 million scheme will require the building of new low-lying walls to separate off existing arable land, placed well inland where they are easier to defend against the sea. Otherwise Wallasea will be allowed to flood by breaching old sea defences, to create a new landscape of marshes, islands, shallow saline (salty) lagoons and creeks – closer to what it used to be before human interference. Advantages will include safeguarding existing wildlife; acting as a buffer zone against flooding, making inhabited areas further inland safer; increasing the areas that are useful as natural nurseries for fish; attracting bird species back to Britain; catering for visitors by creating cycleways and footpaths. This is an example of **managed retreat**, of **sustainable development**; working with natural processes instead of trying to fight against them.

ACTIVITIES

1 (a) Describe the methods of coastal management used in (i) the Thames estuary (ii) Essex marshes.
  (b) State how they are different under the headings (i) Type of engineering (ii) Aims (iii) Future sustainability.

2 (a) What is meant by managed retreat and coastline realignment?
  (b) Groups for and against managed coastal retreat: coastal farmers; homeowners in coastal villages; councils in coastal resorts; government and rural councils; environmentalists and wildlife organisations.
    (i) Make two lists of those against and those in favour of managed retreat.
    (ii) Choose one group from each list and explain why their viewpoints are different.

Practice GCSE question

See a Foundation Tier Practice GCSE Question on the weblink www.contentextra.com/aqagcsegeog.

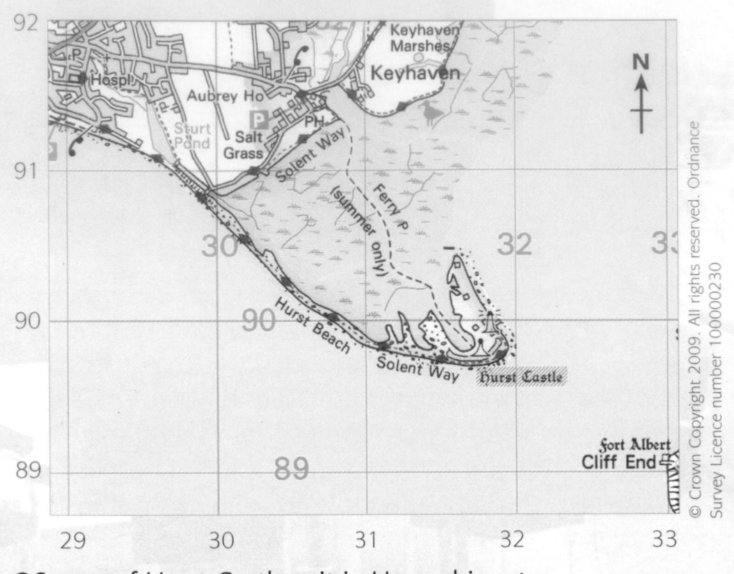

OS map of Hurst Castle spit in Hampshire at a scale of 1: 50 000 (2cm = 1km).

Figure 1 Cliffs at Walton-on-the-Naze in Essex.

7 (a) Study the Ordnance Survey map extract of Hurst Castle spit.
- (i) Give the six-figure grid reference for the lighthouse towards the end of Hurst Castle spit. **(1 mark)**
- (ii) State the direction of the spit from the mainland. **(1 mark)**
- (iii) Is the walking distance along the spit, from the southern end of Sturt pond to the far end of the spit, about 2, 3.5, 5 or 6.5 kilometres? **(1 mark)**
- (iv) Draw a labelled sketch to show the physical features of the spit. **(4 marks)**
- (v) Explain the formation of this type of spit. **(4 marks)**

(b) Study **Figure 1**, which shows cliffs along part of the east coast of England.
- (i) State two pieces of evidence that these cliffs are crumbling. **(2 marks)**
- (ii) Explain why cliffs like these are crumbling and collapsing. **(4 marks)**

(c) For an area of UK coast with eroding and collapsing cliffs, describe some of the methods of coastal management used, explaining why some of them are more sustainable than others. **(8 marks)**

Total: 25 marks

Exam tip
The question for this topic will be Question 7 in Section B in Paper 1.

Improve your GCSE answers

How to recognise coastal landforms from contour patterns and signs on OS maps

Exam tip

- The first thing to look for on any coastal OS map is the **high water mark**, shown by the black line. Only after doing this can you identify landforms.
- The same symbol can be used for different landforms e.g. rock symbols for both cliffs and wave-cut platforms, sand symbols for both beaches and sand dunes.
- What matters is location relative to the high water mark – cliffs and sand dunes are on the land side of the high water mark; wave-cut platforms and beaches are on the sea side.

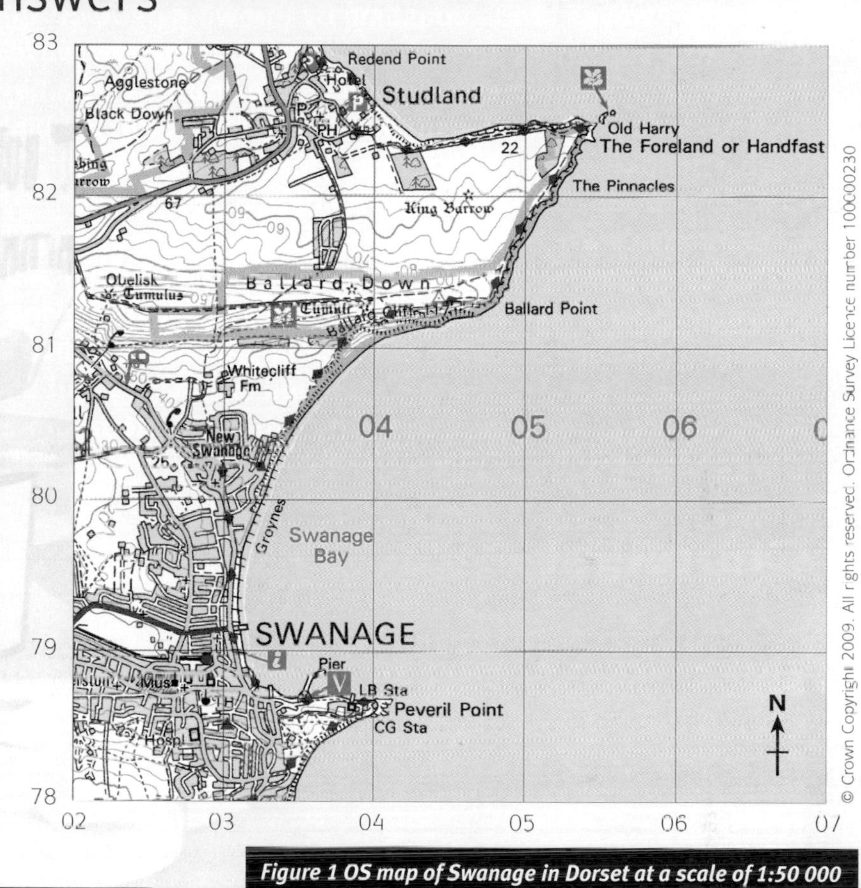

Figure 1 OS map of Swanage in Dorset at a scale of 1:50 000 (2cm = 1km).

Landforms of erosion

In **Figure 1** you should be able to recognise these landforms of erosion.

- **Headland and bay** – identify what is different about relief (steepness and height of the land) for the **headland** which ends at The Foreland, and for the **bay** to the south in which Swanage is located.
- **Cliffs** of different types – notice how cliffs are shown using two different signs.
  1. The Foreland to Ballard Point ... rock sign used, showing vertical cliffs (see **Figure 5**, page 121).
  2. South of Ballard Point ... steep slope sign used, with more than one line, and some lines placed further inland, showing cliffs that are not completely vertical, looking more like the cliffs shown in **Figure 1** on page 130.

Contours that reach the signs (and disappear into them) give you cliff heights. The contours tell you that these cliffs increase in height southwards from The Foreland towards Ballard Point. What is the highest point?

- **Stacks** – Old Harry Rocks; notice the black line around them, indicating that the rocks are tall enough to remain permanently above high tide level.

Landforms of deposition

In **Figure 1**:

- **Beach** in Swanage Bay; the widest area of sand is towards the south in the shelter of Peveril Point.

In **Figure 2** on page 123:

- **Beach** in square 2991.
- **Spit** at Hurst Castle; notice the signs used for pebbles and shingle, as well as sand, on the spit.
- **Salt marsh** filling most of the shallow water behind the spit, between the spit and the mainland, which is sheltered from strong currents and tides. How do you know that all of this marsh is covered by seawater at high tide?

Human management

In both **Figures 1** and **2**:

- Groynes to try to maintain beaches.

ExamCafé

REVISION

Key terms from the specification

Exam Café

Abrasion – waves erode coastline by throwing pebbles against cliff faces
Arch – rocky opening through a headland formed by wave erosion
Bar – ridge of sand or shingle across the entrance to a bay or river mouth
Beach – sloping area of sand and shingle between the high and low water marks
Cave – hollow at the bottom of a cliff eroded by waves
Cliff – steep rock outcrop along a coast
Constructive wave – gently breaking wave with a strong swash and weak backwash
Destructive wave – powerful wave with a weak swash and strong backwash
Hydraulic power – erosion of rocks by the force of moving water in waves
Managed retreat – abandon defence of present coastline in a controlled manner
Spit – ridge of sand or shingle attached to the land, but ending in open sea
Stack – pillar of rock surrounded by sea, separated from the coastline
Wave-cut platform – gently sloping surface of rock, in front of cliffs, exposed at low tide

Checklist

	Yes	If no – refer to
Do you know the difference between wave erosion, weathering and mass movement?		pages 116–17
Can you state and explain the differences between destructive and constructive waves?		pages 116 and 118
How are sand and shingle transported along the coast of the UK?		page 118
Can you name and recognise five or more landforms of coastal erosion?		pages 120–21
Do you understand where and why beaches form?		pages 122–3
Can you recognise landforms of erosion and deposition from OS maps?		page 131
Can you give examples of methods of hard and soft engineering and state differences between them?		pages 124–5
Do you understand what managed retreat of the coastline means, and why it is becoming preferred government policy along rural sections of coast?		pages 124 and 129

Case study summaries

Cliff collapse	Impacts of coastal flooding	Coastal management	Coastal habitat
Rates of erosion	Economic	Why needed	Environmental characteristics
Reasons for them	Social	Methods used	Habitats and species
Human causes	Political	Costs	Strategies for conservation
Impacts	Environmental	Benefits	Sustainable uses of area

Chapter 8

Population Change

People galore in Mumbai (Bombay), India's big business city. Sometime around 2030 India is expected to overtake China as the world's most populous country. Why?

QUESTIONS

- Why is world population growing so fast?
- What is being done to check growth?
- What issues arise from Europe's ageing and greying population?
- Why is Europe now a continent of immigration?

Global population change

Why is the world's population growing so fast? Are there any signs of rates of increase slowing?

A population may increase or decrease over time. How a population changes depends on the birth rate, the death rate and migration (see the Exam Preparation Box). A population grows if the birth rate is higher than the death rate, i.e. there is a natural increase. However, a few countries have a natural decrease where the death rate is greater than the birth rate. In a few countries, migration can have a large impact on population.

World population growth

The growth of world population over the last 200 years has been spectacular (**Figure 1**) and it has not stopped yet. From 1950 there was a population explosion and the total of 6 billion people on Earth was reached in 1999. **Exponential growth** is the term used to describe such a rapid increase. Although there is some evidence that the rate of growth is at last beginning to slow down, the world's population continues to grow because a great majority of countries have higher birth rates than death rates, leading to a natural increase.

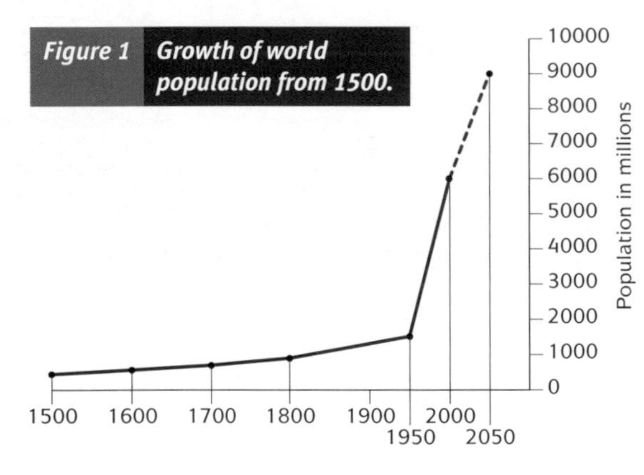

Figure 1 **Growth of world population from 1500.**

	Country	Highest		Country	Lowest
1	Niger	55.2	1	Latvia	7.8
2	Angola	52.3	2	Bulgaria	7.9
3	Somalia	52.1	3	Slovenia	8.3
4	Uganda	50.7	4	Ukraine	8.4
5	Congo	50.2	5	Austria	8.6
6	Liberia	50.0	6	Russia	8.6
7	Mali	49.9	7	Germany	8.7

Figure 2 **Countries with the world's lowest and highest birth rates per 1000 in 2006.**

Birth rates

The average birth rate in the rich, industrialised countries is around 12–13 per 1000; in poorer developing countries it is about 26 or 27 per 1000. There tends to be a general relationship between birth rate and level of economic development – the more economically developed the country, the lower its birth

EXAM PREPARATION

Understanding population terms

Crude birth rate – the number of live births per 1000 population per year.

Crude death rate – the number of deaths per 1000 population per year.

Natural increase – birth rate higher than death rate: birth rate minus death rate.

Natural decrease – death rate higher than birth rate: death rate minus birth rate.

Annual population change – the birth rate minus the death rate plus or minus migration.

Migration – the movement of people either into or out of an area.

Low birth rates in Europe	
Social	• family planning is practised by nearly all couples • women are well educated and career orientated • one or two children is accepted as a normal family size.
Economic	• children are unlikely to contribute to the family income • the average cost of bringing up a child in the UK in 2005 was over £60 000.
Political	• governments support and finance family planning.
High birth rates in Africa and the Middle East	
Social	• family planning is not widely used, especially among the poor • many women receive little formal education and marry young • socially, families of five or more children are considered to be quite normal.
Economic	• children are expected to work to supplement the family income • one child who does well may lift the family out of poverty.
Political	• some governments and religions do not approve of birth control • family-planning clinics are not always available, especially in rural areas.

Figure 3 **Reasons for low birth rates in Europe and high birth rates in Africa and the Middle East.**

rate (page 197). **Figure 2** lists the countries with extremes of high and low birth rates in 2006. It is dominated by countries from only two continents – which are they? All the countries in column 1 have average incomes per head per year under US$ 2500, compared with over US$ 20 000 in the three richest countries in the lists (Germany, Austria and Slovenia). In fact, birth rates in all European countries are low, not just in those countries named in **Figure 2**. In contrast, many countries in Africa and the Middle East have birth rates well over 40 per 1000. The social, economic and political reasons for these differences are given in **Figure 3**.

Death rates

Unlike birth rates, death rates are similar between rich and poor countries; the world average for both is between 9 and 10. During the second half of the twentieth century, death rates fell everywhere, due to the spread of medical knowledge and improvements in primary and secondary healthcare. Primary healthcare is preventing disease, by immunisation for example; secondary healthcare is treatment of illnesses by doctors and nurses. Countries with death rates above 20 per 1000 are now quite exceptional (**Figure 4**). The countries named in **Figure 4** are all from one continent: Africa. Most are countries in southern Africa badly affected by the spread of HIV/Aids, unable to afford the anti-viral drugs. Sierra Leone was war-torn for many years and Zimbabwe is in economic meltdown with severe food shortages. Here the trend towards lower death rates has been reversed.

Natural increase and decrease

The wider the gap between high birth rates and low death rates, the greater the size of the natural increase. Angola features in both **Figures 3** and **4**, which makes it possible to calculate its natural increase. At 31.2 per 1000 (or 3.12 per cent) it is still high, despite Angola's higher than average death rate. Poor developing countries, with their high birth rates and low death rates, are overwhelmingly responsible for continuing world population growth.

A higher death rate than birth rate, resulting in a natural decrease of population, is something new. However, it is a trend that is on the rise in European countries (**Figure 5**). Birth rates are falling as social trends change in some countries in northern and western Europe. People are marrying at an older age and women are delaying starting a family until they have established themselves in a career. Other couples are foregoing children in favour of a higher standard of living. In Russia and the formerly communist eastern European countries, many people are pessimistic about the new economies and their future circumstances.

Death rates are already low, and cannot be expected to fall much further. You need to remember that the death rate is a ratio per 1000 of the population. As a result of an increasing proportion of elderly people within the population, there are many people per 1000 reaching the natural age for death.

	Country	Death rate
1	Swaziland	31.2
2	Botswana	28.4
3	Lesotho	26.4
4	Zimbabwe	23.0
5	Sierra Leone	22.5
6	Central African Rep.	21.8
7	Zambia	21.2
8	Angola	21.1
9	Equatorial Guinea	20.9
10	South Africa	20.6

Figure 4 *Countries with the world's highest death rates per 1000 in 2006.*

Country	Birth rate (per 1000)	Death rate (per 1000)	Natural change (per 1000)	%
Germany	8.7	10.7	–2.0	–0.20
Italy	8.8	10.6	–1.8	–0.18
Austria	8.6	9.8	–1.2	–0.12

Figure 5 *Examples of countries with a natural decrease in population.*

ACTIVITIES

1 Total world population (billions): 1804 1; 1922 2; 1959 3; 1974 4; 1987 5; 1999 6. Estimates for the future (billions): 2013 7; 2028 8; 2048 9.
   (a) State the number of years taken for each population increase of one billion (known and estimated).
   (b) Describe how the data shows that world population growth has been exponential.
   (c) What is the evidence that world population growth is expected to slow in the future?

2 Data for five countries in 2006.

	Birth rate (per 1000)	Death rate (per 1000)	Income per head (US$)
UK	11.0	10.2	35,760
Hungary	8.8	12.9	10,270
Russia	8.6	16.0	4,080
China	14.5	7.1	1,470
Nigeria	39.1	18.4	570

   (a) For each country, calculate the rate of natural population change in 2006.
   (b) Why do rates of population change vary so greatly between countries?

The Demographic Transition Model and changes in population structure

How does the relationship between birth and death rates change over time? What differences in population structure occur?

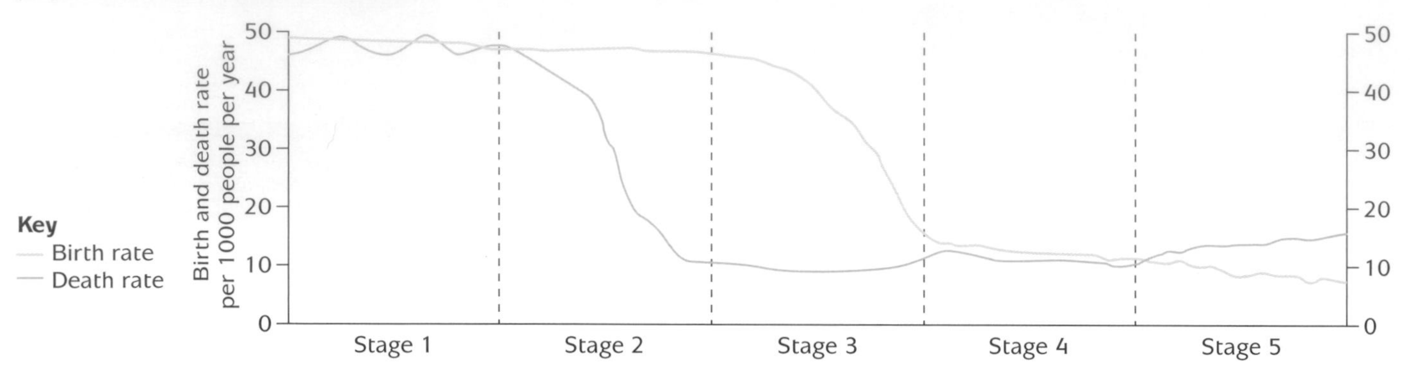

Stage	1	2	3	4	5
Birth rate	High	High	Decreasing	Low	Low
Death rate	High	Decreasing	Low	Low	Low
Natural increase	Low	Becoming high	High becoming low	Low	Natural decrease
Countries	None – all countries have progressed into another stage.	Poor countries with low levels of economic development, many of them in Africa.	Poor countries with improving levels of economic development, many in south and east Asia, Mexico, Brazil.	Rich countries in North America, Australasia, Japan and many European countries.	A few rich countries with very low birth rates, mainly in Europe.

Figure 1 ***The Demographic Transitional Model.***

The Demographic Transition Model (DTM) in **Figure 1** is used to show how countries pass through different phases of population growth over time. For many years there were only four stages, but the recent emergence of countries with a natural decrease of population created the need for a stage 5. Close relationships between birth and death rates are visible in stages 1, 4 and 5, but not in stages 2 and 3, where birth rates are noticeably higher than death rates. The majority of countries are currently passing through these two stages, which explains the large growth in world population.

In many ways the critical stage is stage 2. Death rates fall significantly, mainly due to improved medical care but also as a result of cleaner water supplies and increased food output. Birth rates remain high because the economic, social and political reasons for high birth rates referred to on pages 134–5 still apply. More children means more workers in the fields, children can look after their parents in old age, and some governments either cannot afford or are unwilling – for religious or other reasons – to support family planning. The adoption of improved medical care is always faster than that of contraception.

When birth rates start to fall in stage 3 it is a major change. The underlying reason is the introduction of family planning and birth-control policies, but economic growth is a great help in financing the policies, improving education and changing social attitudes towards large families. Economic development is accompanied by more people living in cities, where the costs of bringing up children are higher; on farms children are workers. Improved educational opportunities for women make a big difference as well; more women entering higher education, marrying later and pursuing careers greatly reduces average family sizes.

Population structure

Population structure is the composition of a country's population by age and sex. Data for these, usually collected by a **census**, is shown in a **population pyramid**. In the pyramids in **Figure 2**, horizontal bars are used to show the numbers of males and females in each five-year age group (0–4, 5–9, etc.) on either side of the central dividing line. Sometimes the actual numbers of people (instead of

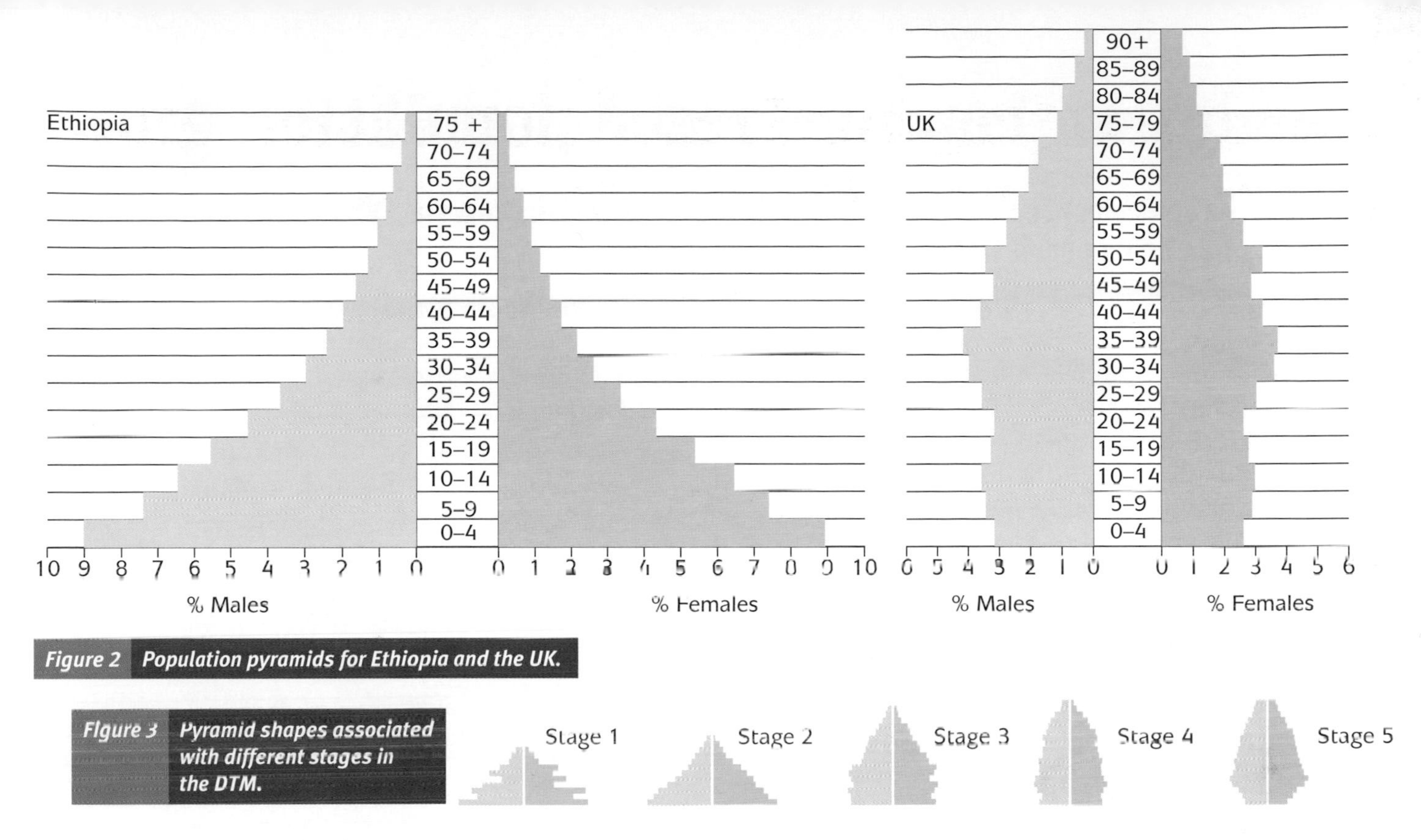

Figure 2 Population pyramids for Ethiopia and the UK.

Figure 3 Pyramid shapes associated with different stages in the DTM.

percentages) are used when constructing pyramids. Pyramids give a good visual impression of the differences in population make-up by age between countries; striking differences are shown here between a poor developing country (Ethiopia) and a rich developed country (the UK).

The pyramid for Ethiopia displays many of the characteristic features of a less economically developed country, particularly the wide base showing a population structure dominated by young people, due to high birth rates. The graph has an almost perfect pyramid shape, progressively tapering towards a narrow top, with few people above the age of 65. The pyramid for the UK is taller and the top is more pronounced, showing significant numbers above the age of 65. The UK's birth rate is low and the narrow base shows this. It is the middle-aged groups that are dominant in pyramids for developed countries. From their shapes it is possible to suggest stages in the Demographic Transition Model – Ethiopia is definitely in stage 2, while the UK has reached stage 4. In summary, population pyramids for developing countries are wider at the base, narrower at the top and less tall than those for developed countries.

Figure 4 Dependants in the UK.

It is customary to subdivide the structure of a country's population into three age groups, namely young (0–14), middle-aged (15–64) and old (65 and above). The middle-aged are distinguished from the other two as the working or **independent population**; they are the group in society that works, earns money, contributes to pensions and pays income taxes. Young and old have in common that they are **dependants**; although some of them work, the majority depend upon services such as education and healthcare, paid for by taxes collected from the working population. The **dependency ratio** is the ratio between the dependent and independent populations.

ACTIVITIES

1 (a) Draw a diagram of the Demographic Transition Model to show all five stages.
  (b) Use different shading for the areas that show natural increase and natural decrease on your diagram.

2 State the key change between each stage and explain why it happens:
(a) stage 1 to stage 2 (b) stage 2 to stage 3 (c) stage 3 to stage 4 (d) stage 4 to stage 5.

3 (a) Draw simple summary sketches of the population pyramids for Ethiopia and the UK, and label the differences between them.
  (b) Describe what the two pyramids show.

4 (a) State likely stages in the DTM for each of the five countries in Activity 2 on page 135.
  (b) Give a brief justification for your choices.

Population issues – rapid population growth

What are the consequences of rapid and continuous population growth? What can be done to lower birth rates and bring population growth under control?

Population problems are very different in poor developing and economically developed countries; in fact, they are exactly opposite. The first group of countries is trying to provide for young and growing populations resulting from many years of rapid and persistent population increase, while the second group is trying to come to terms with **ageing populations** (an increasing proportion of old people) and a reduction in numbers of people of working age (pages 142–3).

Developing countries – implications of more and more people

Crippling rates of population growth have led to many problems – economic, social, political and environmental (**Figure 1**). The environmental problems, resulting from the persistent population pressure to produce more, are making sustainable development ever more difficult to achieve. The overall result is low standards of living, poverty and poor quality of life; all increase vulnerability to natural and human disasters. One thing leads to another.

Many of the *environmental* problems are most noticeable in rural areas. As population growth has increased demand for food and natural resources such as fuel wood and water, farmers have responded by using the land more intensively, in some places to the point where the land and soil have been destroyed. The general term for this damage is **environmental degradation**. One example is **desertification**; a summary of its causes is given in **Figure 2**. This occurs where deserts spread and engulf areas that formerly carried surface vegetation cover and farming settlements. It is a process whereby land is turned into a desert as a result of human activities, although climatic hazards, such as low rainfall and drought, can make it worse. Overgrazing, over-cultivation and overuse of irrigation water leading to salinisation (soils becoming more salty) create land surfaces unable to support vegetation. In other words, future farm output is threatened, lowering the likelihood of sustainable development.

Areas most at risk are semi-arid regions where rainfall is concentrated in one season and the amount that arrives is very variable from year to year. When two or three dry years occur together, human pressures on the natural vegetation increase. One of the worst affected areas is the Sahel on the southern edge of the Sahara desert. Countries located here have some of the highest birth rates in the world (for example, Niger 55, Somalia 52, Mali 50 and Chad 48). The relentless rise in numbers means that subsistence farmers can no longer accumulate a food surplus in wet years to see them through the dry years. In dry years, over-cultivation, overgrazing and further deforestation for fuel wood start the train of events that ends with desertification.

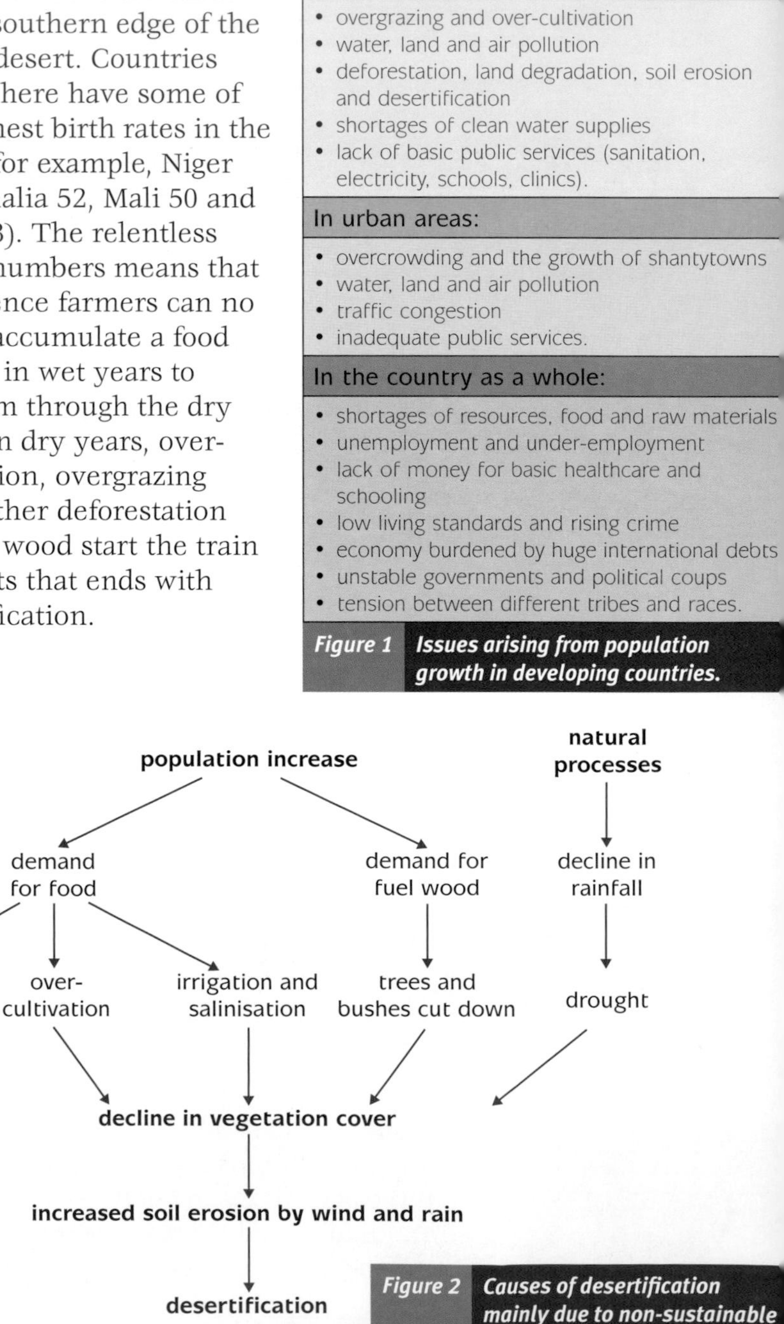

In rural areas:
• overgrazing and over-cultivation
• water, land and air pollution
• deforestation, land degradation, soil erosion and desertification
• shortages of clean water supplies
• lack of basic public services (sanitation, electricity, schools, clinics).
In urban areas:
• overcrowding and the growth of shantytowns
• water, land and air pollution
• traffic congestion
• inadequate public services.
In the country as a whole:
• shortages of resources, food and raw materials
• unemployment and under-employment
• lack of money for basic healthcare and schooling
• low living standards and rising crime
• economy burdened by huge international debts
• unstable governments and political coups
• tension between different tribes and races.
population increase
natural processes
demand for food
demand for fuel wood
decline in rainfall
overgrazing
over-cultivation
irrigation and salinisation
trees and bushes cut down
drought
decline in vegetation cover
increased soil erosion by wind and rain
desertification

Figure 1 Issues arising from population growth in developing countries.

Figure 2 Causes of desertification mainly due to non-sustainable human activities.

The poverty in rural areas is transferred to the urban areas by migration. It is in urban areas that *socio-economic* problems such as overcrowding, poor housing, unemployment and inadequate public services are most concentrated. The *political* problem for governments and city authorities is how to plan and pay for public services (health, education), public utilities (clean water supply, sanitation, electricity) and housing. The *environmental* consequences of population growth in urban areas without effective pollution controls are alarming. Air pollution is at dangerous levels for human health in many big cities, from traffic congestion on roads full of old trucks and cars and unsupervised factory emissions. Water pollution is widespread because rivers are used for waste and litter disposal. Sprawling shantytowns lead to removal of vegetation on the urban edges, while inside cities they climb up hillsides, increasing the risk of landslides.

Figure 3 *Self-help housing in Mumbai (Bombay), the type of housing in which more than half of the city's 18 million inhabitants live.*

Controlling population growth

Many governments in developing countries now recognise high birth rates as a problem, and have family planning and population policies in place (**Figure 4**). The policies vary from persuasion and incentives, as in Sri Lanka, to passing strict laws reinforced by severe punishment, as in China. Some countries, such as Iraq and Saudi Arabia, have shown little interest in controlling their populations. In most cases this is due to religious beliefs. In Muslim and Catholic countries religious teaching opposes any form of contraception.

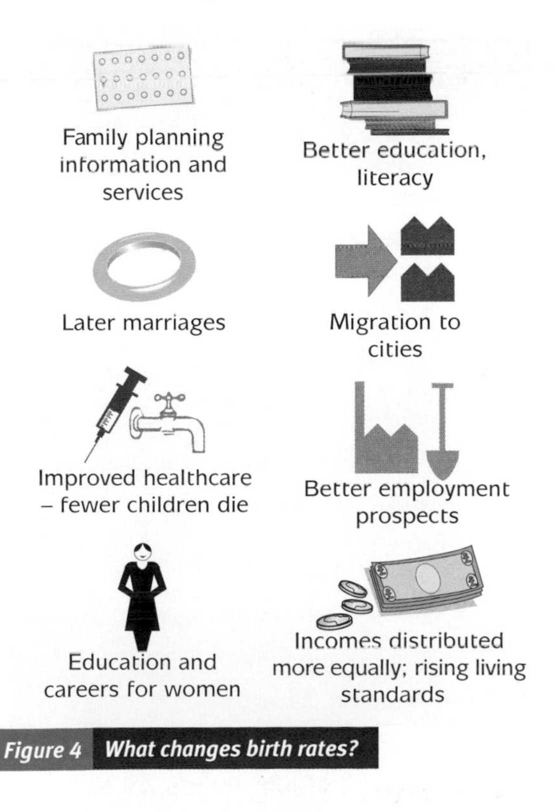

Figure 4 *What changes birth rates?*

ACTIVITIES

1 Make a table entitled 'Problems of high population growth', using the headings Economic, Social, Political and Environmental. Fill it using information from **Figure 1** and the text.

2 (a) Explain how non-sustainable human activities cause desertification.
(b) Which do you consider to be the more important cause of desertification – physical or human? Explain your answer.

3 (a) (i) Make a frame and draw a labelled sketch of the housing area in **Figure 3**.
(ii) Add labels for the urban problems likely to exist here.
(b) Explain how population growth leads to urban problems like these.

4

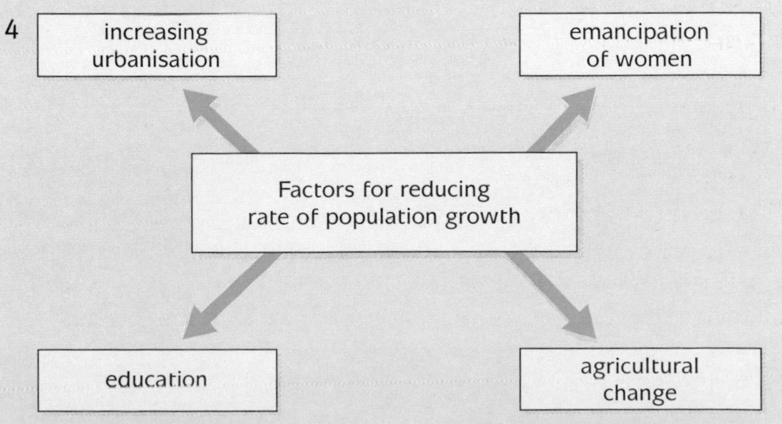

Choose three of these factors. Using pages 136–7 and **Figure 4** on this page, explain how each one helps to lower birth rates from high to low.

Population policies for reducing growth

What strategies do governments use to reduce rapid population growth? Why are they more effective in some countries than in others?

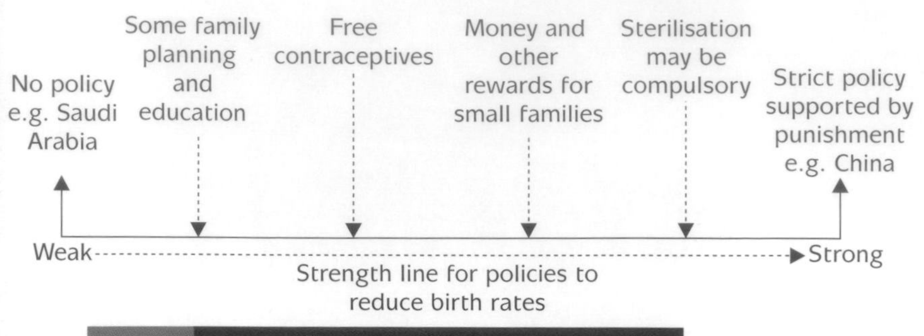

Figure 1 **Strength line for national population policies.**

Population policy in this context means a government strategy for reducing birth rates. There are great variations in strength of policy and implementation between countries (**Figure 1**). China (page 141) is at the top end of the scale for the ruthless and relentless manner in which it has implemented its 'one child' policy since the 1980s. Countries without elected governments, like China, can impose an unpopular policy and make it work, but it is different in a democracy. An earlier Indian government forced through a massively unpopular programme of compulsory sterilisation, and was voted out at the next election. The present voluntary programme does not work well in backward rural states where social and religious traditions remain strong. Between now and 2050 India is expected to overtake China as the world's most populous country (**Figure 4**).

Thailand has a successful and effective family-planning programme, which uses a mixture of media, economic incentives and community involvement to increase contraceptive use. The immediate benefits of smaller families are stressed; the key message links family planning and low family size to high standards of living. To get everyone involved, even in remote rural areas, fun events such as birth-control carnivals are organised. In Thailand the message is carried everywhere, not just left in the cities. Farmers who have registered for family planning are given financial benefits, such as above-market prices for their crops and reduced transport costs to market. There is a big difference in Islamic Pakistan, which does not have a well-promoted population policy and where there is constant political rivalry with India, its much more populous neighbour.

In most African countries, population policies are either non-existent or given a low priority. This is due to a mixture of financial, political and cultural factors:

1. Poverty – many countries are too poor to provide the necessary health infrastructure.
2. Political instability, wars and conflicts, and widespread government corruption.
3. Islam (the dominant faith in the north) and tribes of subsistence farmers view children as assets.

FURTHER RESEARCH

Find out more about Niger's population on the weblink www.contentextra.com/aqagcsegeog.

Niger's national action plan for population – too little, too late

Niger is a country of young people. Half the population is under 15. Its population will double by 2025. At last, someone in the government has a population plan – to bring the average number of children per woman down from the current 7.1 to 5.0 by 2015.

All previous ideas for stopping the population explosion have been wrecked by male domination, lack of education for women, poverty and repeated struggles against drought. Islamic clerics rail against the evils of contraception. The fact that 60 per cent of marriages involve girls under 15 does not help.

When some villagers were asked in a recent survey 'How many children would you like to have?', women said nine and men twelve. But some families said they would like 30 or 40 children. This is a society that is pro-children. Subsistence farmers, making up 75 per cent of Niger's people, see the worth of children in a different way from city folk.

Only 5 per cent of women in Niger use any form of contraception, mainly because men forbid it. In the capital, Niamey, there are clinics where contraceptives are dispensed free. One mother of ten said that she had been taking the pill without her husband knowing, but, when he found out, he forced her to stop. What hope is there for the action plan target of 10 per cent of women using contraception by 2015?

Figure 2 **Newspaper report (December 2007).**

INFORMATION

Fact file for Niger

Population: 13 million

Religion: 80% Islamic

Income: US$ 870 per head

Birth rate: 55 per 1000

Death rate: 20 per 1000

Infant mortality rate: 145 per 1000

Population structure:
- Under 15 50%
- Over 65 2%

Living in urban areas: 24%

China's 'one child' policy

In the 1980s China's population was already over 1000 million. The government decided that the existing 'two child' policy was not sufficient to reduce population growth. China's population would continue to grow at least until 2025, by which time there would be 1.8 billion Chinese. The government introduced the 'one child' policy in the hope that the population would stabilise at about 1.2 billion early in the twenty-first century. **Figure 3** shows an example of the advertising information used by the Chinese government. The policy of 'one couple, one child' is very strict. It is virtually illegal to have more than one child and families are criticised and fined; forced abortions and sterilisations have also been reported.

WHY HAVE ONLY ONE CHILD?

For you with one child:
Free education for your one child.
Family allowances, priority housing and pension benefits.

For those with two children:
No free education, no allowances and no pension benefits.
Payment of a fine to the state from earnings.

To help you
Women must be 20 years old before they marry.
Men must be 22 years old before they marry.
Couples must have permission to marry and have a child.
Family planning help is available at work.

REMEMBER: One child means happiness.

Figure 3 *Part of the Chinese government's advertising campaign.*

Among ordinary Chinese, the policy is desperately unpopular; the Chinese love children. City couples find it easier to abide by the law, but two-thirds of China's population are peasant farmers living in rural areas. Chinese couples in the countryside want large families to help them work in the fields and to look after them in their old age. At the same time, Chinese culture has always held boys in higher esteem than girls, and there have been reports of infanticide where girl babies have been killed by couples who want a son. There are also concerns that the country is breeding a society of spoilt children, mostly boys. Young men are having difficulty in finding partners because of the shortage of women. Such concerns have forced the government to relax the rules in rural areas to allow up to two children.

For the first time, in 2008 there was a suggestion that China might have to scrap its one child policy because of fears of a demographic nightmare from an ageing population. As in many European countries (pages 142–3), the economic time bomb is now ticking – too few young people to support the increasing millions of old people. China will soon suffer a severe labour shortage. The government is faced with a dilemma; if it increases the limit from one to two children for all Chinese, this will mean 400 million more people over the next 25 years.

GradeStudio

1 Study **Figure 4**.
   a Draw a graph to show total populations in China and India for 2006 and estimated for 2050. (3 marks)
   b Calculate the rate of natural increase (per 1000) in China and India for 2006. (2 marks)
   c Does the data for China, when compared with that for the other countries, suggest that the one child policy has been effective? (3 marks)

2 Using examples of countries to support your answers, explain these statements.
   a Social and cultural factors can limit the effectiveness of population policies. (4 marks)
   b Governments have a big influence on the effectiveness of population policies. (4 marks)

Exam tip

When asked for an *example*, naming one will be given some credit. But you will get even more credit for giving information about it.

	Total population 2006 (millions)	Birth rate per 1000 in 2006	Death rate per 1000 in 2006	Average annual population growth 2000–2005 (%)	Estimated population in 2050 (millions)
China	1313	14.5	7.1	0.73	1409
India	1031	23.8	8.3	1.51	1658
Pakistan	157	35.9	7.5	2.44	292
Thailand	64	17.3	7.2	1.01	77

Figure 4 ***Population data for four Asian countries.***

Population issues – ageing populations

How are ageing populations different from young populations? Why are governments worried about them? What can they do?

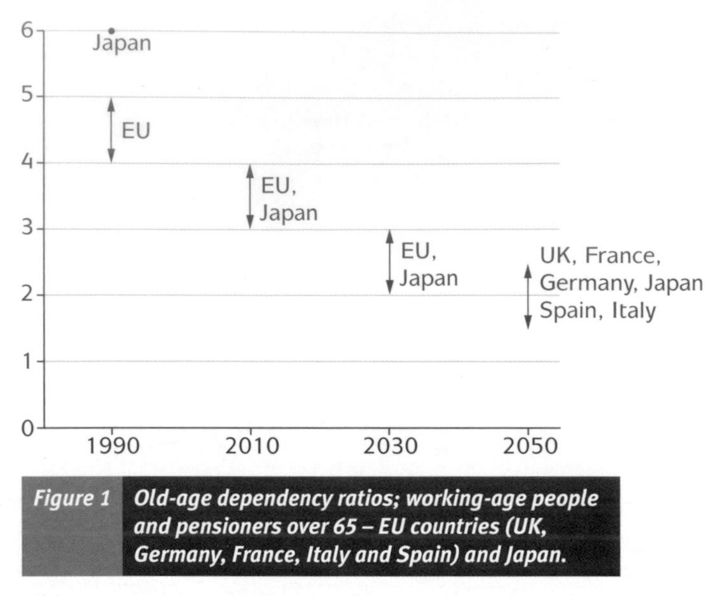

Figure 1 *Old-age dependency ratios; working-age people and pensioners over 65 – EU countries (UK, Germany, France, Italy and Spain) and Japan.*

The rise of the 'grey population' (as ageing populations are often referred to in the media) is most noticeable in Europe and Japan, where birth rates are low, and in some countries below the death rate, leading to natural decrease (**Figure 5**, page 135). The countries named here have reached stage 5 in the Demographic Transition Model (pages 136–7). This is leading to big falls in dependency ratios between working populations and pensioners over 65 (**Figure 1**). In 1990, there were between four and five people of working age to support each old person over 65 in many European Union (EU) countries, and as many as six in Japan. Now the typical ratio in EU countries is 3–4:1. It is predicted to fall to 2–3:1 by 2030, and to below 2:1 in Italy and Spain by 2050. This has been described by some as a ticking demographic time-bomb.

Changes in population structure and its economic implications

Apart from a few countries ravaged by HIV/Aids and devastated by wars (page 135), human life expectancy is increasing due to continued improvements in medical knowledge and the discovery of new drugs. Not only are most drug companies based in the rich countries of North America and Europe, but people living in developed countries can also afford to buy the new drugs and enjoy higher levels of nutrition than in the developing world. Life expectancies at birth in developed countries, already high, are set to soar – for men from 71 in 2000 to 80 by 2050, and for women from 78 to 85. This is causing big changes in population structure. **Figure 2** shows expected changes in EU population by 2050. Notice the thinning of the working age groups, especially between 30 and 50, which will no longer contain the biggest age groups. This is a ticking economic time-bomb.

As people get older, their need for state-funded healthcare increases; and most European countries provide generous pension schemes. But with fewer working people, where will the money come from? Health services and pensions are paid for by taxes, most of which come from the working population. At the same time as more money is needed for the elderly, the population of working age is shrinking. European governments have just begun to wake up to the problem, and are increasingly worried about their continued ability to finance state pensions. As long as birth rates remain low, there are not going to be enough workers to generate all the money needed. Needless to say, telling today's workers that they must pay more in taxes and into pension funds to support the elderly is not popular – and not a vote-winner for any elected government!

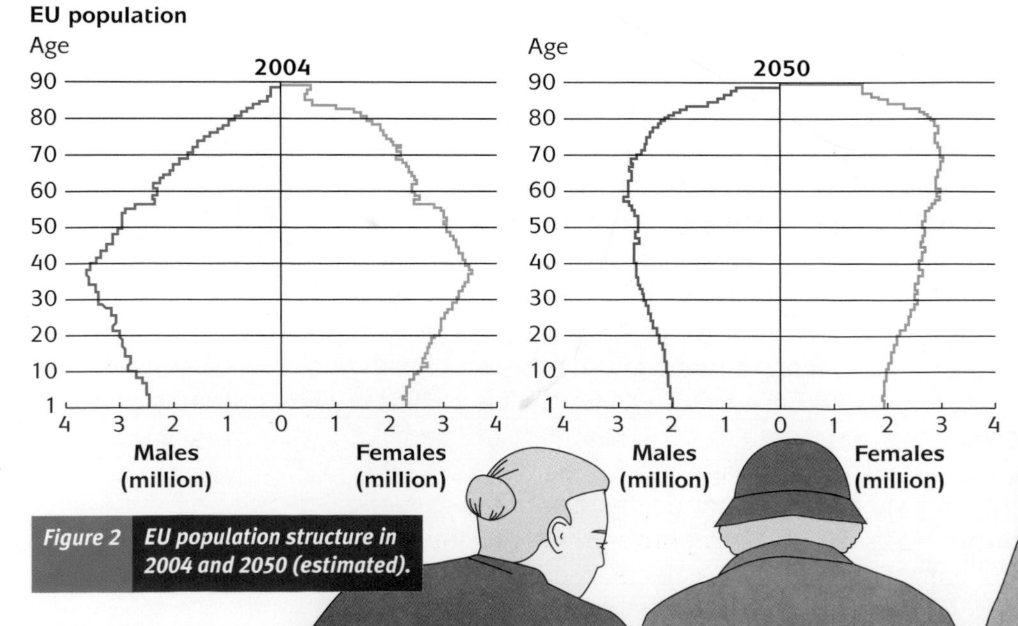

Figure 2 *EU population structure in 2004 and 2050 (estimated).*

Coping with ageing populations

The future problems for EU governments can be summarised like this:

1 Their populations aged 65 or more will rise by 75 per cent by 2050, leaving about two instead of four workers for every pensioner.

2 Funding for pensions, healthcare and long-term care will need an extra 4–8 per cent of their total national budgets by 2050.

3 Their labour forces are expected to drop by 16 per cent by 2050.

What is being done? Retirement ages are already going up, along with the age when governments begin to pay state pensions. European countries that have already raised retirement ages include Iceland (70 years), Norway and Denmark (67); others are to follow soon. Accompanying this are active policies to promote the continued employment of older people.

In response to plunging birth rates, Germany in 2006 became the latest of the EU countries to offer incentives to couples to have more children, including tax breaks and up to €1800 a month for parents to take time off work. However, schemes to encourage couples to have more children have not worked well in the past, and most governments accept that there are limits to what they can do to influence highly personal decisions, like having a child.

Many believe that the only way for Europe to maintain its future working population will be through immigration. There is no shortage of people willing to work in Europe, indeed many countries already depend upon foreign-born workers to plug their skills gaps and fill job vacancies (pages 146–7). But this policy is very unpopular with the public. Therefore, increasing the retirement age, and forcing people to make higher pension contributions during their working lives, seem to be the only options for governments, though neither of these is exactly popular. Bumpy times lie ahead for European governments as they grapple with the pension problem, particularly because the growing 'grey vote' is significant at election time.

Ageing populations are spreading

Some Asian countries are 'greying' fast. This is a sign of economic growth and success in countries such as Singapore, South Korea and China, but it is a tricky issue for the future. Should China relax its 'one child' policy (page 141)? Singapore has the look and feel of an economically developed western country (**Figure 3**). Singapore's previous 'Stop at two' population policy was too successful, because it brought fertility rates down below the replacement level of 2.1 children per woman. Now its national population policy includes tax breaks that are progressively increased for couples with three or more children to try to arrest falling birth rates.

Figure 3 **The modern and dynamic business centre of Singapore.**

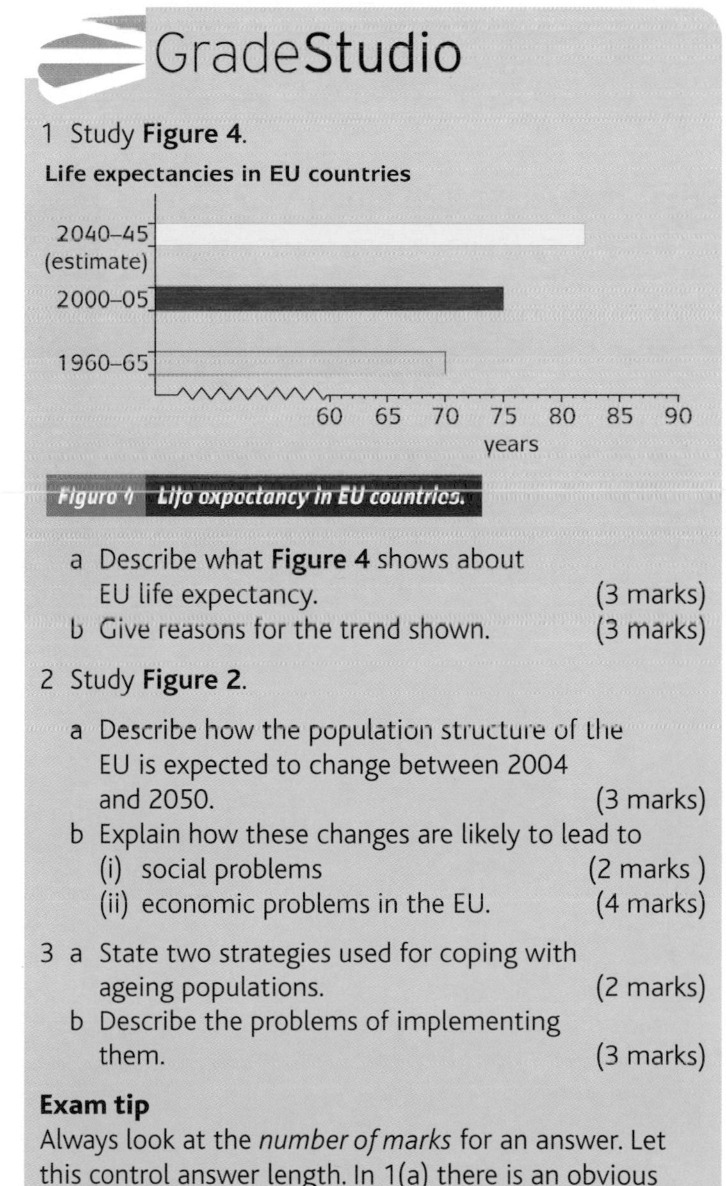

GradeStudio

1 Study **Figure 4**.

Life expectancies in EU countries

Figure 4 **Life expectancy in EU countries.**

a Describe what **Figure 4** shows about EU life expectancy. (3 marks)

b Give reasons for the trend shown. (3 marks)

2 Study **Figure 2**.

a Describe how the population structure of the EU is expected to change between 2004 and 2050. (3 marks)

b Explain how these changes are likely to lead to
(i) social problems (2 marks)
(ii) economic problems in the EU. (4 marks)

3 a State two strategies used for coping with ageing populations. (2 marks)

b Describe the problems of implementing them. (3 marks)

Exam tip

Always look at the *number of marks* for an answer. Let this control answer length. In 1(a) there is an obvious 1-mark answer. But what is needed for 3 marks?

Migration – causes, types and impacts

Why do people move? When is migration beneficial to all involved? And when is it disastrous?

Migration is the movement of people to live in a different place, either within the same country (**internal migration**) or to another country (**international migration**). Moving out of a country is **emigration**. All people moving into a country are **immigrants**; those without documents for entry, such as valid passports and visas, are classified as **illegal immigrants**. Some of them slip across borders without being noticed by border guards, for example Mexicans into the USA. Others detained at entry points by customs officers are **asylum seekers**; they need to make a case for staying, often using the argument that it would be unsafe for them to return to their home country.

What makes people move?

In most decisions more than one factor is involved. Often these factors are a mixture of push and pull (**Figure 1**). **Push factors** are people's dislikes about where they live – the disadvantages of living there. There may be no work, or the work that exists may be badly paid, or available for only part of the year. This is an example of an economic factor. Public services and utilities such as schools, hospitals, electricity and clean water supply may not be widely available. These are more social factors affecting people's quality of life. A natural disaster, such as an earthquake or flood, might force people to leave immediately: environmental factors like these can be so strong that people do not even think about staying.

Pull factors are the attractions of the place people are moving to – the advantages of moving there. The place may offer physical advantages such as a wetter climate, more reliable rainfall and better soils. Or it could offer economic attractions, such as job opportunities with good prospects of improved standards of living. Attractions can also be predominantly social, like moving closer to family and friends.

Rural areas *Push factors*	**Urban areas** *Pull factors*
• poverty	• better-paid jobs
• work only in farming	• work in factories, offices and shops
• land shortages, overuse of farmland and drought causing food shortages and famine	• reliable food supplies
• lack of services, shortage of clean water	• schools, hospitals, safe water supply and electricity
• remoteness – dirt track links only	• focus of roads, mainly paved roads
• little hope of change and improvement; old and traditional ways of life	• always changing; new skyscrapers and road systems; proper shops; dynamic feel to the place

Farming village in the mountains of Peru

Business zone in Lima, the capital city

Figure 1 ***Push and pull factors for rural to urban migration.***

Although in general there have been great improvements in communications and transport in the world, there are still **obstacles** to the free movement of people. Between countries there are border controls: visas and entry permits are often needed. Within countries there may be roadblocks and political or military controls between different regions, especially during times of civil strife and war. For some people, social obstacles to migration are strong, such as leaving family members behind and fear of the unknown; however, the strength and importance of these varies greatly from person to person.

Types of migration

A simple distinction is between forced and voluntary migrations (**Figure 2**). **Voluntary migration** is when a person makes the decision to move. The decision is usually made after weighing up the advantages and disadvantages of moving. It is likely that both push and pull factors will be involved, but the relative strength of these varies greatly. The most common type of international migration in the world today is **economic migration**; basically this is movement for work, usually from poor to rich countries. **Economic migrants** are seeking the better life that a higher income brings. The push factors for this are no work and low standards of living; the powerful pull factor is the amount of money that can be earned by working in a developed country compared with the home country. Even if a person is in work in a developing country, the higher wages in developed countries are still an attraction. This suggests that the pull factors are often stronger than the push factors for economic migrants, such as African and Asian immigrants into the EU (pages 146–7).

Types of migration

- **Voluntary**
  - **Reasons**
    - *Economic*
      - for work and improved standards of living
    - *Social*
      - joining up with relatives or friends
      - for retirement
  - **Examples**
    - *National*
      - rural to urban (mainly in developing countries)
      - rural to rural (new land for settlement)
      - urban to rural (mainly in developed countries)
    - *International*
      - from developing to developed
- **Forced**
  - **Reasons**
    - *Physical*
      - after a natural disaster (earthquake, volcano, flood, drought, cyclone)
    - *Human*
      - after a human disaster (war, revolution)
  - **Examples**
    - *National*
      - movement to camps and temporary shelters
    - *International*
      - refugees to other countries

Figure 2 Types of migration.

Forced migration is compulsory migration and people have little choice about moving. Forced migrants who move to another country are **refugees**. Both physical and human factors can cause this movement. The major volcanic eruption in the Caribbean island of Montserrat (pages 16–17) forced out over half this island country's population to neighbouring islands, the UK and USA. War and persecution of people of different ethnic groups are the most important human factors in forced migration. Since 2000 most refugees have come from just two areas of the world – western Asia (especially Afghanistan and Iraq) and sub-Saharan countries in Africa (especially Sudan, Burundi, Congo and Somalia); upheavals caused by armed conflicts and civil wars have been made worse in Africa by physical problems such as drought, which has led to widespread famine and mass movements of people. The scene shown in **Figure 3** has become depressingly frequent in recent years.

Figure 3 African refugees, dependent on aid for survival.

Effects of migration

Movements of people can arouse strong feelings – among families and friends, in the national population of countries receiving large numbers of immigrants and among politicians and city officials trying to provide for large numbers of new arrivals. As with most things involving people, there are both positive and negative impacts from migration. Advantages and disadvantages for both losing countries (from which people migrate) and the receiving countries (into which people move) are included in the summary in **Figure 4** for migrations that involve cross-border movement.

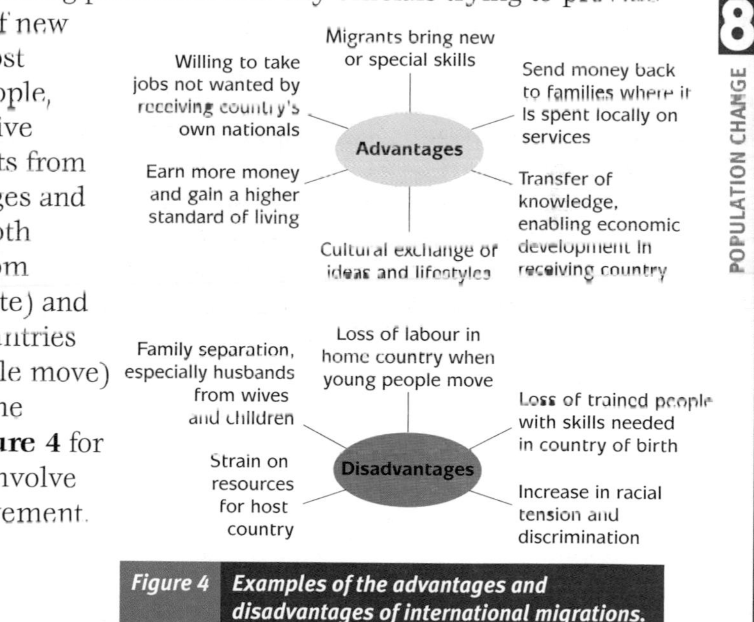
Advantages
Migrants bring new or special skills
Willing to take jobs not wanted by receiving country's own nationals
Send money back to families where it is spent locally on services
Earn more money and gain a higher standard of living
Transfer of knowledge, enabling economic development in receiving country
Cultural exchange of ideas and lifestyles
Disadvantages
Loss of labour in home country when young people move
Family separation, especially husbands from wives and children
Loss of trained people with skills needed in country of birth
Strain on resources for host country
Increase in racial tension and discrimination

Figure 4 Examples of the advantages and disadvantages of international migrations.

ACTIVITIES

1. Briefly state the differences between the following pairs:
   (a) international and internal migrations
   (b) immigration and emigration
   (c) push and pull factors for migration
   (d) voluntary and forced migrations.
2. Use **Figure 4**. Rearrange the advantages and disadvantages in two lists: (a) For the losing country (b) For the receiving country.
3. Draw two spider diagrams to show advantages and disadvantages of rural to urban migration within developing countries.

Movements of people within and into the EU

Why do all EU countries have significant numbers of foreigners working in them? What are the impacts (good and bad) in countries in the front line for immigration such as Spain and Italy?

The EU allows and promotes the free movement of people, goods and services between its 27 member countries. Foreign-born workers make up a third of the total workforce in Luxembourg and between 10 and 15 per cent in the bigger economies, such as the UK, Germany and France. Everyone with a British passport is free to go and work in any EU country.

For many centuries, Europeans (Spanish, Portuguese, British, French and Dutch in particular) settled the world and established colonies in the 'new' continents of North and South America, Australasia and Africa. Now the direction of flow has been reversed as people migrate from poor to rich countries. During the 1990s, Europe became a continent of immigration, now housing up to 35 million non-European-born immigrants. Most of them are economic migrants, moving for work.

They fall into two main categories, either highly skilled or unskilled, with few in between. The former plug vital skills gaps; for example, could the health service in the UK operate without doctors and dentists, already qualified in their country of birth, arriving from India and other Asian countries? The latter (unskilled) mainly fill vacancies in the service sector, such as transport, food and drink, retail and public services, jobs that home-born people are unwilling to take – often citing low pay or unsociable hours as reasons. The wealth gap between Europe and developing world countries ensures a constant supply of willing migrants; wages for unskilled jobs can be 10–15 times higher in Europe.

In every country, the actual pattern of movements and flows in and out is incredibly complex. Take the example of the UK (**Figure 1**). The first point to remember, which newspapers often forget when talking about immigration, is that UK migration is not just one way.

The map highlights certain key points:

- that *nearness* matters – movement (both ways) is high with near neighbours like Germany and France
- that *history* is important – links are strong with the 'old colonies', such as Australia, New Zealand, South Africa, India and Pakistan
- that out-migration for *retirement* is now significant, most noticeable in the numbers to Spain, but also a factor in migration to France and the USA.

The enlargement of the EU in 2004 saw a big rise in UK arrivals from eastern European countries. Poland was the number one supplier of immigrants in 2005 with 49 000, just ahead of India with 47 000. The difference between these two groups is that it is unlikely that many of the eastern European arrivals will choose to settle in Britain permanently. This was confirmed in the economic downturn in 2008, when large numbers of Polish people left the UK – not only due to the contracting job market, but also because of the pound weakening against other currencies.

Immigrants from outside the EU

In 2005, just two countries accounted for over 60 per cent of new EU arrivals, Spain and Italy. This statistic is mainly explained by their locations on the Mediterranean frontier, the great wealth divide between rich Europe to the north and poverty-stricken Africa to the south. These countries act as entry points into the EU. In addition, Spain

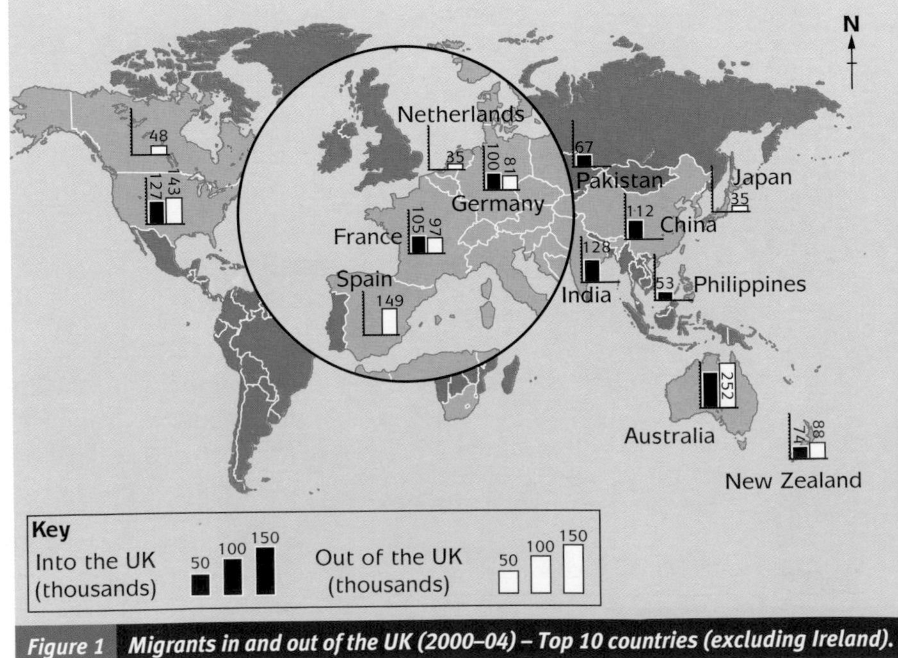

Figure 1 Migrants in and out of the UK (2000–04) – Top 10 countries (excluding Ireland).

EXAMPLE: Illegal immigrants (refugees and others) into Italy

Source regions of illegal migrants

- Top four countries, 2005

1	Romania	95 000
2	Albania	66 000
3	Morocco	60 000
4	Ukraine	40 000

These four countries accounted for half the total illegal migrants coming into the EU.

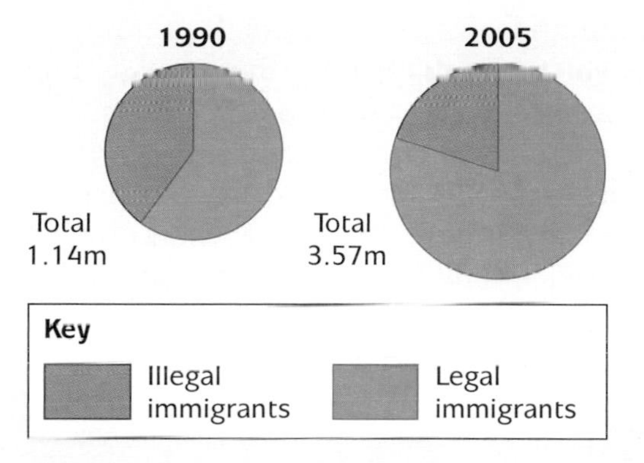

Employment

- On farms, picking oranges, olives, etc.
- In fruit-canning plants.
- Agriculture depends on them.

Refugee migrants from Africa

- Many smuggled in on boats; 11 000 landed in a six-month period in 2006.
- Many more drowned when overloaded small boats sank.
- Normal route is: various African countries – Libya – Italian-owned Lampedusa island off the African coast – southern Italy.
- Gather daily in groups for work; without papers, at mercy of gang masters.
- Sleep rough or in overcrowded, overpriced rooms in run-down property, described as 'living in inhuman conditions'

Government response

- Migrants detained on arrival and most issued with deportation orders.
- Orders not enforced; once free, they make for agricultural areas in the south.

The attitude is that it is a European problem that Italy has become the EU frontier for refugees and other migrants from Africa. The south's agricultural economy, based on fruit, would collapse without plenty of cheap labour at peak times. African migrants complain of exploitation, about not being paid for work done and about violence (beatings and shootings) against them by Italian youths, while the police do nothing. Without legal papers – what can the migrants do? Nor can they get back home.

Figure 2 *Illegal immigrants (refugees and others) into Italy.*

(like Britain) has colonial links in Africa and the Americas, while Italy is next to the old communist bloc countries of eastern Europe with significantly lower levels of economic development. Refugees from Africa, usually trafficked by smugglers into the EU, are included in the totals for illegal immigrants. Most claim to be asylum seekers, but it often turns out that they are not from one of Africa's many conflict zones; in reality, many of them are economic migrants without EU entry papers.

In Spain the public mood towards immigrants, whether legal or illegal, changed in 2008, as economic gloom replaced boom (**Figure 3**). When the jobless total was rising by 3000 per day in 2008, the government launched a plan to pay jobless immigrants their unemployment benefit in advance, provided they went home and promised not to return to Spain for three years. But only 10 000 of the 150 000 eligible were expected to take up the offer.

1998–2007	2008
• Construction, tourism and wider economy thriving • Cheap labour for building sites, hotels and farms • Immigrants needed and welcomed; they increased from 3% of population in 1998 to 12% in 2008 • Allowed use of public services (hospitals, schools, etc.)	• Economy slowing down towards recession • Rising unemployment (11% of workforce) • Pressure on politicians to cut visas for migrant workers • Social unrest in housing areas as immigrants are blamed for rises in crime, queues in surgeries and shortage of nursery places(hospitals, schools, etc.)

Figure 3 *Immigrants in Spain (legal and illegal), in 1998–2007 and 2008.*

ACTIVITIES

1 From **Figure 1**, for the five years 2000–04:
   (a) name the top three source countries of immigrants to the UK, and state what is similar and different about them
   (b) name the top three destination countries of emigrants from the UK, and state what is similar and different about them.

2 (a) (i) Draw two spider diagrams to show the negative and positive impacts of population movements within and into the EU.
   (ii) Use different colours or shading to highlight economic and social impacts.
   (b) Choose the greatest advantage and disadvantage of migration in the EU (in your view) and explain your choices.

Ageing population in the UK

The demographic trends in the UK are clear (**Figure 1**). In early 2008 the Office for National Statistics (ONS) announced that people of pensionable age (at present 60 for women and 65 for men) outnumbered those of school age under 16 for the first time – 11.58 million compared with 11.52 million. The gap is certain to widen in future years. There are now 2.7 million people in the UK over 80; indeed the over-80s is the fastest growing age group. This compares with less than half a million in the census of 1951.

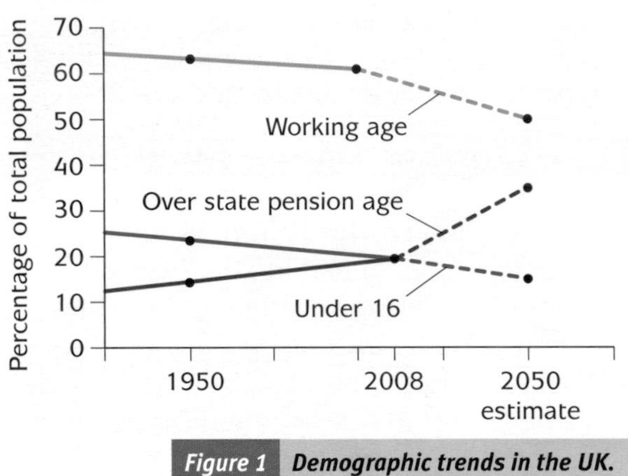

Figure 1 **Demographic trends in the UK.**

The boom in the UK pensioner population poses a tough set of problems for the UK government. People are drawing their pensions for longer. Living longer increases the need for healthcare and social services, for more places in nursing and care homes and in warden-assisted flats and bungalows, and for home services such as daily care and meals on wheels. The cost of maintaining current levels of care services, often regarded as inadequate, is set to double to £24 billion by 2026.

What is the UK government doing?

The UK government has no money of its own to fund these services. All it has is tax revenue raised from voters. Its options are limited to raising taxes, reducing spending on state pensions and cutting costs. **Figure 2** suggests that something needs to happen soon. A start has been made on tackling the pensions issue, with a phased programme underway to increase the retirement (and pension) age for a woman from 60 to 65, the same as a man's. The state pension age will be raised to 68 for everyone by 2050, accompanied by reforms of the whole pension system. As for cutting costs, the NHS restricts access to drugs for diseases of old age, such as Alzheimer's, amid fears that their cost could cripple the health service. The Health Ministry admitted in 2008 that there will be a £6 billion black hole in funding care for older people in future years, without knowing how it will be filled.

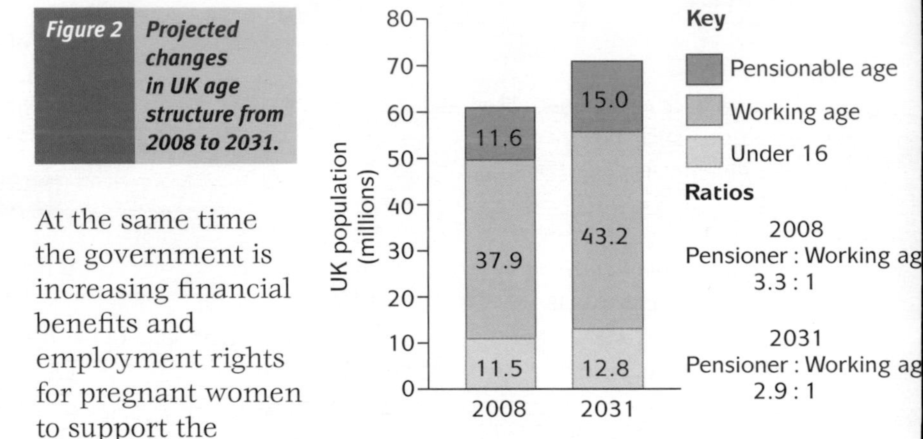

Figure 2 ***Projected changes in UK age structure from 2008 to 2031.***

At the same time the government is increasing financial benefits and employment rights for pregnant women to support the birth rate. To offset worker shortages and keep the economy growing and tax revenues coming in, the government has welcomed migrant workers into the UK; 2007 saw a record number (600 000) moving in and staying here for a year or more. Many came from the new EU countries of eastern Europe. Certain industries, notably construction and farming for vegetables and fruit, depend greatly upon them.

Are there any benefits of an ageing population?

There is always a positive side. Some pensioners (but by no means all) are well off and determined to 'spend the kid's inheritance' (dubbed SKIHERs). They pour money into the leisure sector at off-peak times of the year; they extend the holiday season in coastal resorts. Some manufacturing companies have built up successful businesses by targeting elderly people for the bulk of their sales, such as the makers of chair lifts and mobility vehicles. Many in good health are happy to continue working beyond retirement age; new anti-ageism employment laws will make this easier in the UK. To date only a few companies, notably B&Q, have openly recruited retired people. And, of course, pensioners still pay tax on income and savings, just like everyone else.

Migration issues in the UK

Across the UK the proportion of overseas-born residents is 7.5 per cent, but in London one in four was born abroad, and in some parts of the capital one in two. However, London has long been a multinational and multiracial city. The significant rise in the foreign-born workforce during the ten years of economic growth from 1997 (**Figure 3**) has been noticed much more in rural areas with little previous experience of immigration. One place that has felt a big impact is the port town of Boston in Lincolnshire, where foreign workers were estimated to make up a quarter of the population in 2008. It is surrounded by rich farmland, specialising in intensive vegetable growing, now absolutely dependent on migrant labour for harvesting crops. Most of the workers are from eastern Europe, especially Poland and the Czech republic. The scale and pace of the change increased the impact.

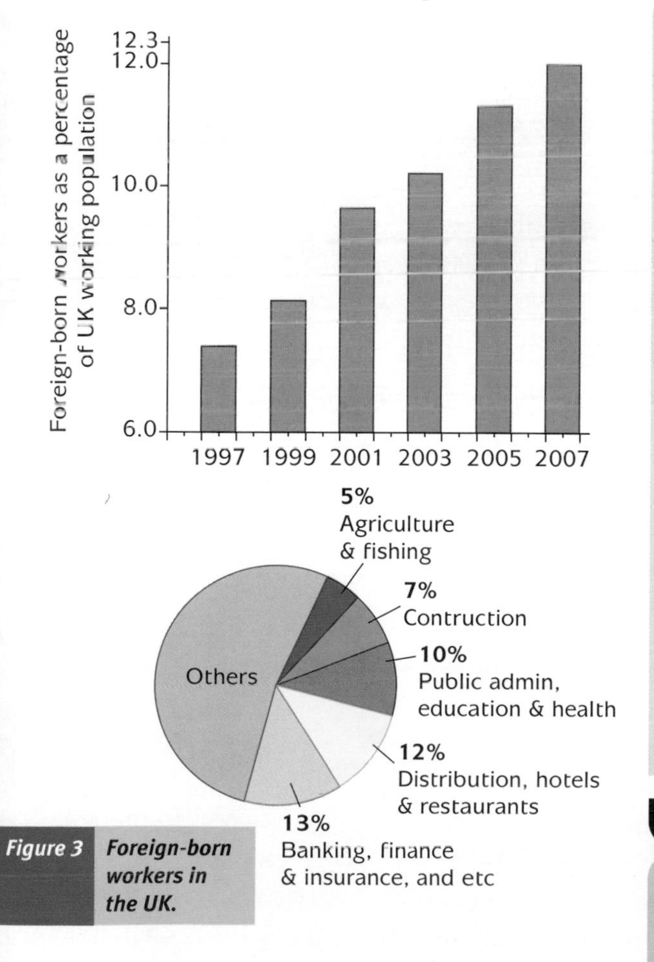

Foreign-born workers as a percentage of UK working population
12.3
12.0
10.0
8.0
6.0
1997
1999
2001
2003
2005
2007
5%
Agriculture & fishing
7%
Contruction
10%
Public admin, education & health
12%
Distribution, hotels & restaurants
13%
Banking, finance & insurance, and etc
Others

Figure 3 Foreign-born workers in the UK.

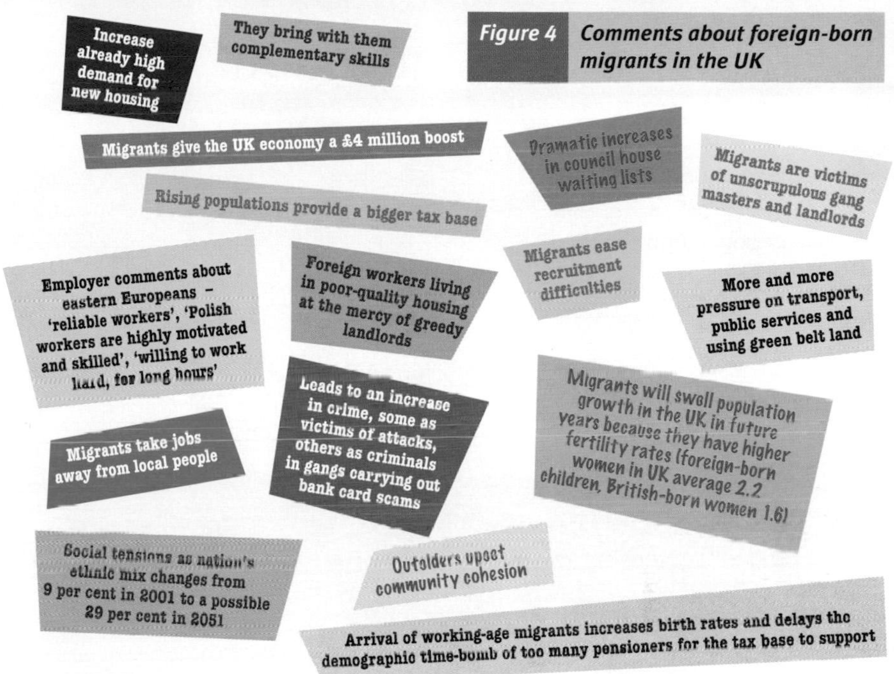

Increase already high demand for new housing
They bring with them complementary skills
Migrants give the UK economy a £4 million boost
Dramatic increases in council house waiting lists
Migrants are victims of unscrupulous gang masters and landlords
Rising populations provide a bigger tax base
Migrants ease recruitment difficulties
Employer comments about eastern Europeans – 'reliable workers', 'Polish workers are highly motivated and skilled', 'willing to work hard, for long hours'
Foreign workers living in poor-quality housing at the mercy of greedy landlords
More and more pressure on transport, public services and using green belt land
Leads to an increase in crime, some as victims of attacks, others as criminals in gangs carrying out bank card scams
Migrants will swell population growth in the UK in future years because they have higher fertility rates (foreign-born women in UK average 2.2 children, British-born women 1.6)
Migrants take jobs away from local people
Social tensions as nation's ethnic mix changes from 9 per cent in 2001 to a possible 29 per cent in 2051
Outsiders upset community cohesion
Arrival of working-age migrants increases birth rates and delays the demographic time-bomb of too many pensioners for the tax base to support

Figure 4 Comments about foreign-born migrants in the UK

ACTIVITIES

1 Make case study notes for 'The problems and strategies in one EU country with an ageing population'.

2 (a) Give two reasons why the UK government supported in-migration of foreign workers from 1997.

(b) Average wages in the work sectors in **Figure 3** vary greatly.

(i) Name examples of sectors with pay likely to be higher and lower than average wages.

(ii) How and why will the nationalities of their foreign workers be different?

3 (a) Make a table for impacts of in-migration into the UK, with these headings for its four columns.

Positive impacts		Negative impacts	
Economic	**Social**	**Economic**	**Social**

Use **Figure 4** to fill it in. (You can add other valid impacts known to you.)

(b) One verdict on UK in-migration is 'Economic advantage, social disadvantage'. How great is the evidence for this?

4 Explain why there are many different views about in-migration among people and groups already living in the UK.

FURTHER RESEARCH

Find out more about population change in the UK, as well as Italy's asylum seekers, on the weblink www.contentextra.com/aqagcsegeog.

GradeStudio

Practice GCSE question

See a Foundation Tier Practice GCSE Question on the weblink www.contentextra.com/aqagcsegeog.

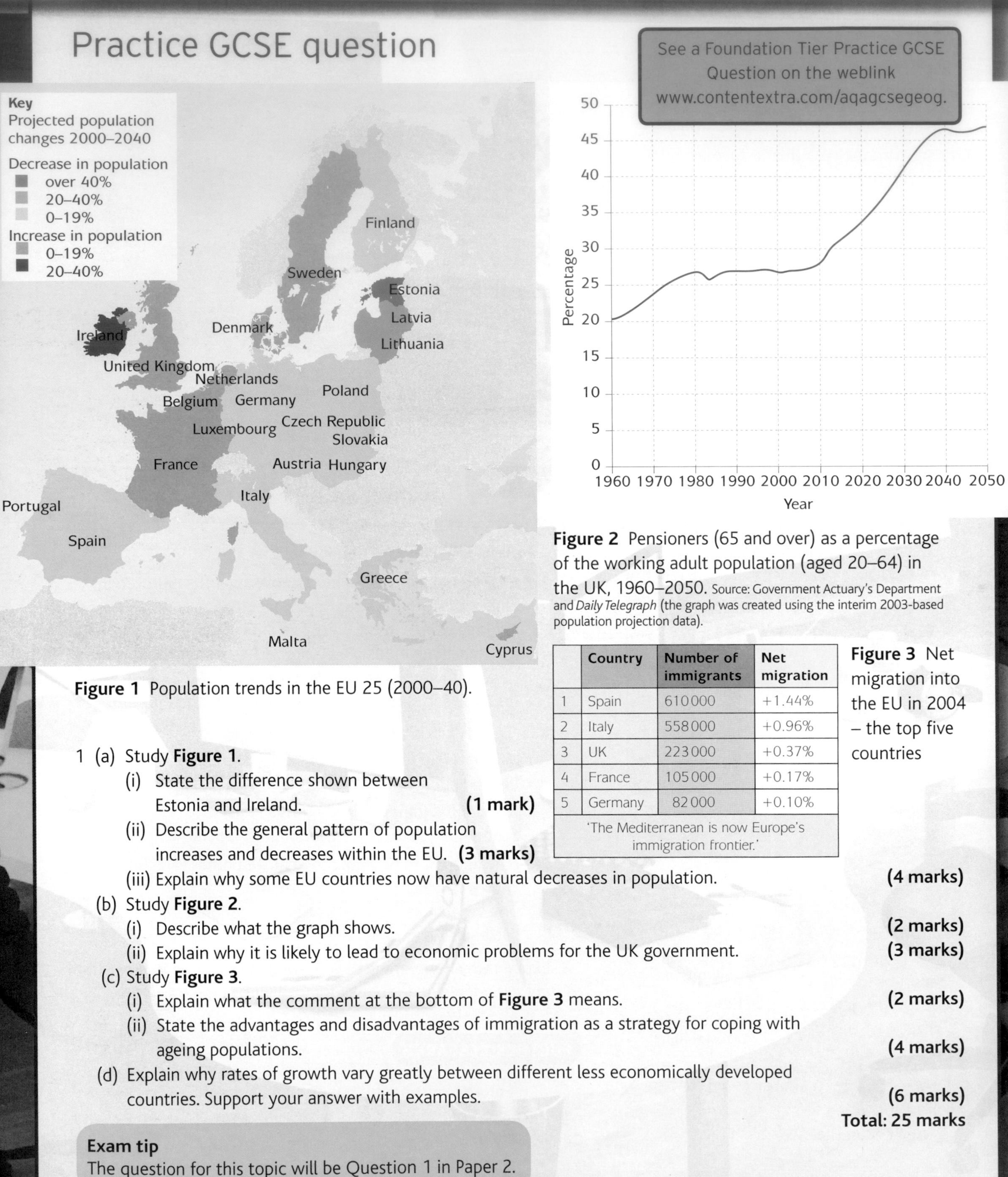

Figure 1 Population trends in the EU 25 (2000–40).

Figure 2 Pensioners (65 and over) as a percentage of the working adult population (aged 20–64) in the UK, 1960–2050. Source: Government Actuary's Department and *Daily Telegraph* (the graph was created using the interim 2003-based population projection data).

	Country	Number of immigrants	Net migration
1	Spain	610 000	+1.44%
2	Italy	558 000	+0.96%
3	UK	223 000	+0.37%
4	France	105 000	+0.17%
5	Germany	82 000	+0.10%
'The Mediterranean is now Europe's immigration frontier.'			

Figure 3 Net migration into the EU in 2004 – the top five countries

1 (a) Study **Figure 1**.
  (i) State the difference shown between Estonia and Ireland. **(1 mark)**
  (ii) Describe the general pattern of population increases and decreases within the EU. **(3 marks)**
  (iii) Explain why some EU countries now have natural decreases in population. **(4 marks)**

(b) Study **Figure 2**.
  (i) Describe what the graph shows. **(2 marks)**
  (ii) Explain why it is likely to lead to economic problems for the UK government. **(3 marks)**

(c) Study **Figure 3**.
  (i) Explain what the comment at the bottom of **Figure 3** means. **(2 marks)**
  (ii) State the advantages and disadvantages of immigration as a strategy for coping with ageing populations. **(4 marks)**

(d) Explain why rates of growth vary greatly between different less economically developed countries. Support your answer with examples. **(6 marks)**

Total: 25 marks

Exam tip
The question for this topic will be Question 1 in Paper 2.

Improve your GCSE answers

Interpreting population pyramids

A Where to look and what to look for

1. Look first at the base – is it wide or narrow?
   This indicates whether birth rates are high or low for the date shown.
2. Look at the overall shape – is it a true pyramid, is the shape more straight up and down, or is it overhanging the base?
   This suggests the country's stage in the DTM (pages 136–7). A true pyramid indicates stage 2, whereas an upright graph is more likely to be stage 4. Overhangs or bulges of older age groups suggest it could be stage 5.
3. Look at the top – are there significant percentages above 65 years old?
   This indicates a long life expectancy, one of the characteristics of an ageing population.
4. Observe the relative differences between:
   - the young (ages 0–14)
   - middle-aged (ages 15–64)
   - old (aged 65 and above).

   *From these, the dependency ratio can be worked out (at least approximately), but only if you are required to do it. The task is quicker and easier when percentages are used (as in **Figure 2**, page 137) instead of population numbers (**Figure 1**).*

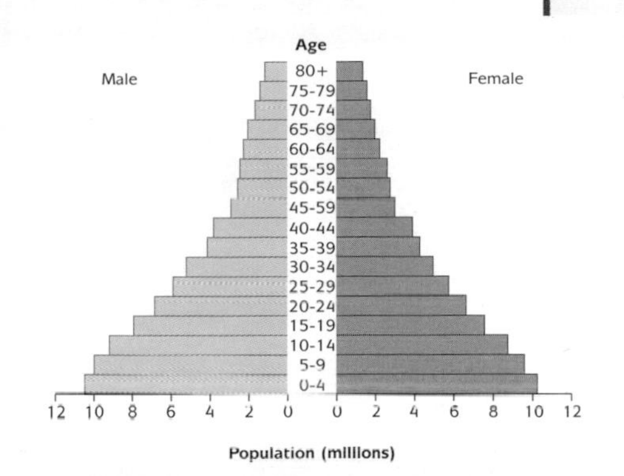

Figure 1 Pakistan – population structure (2000).

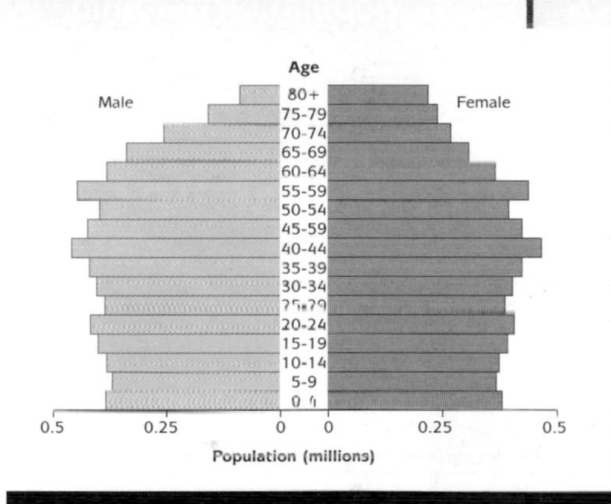

Figure 2 UK – population structure (2006).

Exam question: Describe the population structure of Pakistan shown in **Figure 1**. Answer this following the guidance given above.

Exam question: State three differences between the population structure of the UK (**Figure 2**) and that of Pakistan (**Figure 1**).

Answer this also by following the guidance on what to look for. Make sure that your differences are stated positively in a two-sided way.

Answer 1: The base of the pyramid in the UK is narrow with fewer young children than some older age groups.

Comment: A one-sided answer.

Answer 2: The base of the pyramid in the UK is narrower with fewer young children than some older age groups, whereas in Pakistan the 0–4 age group is the largest.

Comment: A two-sided answer ... notice *narrower* instead of narrow and *whereas* are used, and a statement is made about what the Pakistan graph shows.

B Population pyramids for areas or cities (not countries)

Pyramids might be the same or similar as for the country, but they can be quite different. Look at **Figure 3**. It shows population structure for an African city. It shows the results of rural to urban migration – young workers, especially men in Africa, are more likely to migrate than other age groups. The pyramid bulges for age groups between 20 and 35. What appears to be the effect of the presence of so many young workers on the city's birth rate?

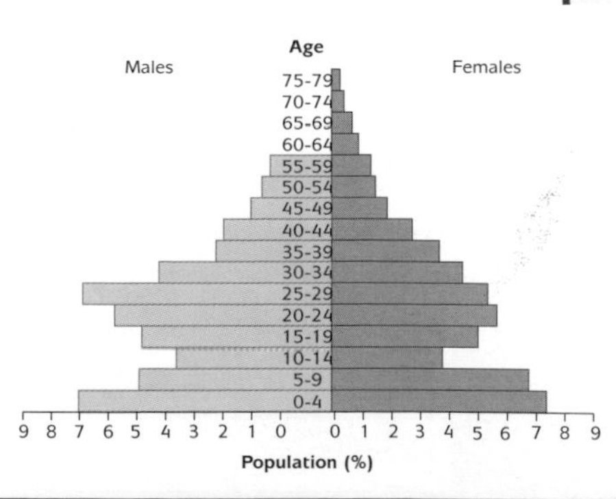

Figure 3 Population structure of an African city.

ExamCafé

REVISION

Key terms from the specification

Exam Café

Ageing population – increasing percentage of old people (aged 65 and over)
Birth rate – number of live births per 1000 population per year
Death rate – number of deaths per 1000 population per year
Dependency ratio – relationship between people of working and non-working ages
Immigration – movement of people into a country from another country
Life expectancy – average number of years that a new-born child can expect to live
Migration – movement of people either into or out of an area
Natural decrease – death rate higher than birth rate, declining population
Natural increase – birth rate higher than death rate, growing population
Population policy – national plan for population change (either to lower or increase birth rates)
Population structure – the make-up (age and sex) of a population, usually shown in a population pyramid
Pull and push factors – circumstances that attract or drive people to migrate
Refugee – person forced to flee from their country or place of residence

Checklist

	Yes	If no – refer to
Can you explain why the world's population is growing so fast?		page 134
Do you understand how to work out, using birth and death rates, whether a country has a natural increase or decrease of population?		pages 134–5
Do you understand the differences between stages 2, 3, 4 and 5 in the Demographic Transition Model?		pages 136–7
Why are high birth rates and rates of population growth more associated with poor developing countries than with those that are economically developed?		pages 138–9
Can you explain why some developing countries, like China, now have much lower birth rates than others, such as Niger?		pages 140–41
What is meant by an ageing population and what economic problems occur?		pages 142–3
Can you give examples of push and pull factors for migration?		pages 144–5
Can you name some of the good (positive) and bad (negative) effects of population movements into and within the EU?		pages 146–7
Do you feel comfortable describing and using population pyramids?		page 151

Case study summaries

China's population policy
Strategies (methods) used
Results
Drawbacks and problems
How effective overall?

EU country with ageing population
Problems caused
Strategies for coping with problems
Likely effectiveness?

Chapter 9

Changing Urban Environments

Overcrowded slum housing in Mumbai (Bombay), the type that houses more than 8 million inhabitants, over half the city's population. Some is a lot worse than the housing shown here.

QUESTIONS

- **Why are the world's big cities growing so fast?**
- **What changes can be observed between city centres and edges of built-up areas in British cities?**
- **Are squatter settlements in developing world cities places of hope or despair?**
- **How can urban living be made more sustainable?**

Urbanisation

Why has the human race become more urban than rural? Why is Asia home to more and more of the world's big cities?

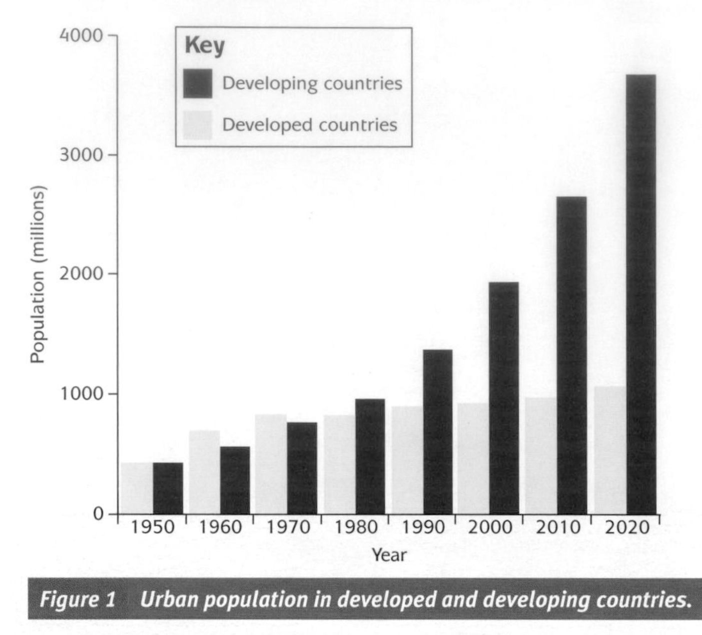

Figure 1 Urban population in developed and developing countries.

Urbanisation is an increase in the percentage of people living in urban areas. So explosive has been the rate of urban growth since 1980 that in 2006 the United Nations (UN) announced that the world's people were split half and half between urban and rural (3.26 billion: 3.26 billion). You are now living in a world where, for the first time in human history, more people live in towns and cities than in the countryside.

What makes urban settlements different from rural settlements (from which most urban areas originally grew)?

- The obvious answer is that urban settlements are larger – anything from 5–10 000 people to many millions.
- Their **functions** are different. Function is the settlement's purpose – why it is there and what 'work' it does. Rural settlements usually have only one function such as farming or mining, whereas the functions of urban settlements are many and varied – for example, industry, port, political administration (e.g. capital city), business offices, market and shopping centre, tourism.
- **Land uses** (the way the land surface is used) are different as well. Rural land uses include crops, pasture, woodland, moorland and marsh; urban land uses include houses (which cover the largest space in all urban areas), factories, offices, shops, parks and open spaces (including derelict land).

Figure 1 shows how recent, present and predicted future urban growth is overwhelmingly concentrated in developing countries. What is significant about the change that occurred sometime between 1970 and 1980?

Urban growth in developed countries

Towns and cities in the developed world grew rapidly during the Industrial Revolution. In Britain this was mainly in the nineteenth century. At the time there was an Agricultural Revolution. New farm machinery meant less labour was needed on farms, so people moved to towns where there were plenty of new jobs available in new factories, mines and shipyards. Urbanisation was happening. Growth was quite rapid, about 10 per cent per annum, and there were enough jobs for people. Many mine and factory owners built houses for their workers.

1950				2005			
Rank	**City**	**Continent**	**Population (millions)**	**Rank**	**City**	**Continent**	**Population (millions)**
1	New York	North America	12.3	1	Tokyo	Asia	35.3
2	London	Europe	8.7	2	Mexico City	Latin America	19.0
3	Tokyo	Asia	6.9	3	São Paulo	Latin America	18.3
4	Paris	Europe	5.5	4	New York	North America	18.1
5	Moscow	Europe	5.4	5	Mumbai (Bombay)	Asia	18.0
6	Shanghai	Asia	5.3	6	Delhi	Asia	15.3
7	Rhine-Ruhr	Europe	5.2	7	Kolkata (Calcutta)	Asia	14.3
8	Buenos Aires	Latin America	5.0	8	Buenos Aires	Latin America	13.3
9	Chicago	North America	4.9	9	Jakarta	Asia	13.2
10	Calcutta	Asia	4.4	10	Shanghai	Asia	12.7

Figure 2 The world's top ten most populated cities in 1950 and 2005.

Towns and cities continued to grow into the twentieth century as a result of the push and pull factors referred to on page 144, causing rural depopulation, particularly from remote areas. As a result, almost 90 per cent of the UK population is now classified as urban; it is almost as urbanised as it can be. If anything, in the UK and in some other developed countries, notably the USA, the movement is the other way. Big cities in particular are witnessing a loss of population as wealthy commuters and retired people are moving out into the country to live in villages and small market towns (pages 174–5).

Urban growth in developing countries

Persistent urban growth in developing countries has led to a change in the distribution of the world's largest cities. How does **Figure 2** suggest higher rates of urbanisation in continents of the developing world? By 2005 there were 20 **megacities** (urban areas of more than 10 million inhabitants) in the world, more than half of them in Asia. In the 50 years between 1950 and 2000, the mean latitude of **millionaire cities** (urban areas of more than 1 million inhabitants) changed from about 40° to 30°; an average move of 10° towards the Equator confirms the faster urban growth rates in developing countries, which are concentrated in the tropics on the southern side of the North South line that separates the developed from the developing world.

What are the causes of these phenomenal rates of growth in developing countries, responsible for what some call hyper-urbanisation? Three main ones can be identified:

1 Rural to urban migration, as a result of both push and pull factors, like the ones named and shown in **Figure 1**, page 144.

2 High fertility rates (large numbers of children per woman) and high rates of natural increase (birth rates higher than death rates) are population characteristics of developing countries (pages 134–9). City population structures are dominated by young people under 25 of child-bearing and child-producing ages. This will almost certainly ensure that population growth will continue for many years, even without more new arrivals from rural areas.

3 The urban areas are dynamic places, growth points, where industry and all other modern economic activities are concentrated. This makes them powerfully attractive to young people looking for work. Look at **Figure 3** to see the benefits that urban growth can bring in developing countries.

Benefits

Economy
- Big cities attract investment from overseas companies, encouraging modernisation.
- More value is added by processing and manufacturing than by exporting raw materials.

People's quality of life
- Essential services such as safe water supply, sanitation and electricity are more likely to be available than in rural areas.
- Often secondary education is only available in cities, opening up more chances of acquiring skills.

People's incomes
- The variety of employment opportunities increases and people have more chance of regular paid work.
- Even self-help work in the informal sector often brings in more money than farming.
- There are more commercial opportunities for farmers to sell produce in city markets.

Opportunities for improvement
- Improvements in shantytowns can be seen as a step on the ladder for future generations.
- Even jobs in the informal sector may enable skills to be acquired, leading to better pay in the future.
- Possibilities exist that are not present in the countryside, even if many will not be able to benefit from them.

Figure 3 Advantages of big cities in developing countries.

FURTHER RESEARCH

For more about the world's growing cities and an interactive urbanisation map go to the weblink www.contentextra.com/aqagcsegeog.

ACTIVITIES

1 (a) Describe what **Figure 1** shows about urban growth (past and projected) in (i) developed and (ii) developing countries.
(b) Give reasons why:
(i) a high percentage (90 per cent) of the population is urban in the UK, but the percentage is not likely to increase much in the future
(ii) a low percentage (under 45 per cent) is urban in developing countries like China and the percentage is likely to keep on increasing.

2 Study **Figure 2.**
(a) (i) Add up the number of cities named for different continents in 1950 and 2005.
(ii) Explain how the results suggest greater urbanisation in developing countries.
(b) (i) On an outline map of the world, using different colours for 1950 and 2005, plot the locations of the cities named.
(ii) Describe how the pattern of urbanisation has changed between the two dates.

Urban morphology in UK and developing world cities

What land use changes can be observed between centre and edge in UK cities? How is urban layout different in cities of the developing world?

Urban areas in the UK tend not to be just a jumble of different land uses. Each type of land use usually clusters together to give distinctive **urban zones** such as the **Central Business District (CBD)**, where shops and offices are concentrated. **Morphology** is the term used to describe the internal structure of a city. Most British cities have a simple three-fold structure: CBD in the centre, **inner city** around it, and **suburbs** filling the rest of the built-up area (**Figure 1**). Where the built-up area touches surrounding countryside, and sometimes pushes urban land uses into it, is the **rural–urban fringe** (pages 174–5).

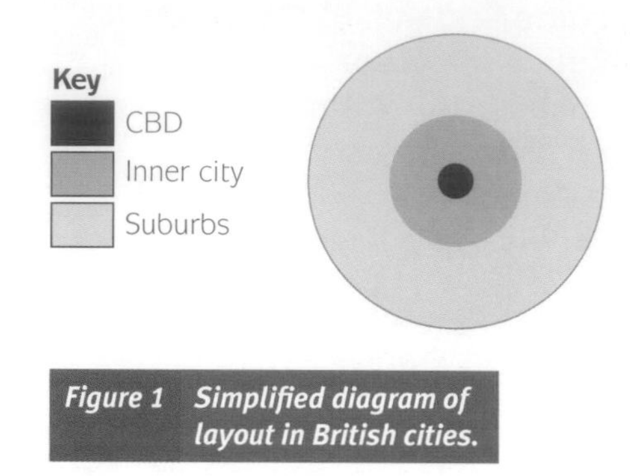

Figure 1 Simplified diagram of layout in British cities.

A British CBD

Characteristics

This is located in the centre of the urban area, around the historical core – in some cities this is a cathedral or a castle. From a distance it is often easy to pick out the CBD by its concentration of skyscrapers and other tall buildings.

Compared with other urban zones, the CBD:

- has the largest offices and shops, including department stores
- has the widest variety of goods on sale
- has high land values, rents and rates (which helps to explain the many tall buildings built close together)
- is the main place of work by day, which leads to traffic congestion (especially in rush hours)
- is the most accessible location where the main roads meet and has the main railway station(s).

The main difference between the CBD and other urban zones is that few people live here.

B Inner city

Characteristics

Located next to the historical core, this is an area of old housing and industry suffering from urban decay. It is a zone of mixed land uses:

- old high-density terraced houses, some three or four storeys high and including basements and attics; these are often let out as flats and badly maintained
- old and sometimes abandoned factories and warehouses
- areas of derelict land around railway sidings, unused docks and canals
- high-rise flats (many built in the 1960s)
- pockets of smart new developments, in many cities around old docks such as the London Docklands, Albert Dock in Liverpool and Salford Quays in Greater Manchester.

The main difference between the inner city and other urban zones is its generally run-down appearance.

Figure 2 CBD in Manchester.

Figure 3 Two views of inner-city Birmingham.

C Suburbs

Characteristics

In most cities these cover the largest area. A suburb is part of the urban area, which has grown outwards from the old centre across what was once countryside. This zone is predominantly residential:

- along the sides of the main roads are inter-war semi-detached houses and small shopping parades
- behind the main roads are more modern housing estates, mainly of semi-detached and detached houses
- some are private estates, others were local-authority built, although many have now been bought by residents
- the houses usually have gardens and garages with areas of open space between them
- the more recent and expensive housing is in the outer suburbs, where the density of housing tends to be lower.

There is less change in this zone than in the other three; the houses are good for many more years and virtually all the land suitable for building has already been used.

Figure 4 Two different styles and types of housing in the suburbs.

Developing world cities

The layout is much less regular. Land uses are less well-segregated for two main reasons. Planning controls are much weaker and explosive growth has meant that newcomers take over any unused land for their homes. People who set up homes on land not owned by themselves are squatters; **squatter settlements** consist of slum housing, which grows into shantytowns wherever there is enough unused land within the city, or outside if not. Therefore the layout shown in **Figure 5** is only a guide, with great variations from city to city. Note that some things are similar to developed world cities, such as a generally circular arrangement, a CBD in the centre surrounded by older housing areas. However, some of the housing is very different (pages 153 and 163). Also, sectors of particular land uses stretching from the centre outwards are more common, such as expensive housing zones or industry following the main highway out.

Figure 5 Simplified diagram of layout of a developing world city.

FURTHER RESEARCH

Investigate functions and land uses for a town or city in your home area. Describe the main differences between the city centre and the edge of the built-up area.

GradeStudio

1 a Describe two characteristics typical of the CBD shown in **Figure 2**. (2 marks)
  b Why is this urban zone given the name CBD? (2 marks)

2 a State what is meant by the *function* of a settlement? (1 marks)
  b What is the evidence for manufacturing industry as one of the functions of Birmingham in **Figure 3**? (1 mark)
  c Describe how the two views in **Figure 3** show that the inner city is a zone of mixed land uses. (3 marks)

3 a Name the two types of house shown in **Figure 4**. (2 marks)
  b Which one is likely to be closest to the edge of the city? Explain your choice. (3 marks)

4 Describe and explain the main similarities and differences in the layout of functional urban zones between settlements in developed and developing countries. (6 marks)

Exam tip

Observation from photographs – go to page 171 so that you know what you are looking for.

Issues in the CBD and inner city in the UK

How are city centres changing to make sure that shoppers keep coming? Why did strategies for inner-city improvement need to change?

The CBD of a city is not static; it is a dynamic area that goes through phases of growth and decline. You can pass through any CBD of a large city and see areas in decay, with closed shops and a run-down appearance, and others that appear lively, smart and successful. Often it is possible to identify a **core** and a **frame** (**Figure 1**). The core houses the large department stores and branches of many national chains of shops and the city offices of big companies; it is the busiest part, but facing increasing competition from out-of-town shopping centres and business parks. The frame has smaller, more specialist shops, some of them struggling to survive because of high land values and rates; notice that the services and land uses here are more mixed.

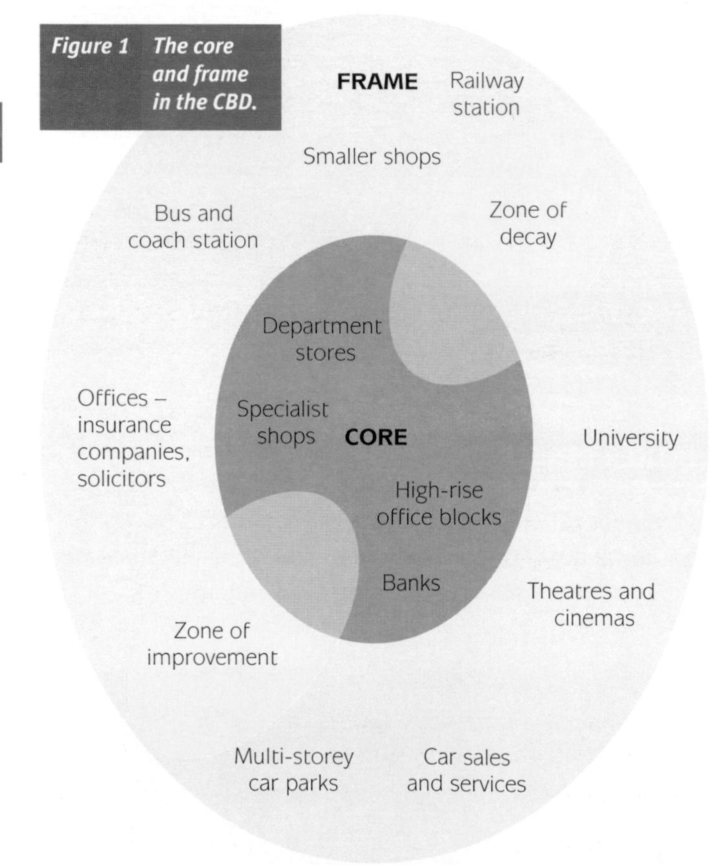

Figure 1 *The core and frame in the CBD.*

Exam tip

Don't be afraid to name and use examples from local urban areas in your answers.

They are just as good to use as the examples in this book.

The big advantage is that you know more about them.

Revitalising CBDs

In order to maintain the CBD's traditional role as the main shopping centre for the surrounding region, most city authorities realised that the shopping experience needed improvement. This has been done in two main ways. One has been to do more to separate shoppers from traffic, by diverting traffic onto inner ring roads and keeping the main shopping high streets for pedestrians only. The other has been to redevelop parts of the shopping core into indoor shopping centres. These give all-year protection from the British weather, as well as providing well-lit and heated environments in which to shop and meet people. Wherever possible, pre-existing big name shops and stores were invited in. The Bull Ring Centre redevelopment, for instance, has revitalised, if not revolutionised, Birmingham city centre's status as the major regional shopping centre for the West Midlands (**Figure 2**). Can you name examples of pedestrianised streets and indoor shopping centres from the local town or city centre most visited by your family?

Figure 2 *Birmingham's Bull Ring redevelopment, which attracted big name London stores such as Selfridges.*

Problems and issues in inner cities

Long ago, workers who could afford it moved out of inner cities into the suburbs and surrounding rural areas, pushed out by the kinds of problems listed in **Figure 3**. Because people tend to buy the most expensive property they can afford, there is **housing segregation** in UK cities – separation according to wealth. House prices and rents were lower in the inner cities and these became the poor housing areas. Among the less well-off groups are members of ethnic minorities, especially recent arrivals in the UK, who find it hardest to obtain well-paid jobs; this is often due to a combination of lack of qualifications and poor English. For obvious reasons, new arrivals in the city are pulled towards relatives, friends and areas of familiar culture for language, religion, food, etc. As the cultural character of the area continues to change, more of the long-established residents move out. This produces **ethnic segregation**, when areas become dominated by particular racial or national groups. In two London boroughs, Newham in east London and Brent in north-west London, more than 50 per cent of residents are from the black and Asian communities – compared with a UK average of 9 per cent for ethnic minority groups.

Environmental problems

- Housing is either old terraces or cheap tower blocks
- Many derelict buildings – factories, warehouses, churches, houses and flats – often vandalised
- Shortage of open space; most of what exists is wasteland

Social problems

- Above-average numbers of pensioners, singe-parent families, ethnic minorities and students
- Poorer than average levels of health, but higher than average levels of drug abuse and crime
- Difficult police–community relations

Economic problems

- Local employment declined as industries and docks closed
- Higher than average rates of unemployment, especially for the young and ethnic minorities
- High cost of land compared with the suburbs
- Low income and widespread poverty

Figure 3 The environmental, social and economic problems of British inner cities.

Poverty and dilapidation are seedbeds for crime, vandalism and drug trafficking. From time to time, racial tension and general discontent have flared up into riots, for example in parts of Leeds, Bradford and Oldham in 2001. Also these areas have inherited problems from bad planning decisions made in the 1960s, when comprehensive redevelopment was fashionable, which led to widespread clearances of terraced houses and their replacement with high-rise flats. Many of these were badly built: residents complained about the cold and damp, broken lifts, open staircases taken over by gangs, criminals and drug dealers. Pensioners and mothers with children felt like prisoners in their uncomfortable homes. Many inner-city landscapes were further blighted by efforts to relieve traffic congestion in city centres, which was tackled by using land in inner cities for building ring roads, flyovers and urban motorways.

Improving inner cities

Since the 1990s the attitude of government and planners have changed. After an unsuccessful redevelopment in Hulme in inner-city Manchester in the late 1960s, which produced highly unpopular, but award-winning, concrete crescents of high-rise flats, the Hulme City Challenge was launched in 1992.

Figure 4 New homes in Hulme.

The involvement of the local people resulted in a landscape that was totally different from what was there before:

- Varied accommodation for large and small families, single parents and first time buyers.
- Friendly, welcoming architecture with different dwelling types, structures, colours of brick and building materials (**Figure 4**).
- An old church redeveloped to provide facilities for dance and music (Zion Arts Centre).
- Hulme Park, which links up with Castlefields (**Figure 2B** on page 160), the first new large park Manchester has seen in the last 60 years; within it are well-equipped play and sports areas as well as open space to give relief from the high urban densities around it.

While the high concentration of individual ethnic groups in certain inner urban areas can make wider integration a challenge, it does make it easier to target these groups' particular health, education and cultural needs.

ACTIVITIES

1 (a) State two differences between shops in the core and frame of a CBD.
  (b) Explain why the land uses shown in the core in **Figure 1** can afford higher rates and rents than those in the frame.

2 Choose one environmental, one social and one economic problem from **Figure 3.** For each one, explain (a) why it exists (b) why it is difficult to overcome.

3 (a) Write about the disadvantages for families living in high-rise flats.
  (b) Describe the improvements shown in **Figure 4.**

4 Using the example of Hulme or an example in your home area, make notes for a case study of inner-city improvement since 1990. Use these headings:
(a) Location (b) Why change was needed (c) Changes made (d) Whether successful or not.

Key issues in British cities – housing and traffic

What are brownfield sites? What are the advantages of using these for new buildings? What more can be done about traffic problems in British towns and cities?

The UK is facing a housing crisis as demand for homes continues to outstrip supply. Several social trends are pulling in the same direction, creating the extra need:

1 Total population is increasing (in 1950, 50 million; now 60 million).

2 Total households are rising (in 1950, 15 million; now 25 million).

3 More people now live alone (in 1950, 10 per cent of households with one adult; now 30 per cent).

The need is not equally distributed throughout the UK. It is greatest in London and south-east England, the main magnet for economic growth in the UK. Here additional pressures from in-migration from other parts of the UK and immigration from the EU and elsewhere are felt most keenly. Government policy is to use **brownfield** sites (areas of previously built-up land available to be built on again) whenever possible, to save the countryside and its **greenfield** sites (open land that has never previously been built on) from further urban sprawl. **Figure 2B** shows a brownfield site while **Figure 1** shows a greenfield site.

Figure 1 ***Land next to a city housing estate in the rural–urban fringe.***

Brownfield sites in inner cities

Inner-city areas have a larger supply of brownfield sites than other urban zones. This is an important consideration when new building land is at a premium and objections to using greenfield sites are increasing. Some of the largest spaces are to be found in abandoned dockland and industrial areas. An important part of many dock-, river- and canal-side redevelopment schemes is the conversion of old warehouses into good-looking modern apartments. At the same time the environment in the area around them is cleaned up and landscaped: Salford Quays and Castlefields in inner-city Manchester are examples (**Figure 2**). The associated movement of wealthy people back into areas of former urban decay is known as inner-city **gentrification**.

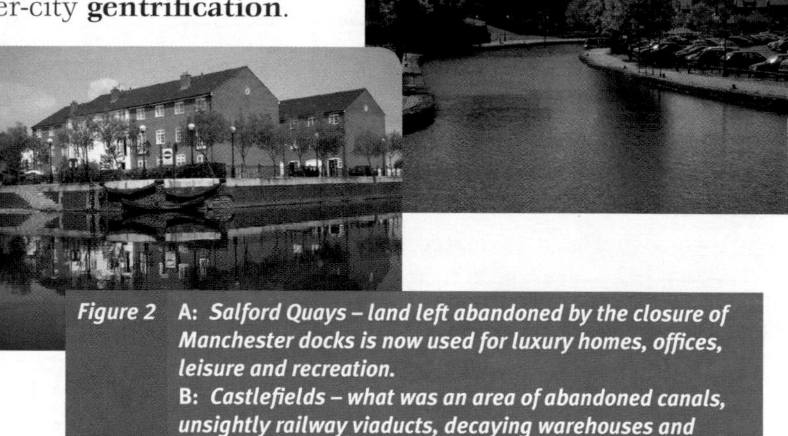

Figure 2 ***A: Salford Quays – land left abandoned by the closure of Manchester docks is now used for luxury homes, offices, leisure and recreation.***
B: Castlefields – what was an area of abandoned canals, unsightly railway viaducts, decaying warehouses and factories has been transformed into a zone of commercial, residential and leisure developments that attracts over 2 million visitors a year.

Unused brownfield sites in city centres are fewer; also buildings of historical and cultural importance (cathedrals and castles, town halls and other public buildings from Victorian times such as galleries and museums, libraries and railway stations), which are listed buildings for preservation, are concentrated in this zone. Some fine old commercial buildings, no longer needed as offices, have been renovated for use as luxury flats and up-market hotels. St Pancras railway station in London is a good example. However, using brownfield sites can still make only a partial contribution to the 3 million new homes that, according to government estimates, will be needed by 2020.

Traffic problems in urban areas

The most congested part of any city is the centre. Its street plan was laid out long before the motor car was invented; the oldest streets are often the narrowest. With little pavement space, the conflicts between the needs of pedestrians and drivers are at their greatest. As this urban zone is unique, with few full-time residents, anyone who works there, goes there to shop or visits its restaurants, cinemas, theatres and clubs needs transport to get there. Congestion peaks during morning and evening rush hours on working days. 'Grid-locked' traffic, stopping, starting and crawling along, causes more **air pollution** than smoothly flowing traffic. Pollutants are trapped in deep narrow streets, explaining why many cities suffer from poor air quality, with related dangers for human health (fatigue, headaches, asthma and bronchitis). There are also the wider global pollution issues that arise from burning fossil fuels (pages 54–5 and 223). Space is in short supply in such a highly built-up area, with high land values; this leads to the next problem – where to park once you are in the centre?

Underground
- Lines serve everywhere in the centre and some of the suburbs.
- The world's first underground railway system.
- Despite its age (and some problems with reliability), it works well most of the time.
- Over a billion passenger journeys are made in a year.

Surface railways
- Most towns in south-east England have a service to London.
- Lines radiate like the spokes in a bike wheel from stations such as Waterloo and Victoria.
- More trains operate to and from London stations in a day than in the whole of Switzerland in a year.

Buses
- A dense network of red buses serves other areas and links up with the stations.

Figure 4 London – an example of a city with an integrated public transport system.

FURTHER RESEARCH

Investigate traffic problems and attempted solutions in your nearest town or city.

Figure 3 'Park and Ride' in Nottingham – one person in many commuter cars compared with up to 70 people on a double decker. Bus lanes lower the 'ride' time into the city centre.

Traffic solutions

For a long time the focus was on trying to accommodate the traffic, especially people and their cars. Widely used attempted solutions included:

- ring roads and bypasses to keep through-traffic out of the centre
- urban motorways and flyovers to help drivers reach destinations faster
- multi-storey and underground car parks
- one-way traffic systems, avoiding pedestrianised high streets.

How many of these are used in your local town or city?

Today the focus has shifted to encouraging people to give up their cars and use public transport instead. The policy is one of 'stick and carrot'. The *stick* is charging for entry into city centres, such as the London Congestion Charge, accompanied by increased charges for parking. Another idea that has just arrived, widely used in American cities, is creating traffic lanes that only cars with two or more people in them can use. The *carrot* is providing an improved, efficient public transport system. Big city authorities in countries all over the world are looking at integrated public transport systems where metros (underground railways), trams (**Figure 2**, page 156) and other surface railways and buses meet at interchange points, each one feeds into the others to cover the whole built-up area. Only in London does a system like this exist in the UK (**Figure 4**); there is even a boat service on the Thames for people living downriver from the centre. 'Park and Ride' operations are common (**Figure 3**). Cleaner buses, which run on natural gas and biofuels, are being progressively introduced.

ACTIVITIES

1 Why is there a housing problem in the UK?
2 (a) (i) Describe the land uses in **Figure 1.**
   (ii) Why is it an example of a greenfield site?
 (b) (i) Describe the land uses in **Figure 2B** on page 160.
   (ii) Why is it an example of a brownfield site?
 (c) (i) Make side-by-side lists of the advantages and disadvantages of greenfield and brownfield sites.
   (ii) Which one is more sustainable? Explain your choice.
3 Make notes for traffic problems in UK cities using these headings: (a) Nature of the problems (b) Causes (c) Impacts (d) Attempted solutions (e) Their effectiveness.

Problems caused by rapid urbanisation in developing world cities

What problems arise from the rapid urbanisation of developing world cities? To what extent are they the same as those of developed world cities? What are the differences?

Non-stop population growth is leading to a multitude of urban problems – without any hope of a breathing space for their effective management. Governments and city authorities are struggling to provide an adequate supply of essential services, such as piped water, sewerage, electricity, health and education, for existing city inhabitants, and yet they are expected to provide even more for the constant flow of new arrivals. More would-be migrants are queuing up in the countryside, ready to move in, undeterred by the great economic and social problems faced by those who have already moved there. The numbers can be frightening – over 8 million a year in China alone.

Figure 1 Traffic congestion, Cairo style.

Environmental problems

Some problems, such as **traffic congestion**, are faced by all big cities irrespective of whether they are located in developed or developing countries (**Figure 1**). However if anything, traffic congestion is more acute in developing countries despite lower rates of car ownership. In Asian cities motorised transport must compete for space with pedestrians, rickshaws, scooters, donkeys and other animals. Congested streets are clogged all day with taxis, buses, lorries and cars, many of them old, with inefficient exhaust systems, and this leads to high levels of air pollution and many health risks, especially asthma and bronchitis. Some cities are notorious for these problems. Mexico City is one: it is sited in a mountain basin where pollutants are easily trapped. Beijing is another: so rapid has been the growth of car ownership there (with vehicles running on low-grade petrol) and industries (burning low-grade coal) that smog reduces visibility to a few hundred metres at times and gives a putrid smell to the air. Doctors are blaming dangerous levels of smog for increases in the incidence of bronchitis, tuberculosis and lung cancer.

Water pollution of rivers and seas is widespread, as they are used as dustbins for human wastes, destroying all traces of plant and fish life. Woodland and **wildlife habitats**, as well as good agricultural land, are destroyed by the sprawl of squatter settlements and industries on the edges of cities. Underground **water supplies** are being used up at alarming rates; Mexico City has sunk over 7 metres during the last century as the aquifer below it has been emptied of water.

Managing environmental problems

In many cities the human problems are so great that there is little time or money to devote to environmental problems; these are viewed as a necessary cost of economic development, and have a low priority. This is no different from the situation in London and the big cities of the Midlands and northern England during Victorian times. When it mattered, as in Beijing during the Olympic Games in 2008, the authorities imposed drastic measures, which included temporary closures of the worst polluting factories and limiting daily entry of cars into the city according to registration numbers. Otherwise, even where regulations exist, they tend to be weakly enforced. Cairo is one example of a city where the authorities are at least trying to manage the mountain of environmental problems. Look ahead to pages 166–7 to read about the methods they are using.

Social problems caused by the growth of squatter settlements

Social problems are mainly to do with housing and the effects on health and family life for those living in squatter and shanty settlements. Most big cities in Africa, Asia and Latin America are surrounded by unplanned, makeshift shantytowns (**Figure 2**), and it is common for 50 per cent or more of the city's population to live in them. They are known by different names in different parts of the world – *favelas* in Brazil, *bidonvilles* in North Africa and *bustees* in Kolkata (Calcutta).

Squatter settlements are sited on any spare land the migrants can find. This includes steep slopes (as in Rio de Janeiro and Lima), swamps and rubbish tips. The areas used are often avoided by others because they are prone to landslides, flooding or industrial pollution. The settlements are illegal. The shacks and shelters are homemade, built from anything the people can find, including bits of wood, sheets of corrugated iron, cardboard, polythene and 5-gallon oil drums. They are a real fire hazard. Typically, there are usually only one or two rooms where the family eats, sleeps and lives. Most shacks lack basic amenities such as electricity, gas, drainage, running water and toilets. In the *bustees* of Kolkata one water tap and one toilet may be shared among 30 people.

Sewage often runs down streets and pollutes the water supply, leading to water-borne diseases such as diarrhoea, typhoid and cholera. Diseases spread quickly because of the high density of housing (**Figure 2**). Also there is often no refuse collection, and any spare space becomes filled with rubbish, another breeding ground for disease. Infant mortality rates are high because babies are the most vulnerable to disease. Healthcare is often too expensive and too far away to access. Many families, and especially the children, suffer from malnutrition. Local shops and stalls, often fly-ridden and dirty, sell a limited range of poor-quality foods that lack the proteins, vitamins and calories needed for a healthy diet. The stress of living in shantytowns leads to frequent breakdown of marriages and increases in crime, mainly theft and robbery. In some cities there are large numbers of 'street children', who have either run away or been abandoned due to family break-up. Some areas are controlled by gangs and 'drug lords' and are no-go areas for the police.

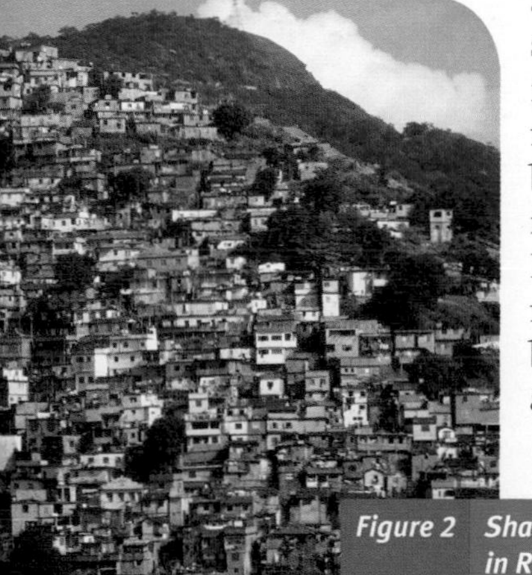

Figure 2 Shantytown on a hillside in Rio de Janeiro, Brazil.

Economic problems

The underlying cause of all the social problems is economic – lack of income. For the newly arrived migrant the main problem is finding work. With few jobs available in the formal sector for non-skilled and illiterate rural people, most are forced to look for work in the **informal sector**, as very low-paid street sellers, shoe shiners, human carriers, waste collectors and domestic servants. Surviving in the city is tough for all newcomers. Even if they find jobs in the city centre or in the industrial zones, travelling there can be expensive. In the meantime, poor shanty dwellers continue to have more and more children, partly in the hope that at least one of the family will get a job, either in the home city or overseas, that will allow the whole family to move out of the slum into a proper residential neighbourhood.

Figure 3 'Working while you work' as market sellers in Ecuador.

ACTIVITIES

1 (a) Name and describe two environmental problems in developing world cities.
  (b) Explain why they are worse than in developed countries.

2 (a) From the photograph on page 153, draw a labelled sketch of a shantytown.
  (b) Describe how some shantytowns are worse than this one.

3 (a) List the health problems of shantytown dwellers.
  (b) Explain (i) why there are so many health problems
  (ii) why infants and children are most at risk.

4 One person's view: 'The simple solution to urban problems such as shantytowns is birth control.'
  (a) Explain this person's view.
  (b) Why is it not the simple solution this person suggests?

Squatter settlements – what are the solutions?

How do squatter settlements change with time? Are they slums of hope or despair?

The worldwide slum problem is too big to ignore, now that an average of 45 per cent of urban populations in developing world cities live in slums (**Figure 1**). It is not as easy now for city authorities to carry on using their old idea of slum management, sending in the bulldozers to demolish and clear everything, thereby making residents homeless again.

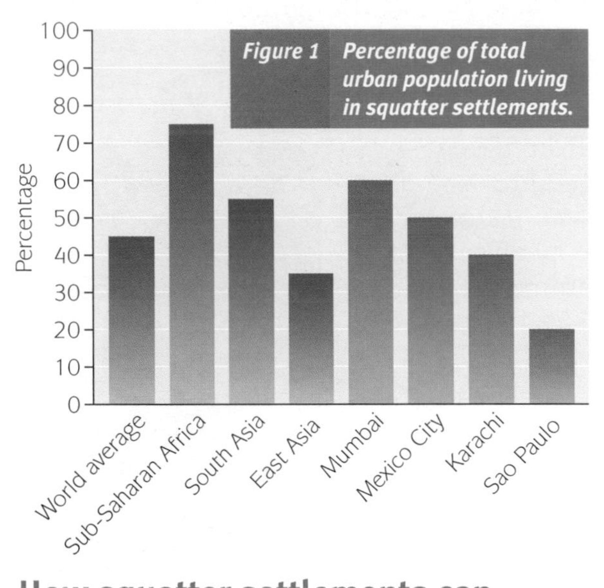

Figure 1 *Percentage of total urban population living in squatter settlements.*

How squatter settlements can change over time

Authorities do not have the money to rehouse all the people. One policy is Aided Self-Help (ASH). A few changes, implemented with the cooperation of city authorities and help from non-governmental organisations (NGOs), can speed up the process of improvement. ASH involves:

- giving slum dwellers legal titles to the land
- connecting them to essential services such as water, electricity and roads
- providing building materials, technical help and loans.

Community participation is vital in self-help schemes because families and neighbours do the work; they are responsible for building houses on allocated plots. Successful schemes encourage and reinforce the community spirit that often exists among people living in difficult circumstances.

Concern is growing among Indian authorities that the commercial progress of Mumbai, India's number one business city, is being held back by its image as a city of slums (**Figure 1** and page 153). This has led to radical plans being put forward for the redevelopment of Dharavi slum, located in central Mumbai. Asia's largest slum, it is a place where 600 000 people live and work. Really Dharavi is two places in one. It is a rabbit warren of two- and three-storey dwellings constructed of concrete, corrugated-iron sheets and cardboard. Hundreds of people share the common toilet and water supplies and its narrow walkways run with human waste. Tiny rooms, often less than 10 square metres in size, can house more than a dozen people from two or three families. Then by day Dharavi is transformed into a workplace, full of human energy, enterprise and skills. Homes double up as workshops and are alive with informal businesses making all sorts of things, including clothes and leather goods. This is why residents are very suspicious of the state's plan for Dharavi's development (**Figure 3**).

The plan

- To rehouse some residents on site in apartment blocks with all modern facilities.
- To move remaining residents to apartments in other parts of the city.
- For the developer to use the rest of the site for commercial developments.

FURTHER RESEARCH

See a video clip of life in the Dharavi slum at the weblink www.contentextra.com/aqagcsegeog.

POOR MIGRANTS FROM RURAL AREAS

↓

SQUATTER SETTLEMENT
Poor shacks

↓

Gain work: have some income to improve houses

↓

SHANTYTOWN
Self-built houses

↓

Regular work
Community action
ASH
Legal titles
Public services provided

↓

LOW INCOME RESIDENTIAL DISTRICT

Figure 2 *How squatter settlements can change over time*

Running a factory in a seven-storey block of flat is impossible. I need to b on the ground floor.

It is really a plan to make money out of Dharavi by selling the land.

It is a way to force us out of the city to the edges where there is no work.

Figure 3 *Residents' views on the Dharavi redevelopment plan.*

São Paulo, Brazil

São Paulo is Brazil's biggest city, with at least 2.5 of its 18 million people living in squatter settlements. It is the major industrial and business centre and is often described as 'the economic engine that pulls the rest of Brazil'.

ASH-type schemes for improvement began in the favelas in the 1990s. Simple single-storey homes with a water tank and indoor bathroom were built with cheap and easily accessible building materials such as breeze-blocks (**Figure 4**). The status of the housing areas improved and some have become lower class/ low-income urban residential districts with their own services. Roads built by the authorities through the settlements, even if not paved, enabled the area to be integrated into the city's bus network. Some families in formal work, with a regular income, have built upwards and added a second storey.

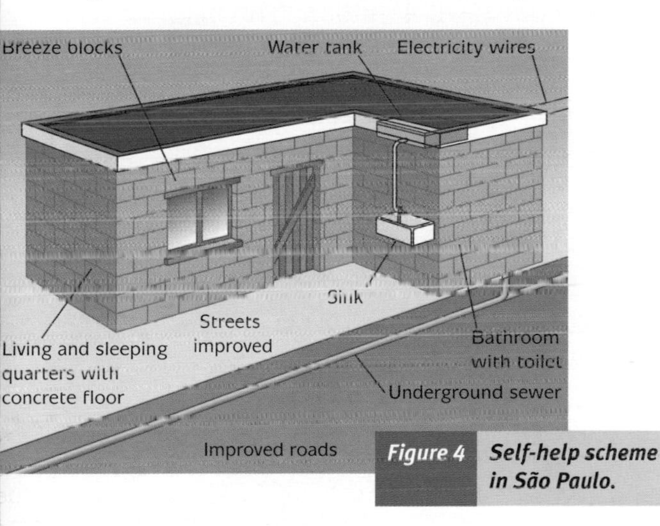

Figure 4 ***Self-help scheme in São Paulo.***

Paraisopolis (which means 'Paradise City') is one of Brazil's largest favelas, home to about 80 000 people (**Figure 5**). In recent years it has been linked into the main service network of São Paulo, upgraded and urbanised. Many areas now have piped water and electricity and schools have been built. This has made a big difference. A market for houses now exists and they are changing hands for between R$15 000 (about £4000) and R$300 000. Many residents have low-skilled jobs. In 2008 the Brazilian chain store, Casas Bahia, which sells furniture and electrical goods, opened its first ever store in a favela. It is a measure of how far Paraisopolis has come because most favelas are strict no-go areas for outside businesses. Until recently, any business in this favela had to pay a bribe to the PCC, a criminal organisation that controlled the slum. Only real economic and social improvements allowed the PCC to be driven out and new businesses to arrive.

Figure 5 ***Favela Paraisopolis in São Paulo, located in the middle of the metropolitan area close to the business centre.***

GradeStudio

1 a State two **push** and two **pull** factors for migration from rural to urban areas in developing countries. (4 marks)

b Explain why most new migrants to the city live in squatter settlements. (3 marks)

c (i) Name two different types of location for squatter settlements in cities. (2 marks)

(ii) What do they have in common? (1 mark)

2 a Describe two differences between squatter settlements and low-income residential areas. (4 marks)

b With reference to a case study, describe how self-help schemes for improving squatter settlements work in developing world cities. (6 marks)

3 State the advantages and disadvantages of self-help schemes compared with larger schemes for redevelopment in developing world cities. (5 marks)

ACTIVITIES

1 Are the squatter settlements in developing world cities 'Slums of despair' or 'Slums of hope'?

(a) Write a paragraph in support of each view.

(b) With which one do you agree most? Explain your view.

CASE STUDY Attempts to solve urban problems in a developing world city

Cairo

Cairo is an example of a big city with all the urban problems referred to on the previous pages (especially traffic congestion, sewage and housing). It is the largest city in Africa, with an official population of about 11 million but an estimated total of at least 16 million. Growth took off in the 1960s and shows no sign of slowing down (**Figure 1**). Despite the usual shortage of funds, the Egyptian authorities have implemented a number of projects to tackle its urban problems.

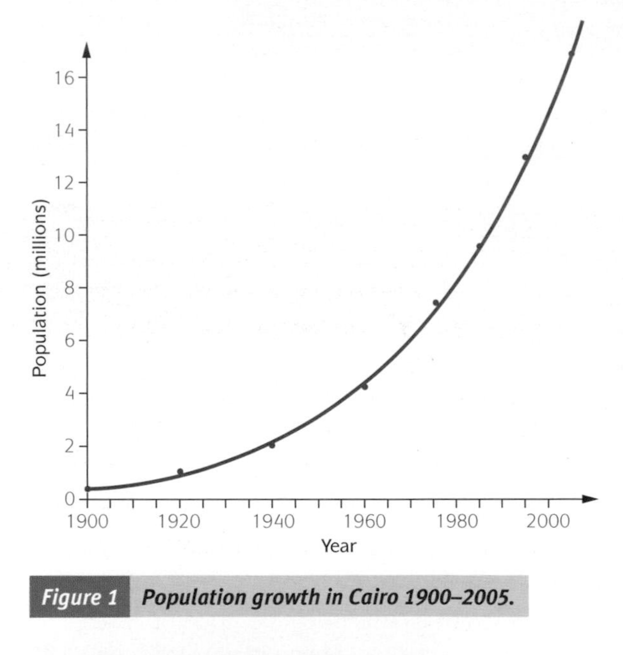

Figure 1 ***Population growth in Cairo 1900–2005.***

Traffic congestion

The scene in **Figure 1** on page 162 was taken after a massive new ring road was built around the city – imagine what it was like before! At least not everyone is forced to face the nightmare of travelling on Cairo's congested and fume-filled roads. The construction of the Cairo metro, the first in Africa, has been a great success (**Figure 2**). Two lines are already open and a third is under consideration. The train services are well organised and quick, and offer air-conditioned comfort; the stations are clean and welcoming with televisions on the platforms to watch while waiting for a train. The metro is used by about 2 million commuters a day.

Figure 2 ***The Cairo metro runs underground in the centre but on the surface in the suburbs.***

Sewage

The Greater Cairo Sewage project provides aid to repair the city's crumbling sewers, some of which date back to 1910, when the population was only half a million. The main work is to extend the sewers into areas currently without any; these are mainly the poor neighbourhoods where sewage on the streets is a common sight. At least Cairo has an efficient system of street collection. The Zabbaleen, with their donkey carts, are the traditional collectors of rubbish from wealthy neighbourhoods (**Figure 3**). Their licence to operate has been officially extended as collectors and recyclers for some of the slum housing areas as well.

Figure 3 ***The Zabbaleen – official refuse collectors in Cairo's slums.***

Housing

As elsewhere in the developing world, this is the biggest problem faced by urban authorities. The nature of the housing problems might be slightly different in Cairo from most other developing world cities – there are no shantytowns as such, but brick-built houses are built illegally on state-owned irrigated green land next to the River Nile, which the government had reserved for food supply. These 'informal' houses now cover 80 per cent of the land area of Greater Cairo. But otherwise the problems of overcrowding and numbers are the same, with an estimated 2–3 million people having set up homes in the 'Cities of the Dead' among the tombs in the cemeteries of Old Cairo and in homemade huts on the rooftops (**Figure 4**). Every Cairo citizen has an average living space of less than 2 square metres; the overall population density is 33 000 people per square kilometre. Urban sprawl, and its accompanying air pollution, have already extended the borders of the city into the desert up to the Great Pyramids and the Sphinx at Giza, which are being badly eaten away.

Figure 4 ***Old Cairo is overcrowded.***

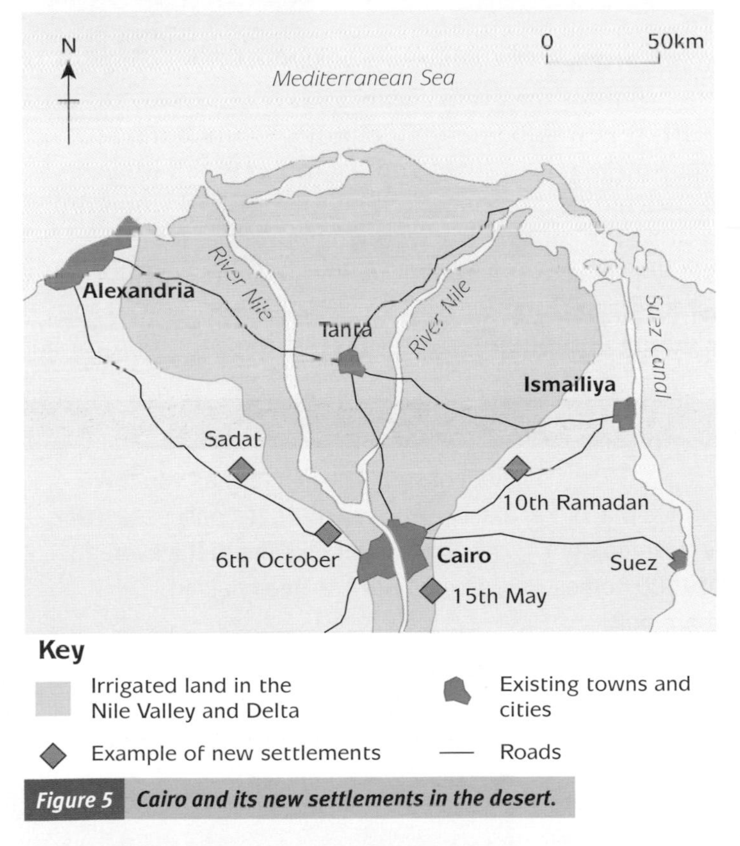

Figure 5 ***Cairo and its new settlements in the desert.***

The long-term plan is for the construction of over 40 new settlements, many of them new towns, capable of housing 15 million people. They are to be located in the desert, away from the fertile irrigated land of the Nile Valley and Delta, which is needed for farming (**Figure 5**). One example is 6th October City, west of Cairo. At first glance, it is not a very attractive place, dominated as it is by multi-storey apartments, many of them identical. However, included in the plans are open spaces and gardens, mosques, junior schools, shopping centres and industrial zones. It was considered essential to provide modern services and work opportunities if people were to be persuaded to leave Cairo. The Mercedes car factory has been established here for several years, and employs about 400 people, which is helping to attract others. As further attractions to new residents, loans and mortgages are offered on favourable terms, while a new super-highway provides a link to Cairo, only 30 minutes away in good traffic conditions. Some of the critics say that the plan favours the middle classes, who can afford the rents. They maintain that the government should not be spending vast sums on new settlements when more than 3 million homes in Cairo are poorly connected to public services. This is another example of a local authority attempting to solve its housing problems.

ACTIVITIES

1 (a) Describe what **Figure 1** shows.

(b)

	Average personal income per year (Egyptian £)	Infant mortality rate (per 1000)
Cairo	3461	50
Rural Egypt	2353	68

(i) Calculate the size of the differences between Cairo and rural Egypt for the figures in the table above.

(ii) Explain why there is likely to be no breathing space in growth to allow the authorities to deal with the already existing problems of Cairo.

2 (a) Describe (i) the housing problems of Cairo (ii) the policies intended to solve them.

(b) Read the following comments from residents of 6th October City. What do they suggest about the success of the housing policy for Cairo so far?

I am pleased to have moved – I couldn't stand the pollution, traffic, crowds, and noise of Cairo.

We were promised factories and work. There are none.

All our good friends still live in Cairo

Apartment rents are too high. I am taking my family back to Cairo. I know we can stay more cheaply on a friend's rooftop.

How can urban living be made sustainable?

Why is urban living not normally associated with sustainable living? What can be done to make urban living more sustainable?

Living in cities and sustainable living are not natural partners. The subsistence way of life of an Indian tribe in the Amazon jungle is truly sustainable (page 72). It has to be. Relying upon what nature provides for all its needs, the tribe cannot afford to over-use and exhaust the natural resources without impacting upon future generations. Urban growth, however, followed economic development, powered by energy from non-renewable fossil fuels. As cities have grown and spread, living places have been separated further from places of work. More 'commuter miles' translates into higher energy consumption. Indeed cities are great centres of consumption, and until recently little of the waste they produced was ever recycled.

Eco-towns

The vision of future sustainable urban living is the **eco-town**. In 2008 the UK government put out a grand plan for ten of these to be built by 2020. What is different about them? The guidelines are:

1 Low energy consumption.

2 **Carbon neutral** or **zero carbon** (this means that the amount of energy taken from the National Grid should be no greater than the amount put back by renewable power sources).

3 Built from recyclable materials.

4 Green spaces cover at least 40 per cent of the urban area.

5 Safe walking routes and cycle lanes built into the planned layout.

Figure 1 *Views from Vauban, Germany, an environmentally friendly town encouraging sustainable living.*

Houses will be built using the latest insulation techniques; they will come with renewable energy technologies including rooftop wind turbines and solar thermal panels. Eco-friendly lifestyles will be encouraged, including rainwater capture for watering gardens, total recycling of waste, farmers' shops for consumption of local produce, and cycle storage points. Walking, cycling and public transport running on green fuels

EXAMPLE: Sustainable urban living

BedZED in Wallington, South London

Only a handful of low carbon developments exist in the UK, of which this is one. BedZED, or to give it its full name Beddington Zero Energy Development, is described as an 'eco-village'. People moved into the first of its 100 homes in 2002. The homes shown in **Figure 2** have heating needs that are only about 10 per cent of the typical UK home's heating requirements; energy and water efficiency are designed in. About 10 per cent of the construction materials were from waste timber, doors, kerb stones and paving, reducing the amount of waste sent to landfill. Also included are community facilities and workspace for 100 people.

The 'village' is well served by public transport – it is located on a bus route, only five minutes walk from a train station and a 15-minute walk to the Tramlink from Croydon to Wimbledon. Car club members use a locally based fleet of vehicles on a 'pay as you drive' basis, which encourages only essential car use. Annual mileage of the average club member has been halved.

Figure 2 *Aerial view of BedZED.*

will be promoted to replace private car use. Planting trees will release carbon dioxide to help offset emissions from essential car use.

The idea of eco-towns is based on continental examples, such as Vauban in Freiburg in southern Germany (**Figure 1**). This example of sustainable urban living is home to more than 5000 people and 600 jobs. Community involvement is encouraged among people wishing to create a flourishing and sustainable neighbourhood.

How can city living in the UK be made more sustainable?

The central areas of British cities have many buildings that reflect their long histories (page 160); their maintenance is expensive, but is more easily justified when they support other economic activities such as tourism (**Figure 3**). In old industrial cities in particular there are areas of derelict land and disused buildings. It makes more sense to improve the environment in these areas, as shown in **Figure 2** on page 160, than to destroy the rural environments and cover areas of grass and woodland with concrete. Private car use is a major source of carbon emissions; improve public transport (availability, efficiency and price) to make it a real alternative to the car. As well as improving the quality of living in the area for everyone, greenery and trees use and store carbon. When a household is recycling fully, everything from organic waste for compost to papers, cardboard, cans, glass and plastics, there is little left for the dustbin and landfill. The more individuals and communities are involved, the greater the chance of a successful uptake of environmentally friendly ways of living. This is necessary because going green is not always easy. Why do builders prefer to develop greenfield rather than brownfield sites? How is the economic recession of 2008 badly affecting recycling? Why does renewable energy still make up such a small part of our total energy consumption (pages 224–5)?

Ways to make urban living in the UK more sustainable

Conserve the historic environment
- Repair and renovate historical buildings and structures
- Once lost they are gone for ever
- Enhance their uses and appeal as visitor attractions

Use brownfield sites for new development
- Use waste land or empty spaces for building first
- New development improves the appearance of these areas
- Reduce the loss of greenfield sites in the countryside

Provide more open spaces and greenery
- Make provision for these in the plans
- Improve the quality of life for urban residents
- Landscape the sides of waterways and railways

Reuse, recycle and reduce waste
- Reducing the amount of waste comes first
- Reuse bottles, plastic bags etc. more than once
- Up to 90% of household waste is recyclable

Improve public transport systems
- Link bus, tram and rail routes at interchanges
- Provide feeder services to housing estates
- Use more environmentally friendly vehicles

Involve communities in local decision-making
- Consult local people instead of imposing plans
- Put people first; ask for and act on their ideas
- Foster the growth of a community spirit

Figure 4 Towards a sustainable future for British cities.

ACTIVITIES

1 (a) What is meant by *carbon neutral* or *zero carbon*?
(b) Describe two ways of trying to make urban living more sustainable to achieve this.
(c) Explain why it is not easy to achieve in UK cities.

2 (a) Study the information about cars on page 161. Describe what is being done to make private cars more environmentally friendly.
(b) How can (i) car clubs (ii) integrated public transport systems as in London (iii) converting warehouses and offices in city centres into homes, help to reduce private car use?
(c) Why are many people reluctant to give up their cars?

3 Make notes for a case study of sustainable living using the headings: (a) Location (b) Main characteristics (c) Features that make it sustainable.

Figure 3 York is one of the UK's most popular tourist towns. Its well-preserved historical remains such as the Roman walls and the Minster are major attractions.

FURTHER RESEARCH

Read about eco-towns at the weblink www.contentextra.com/aqagcsegeog. Follow the links to examples of eco-settlements.

Practice GCSE Question

See a Foundation Tier Practice GCSE Question on the weblink www.contentextra.com/aqagcsegeog.

Rates of urban growth 2005–10	
Africa	3.2%
Asia	2.4%
South America	1.7%
North America	1.3%
Europe	0.1%

Figure 1

Figure 2 Village in rural Peru.

Figure 3 Residential area in Lima.

2 (a) (i) State what is meant by urbanisation. **(2 marks)**

(ii) Study **Figure 1**. Describe how the data shows that rates of urban growth are greater in developing world countries than in developed. **(2 marks)**

(b) Study **Figures 2** and **3**.

(i) Describe the differences in housing and environmental quality between the two areas. **(4 marks)**

(ii) Name and explain two strong pull factors of big cities in the developing world for people living in rural areas. **(4 marks)**

(c) (i) Give two reasons why many squatter settlements are located on the edges of developing world cities. **(2 marks)**

(ii) In UK cities, why are the more expensive housing areas usually found on city edges? **(2 marks)**

(d) (i) State what sustainable urban living means. **(1 mark)**

(ii) With reference to a case study of sustainable living, describe the characteristics that make it sustainable and how successful it is. **(6 marks)**

(iii) Choose one characteristic of sustainable urban living and explain why it is difficult to achieve. **(2 marks)**

Total: 25 marks

Exam tip

The question for this topic will be Question 2 in Paper 2.

Improve your GCSE answers

Observation and its importance in urban studies

A Observation from photographs of UK and developed world cities

What to look for – checklist

Building features
- age
- building materials
- appearance and any distinctive features
- level of maintenance
- rich or poor

About the area
- function/purpose
- building density
- open spaces and their uses
- environmental quality
- likely urban zone

Figure 1 Birmingham 1km from the centre.

Figure 2 Birmingham 7km from the centre.

- Use the lists and check everything (there will always be some that do not apply).
- After this you should be well placed to suggest the different urban zones.
- What observations from the photographs support the answers of inner city for **Figure 1** and outer suburbs for **Figure 2**? You should be able to describe at least four points to support each one.

B Observations from photographs of developing world cities

Figure 3 Housing in Mumbai.

Figure 4 Street scene in Cairo.

- The checklist should work just as well for **Figure 3**, if asked to describe what **Figure 3** shows.
- It is still of some use for **Figure 4**, but it all depends on the question.
- This street scene is more about traffic than buildings and urban zones.

Traffic checklist
- Types of transport
- Infrastructure provided for it (e.g. type of road, quality of surface)
- Density of traffic
- Number of pedestrians
- Traffic problems

C Fieldwork observation in urban areas

Examples based on observation
- An urban transect – land use changes from city centre to edge of built-up area.
- CBD study of shops – types of goods sold, size of shops, number of storeys, pedestrian counts.
- Two or more housing zones – types of houses, house quality, environmental quality of the areas.
- Traffic survey – traffic counts on different streets, at different times of the day and week, parking.

ExamCafé

REVISION

Key terms from the specification

Exam Café

Brownfield site – area of previously built-up land that is available to be built on again
CBD (Central Business District) – urban zone located in the centre, mainly shops and offices
Functional parts (of a settlement) – purpose of that area, e.g. residential, industrial, port area
Industrialisation – growth and increasing importance of manufacturing industry (making goods)
Informal sector – not regular paid employment; unofficial work, often self-help small-scale services such as street sellers and shoe shiners
Inner city – urban zone around the edges of the CBD, quite old
Multicultural – when people from different ethnic, racial or religious backgrounds live together
Segregation (in urban areas) – high concentration of land uses and/or groups of people in certain areas of the city, separate from other uses/people
Squatter settlement – homes on land not owned by the people living there, built illegally
Sustainable city – city with low use of energy and raw materials, replacement by renewables and waste recycling
Urbanisation – increase in the percentage of people living in urban areas

Checklist

	Yes	If no – refer to
Can you explain why big cities are growing faster in developing than in developed countries?		pages 154–5
Can you describe changes in land uses between city centre and city edge in British cities?		page 156
Do you know two ways in which layout is different in developing world cities?		page 157
Can you give information about improvements in an inner-city area of the UK?		page 159
Do you know three ways of reducing the amount of traffic entering city centres?		page 161
Can you describe three environmental problems caused by rapid city growth in the developing world?		page 162
Can you explain how squatter settlements can be improved to become proper residential areas?		pages 164–5
Are you able to give three ways of making urban living more sustainable?		pages 168–9

Case study summaries

Squatter settlement redevelopment
Name and location
Characteristics
Changes and improvements made
What made them possible and the results

Sustainable urban living
Location
Characteristics
What makes it sustainable

Chapter 10

Changing Rural Environments

This is Henley-in-Arden. What shows that it is a rural environment under pressure?

QUESTIONS

- What is a rural environment?
- How is urban living impacting on the rural environment?
- What are the environmental consequences of the intensification of farming?
- Will our future use of the countryside be sustainable?
- What problems are pushing rural farmers in the tropics to migrate to the cities?

The rural–urban fringe is under intense pressure

How is urban sprawl impacting on the countryside? And what can we do about it?

A **rural settlement** is characterised by low population, usually less than 10 000 residents, in the countryside, where farming dominates the landscape. In 2005 there were about 10 million people living in rural England, mostly in large villages and small market towns. It is impossible to say precisely where the line separating rural and urban should be drawn. Over time cities have expanded outwards, along with their influence on the surrounding area. This transitional zone between the countryside and the city is known as the **rural–urban fringe**, where residents live in the countryside but tend to work and socialise in the city.

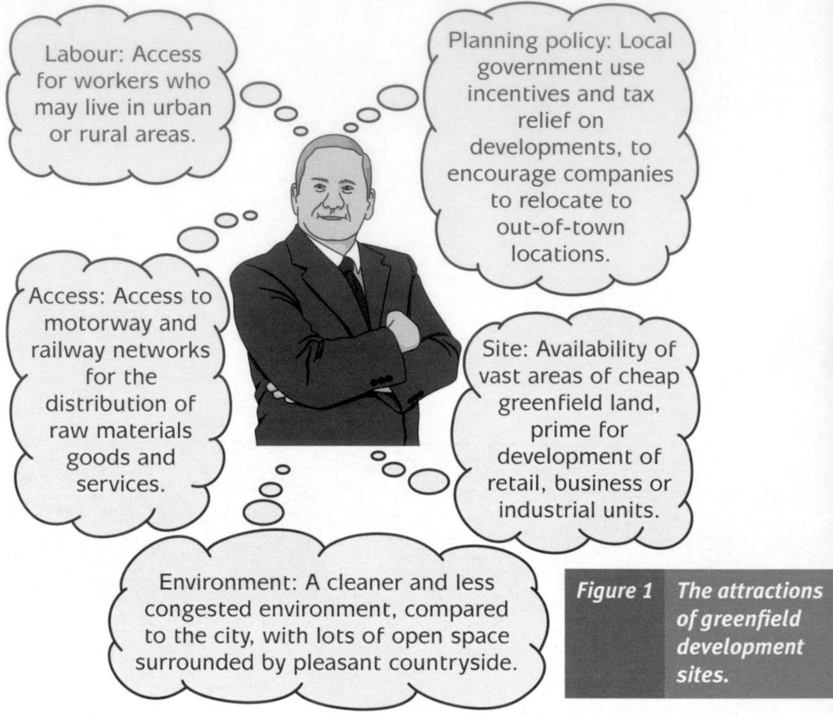

Labour: Access for workers who may live in urban or rural areas.
Planning policy: Local government use incentives and tax relief on developments, to encourage companies to relocate to out-of-town locations.
Access: Access to motorway and railway networks for the distribution of raw materials goods and services.
Site: Availability of vast areas of cheap greenfield land, prime for development of retail, business or industrial units.
Environment: A cleaner and less congested environment, compared to the city, with lots of open space surrounded by pleasant countryside.

Figure 1 ***The attractions of greenfield development sites.***

The causes of urban sprawl

The expansion of cities outwards into the surrounding countryside, or **urban sprawl**, is changing the face of the countryside. The population of the UK is continuing to rise and the number of households is increasing, as people are living in smaller family units. In 2004 there were nearly 3 million more families in Great Britain than in 1961, but there were 8 million more households. Modern technology, including the use of email, video-calling, fast Internet connections and **teleworking**, offers people and businesses greater flexibility as to where work takes place. Over the last 60 years, the development of a 3500-kilometre motorway network has made cities more easily accessible for affluent motorists living in the countryside.

The rural–urban fringe has seen changes, with land being used for housing, business parks, transport routes and recreational land, while the noxious land uses of the city (sewage works, scrap yards and landfill following mineral extraction) are pushed further into the countryside. The development of golf courses, sports pitches, country parks, equestrian centres and other recreational facilities provide a diversity of leisure provision and landscapes for people to escape to and enjoy, while promoting healthy living away from life in the city.

Developers prefer **greenfield** sites (land that has not been built upon) on the edge of urban areas, as they are cheaper and easier to build on than previously developed or **brownfield** sites inside the cities. Other reasons are noted in **Figure 1**. Out-of-town retail outlets, which became popular in the 1990s, are increasing in number as consumer demand increases.

The impact of urban sprawl on the countryside

Urban sprawl is already having a huge impact on the countryside, resulting in longer journeys by car from the edge to the centre of the city, more air pollution, more traffic congestion and gridlock (when use of existing transport networks exceeds their capacity). As development continues, it is the environment that suffers. Wildlife habitats are destroyed and fragmented, leading to a decline in wildlife numbers and diversity. Tranquil areas (**Figure 2**) are places sufficiently far away from development or traffic to be considered unspoilt by urban influences and they have declined by a third in less than 40 years.

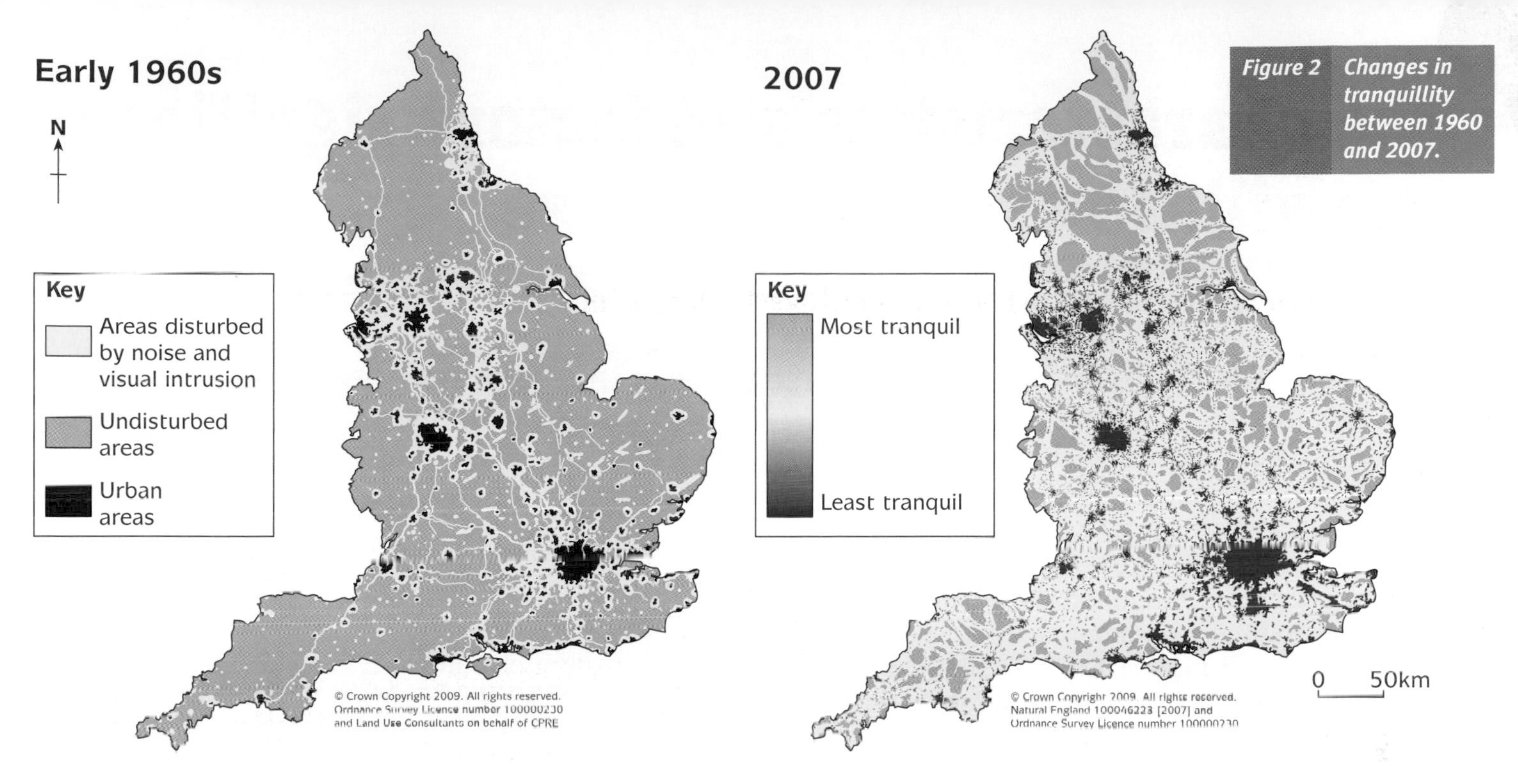

Figure 2 *Changes in tranquillity between 1960 and 2007.*

Responses and strategies for the future

In the UK, planners decided that **green belts** were the best way to stop urban sprawl. Green belts are areas of protected countryside with strict planning controls and they cover 13 per cent of England. Looking at **Figure 3**, you can see green belt areas located around cities and large merged urban areas or **conurbations** (such as Leeds/Bradford). The biggest green belt in the UK was created around London following the 1947 Town and Country Planning Act. In the last 60 years, green belts have been effective in slowing urban sprawl, but they do not stop a city from growing. Development either eats into the green belt or simply leapfrogs the protected land and continues on the higher-quality countryside beyond. The government wants 3 million new homes by 2020, a quarter to be located on green belt land. The Thames Gateway project to the east of London proposes the use of 485 hectares of green belt land around Tilbury. In order for planning authorities to conserve and enhance the quality, diversity and distinctiveness of the countryside, sustainable rural policies need to be put in place. New affordable homes need to be built in key villages, towns or in the urban areas so that people can live and work in the same settlements. In order to be sustainable, transport policies must encourage more people to walk, cycle or use improved public transport systems, and support a reduction in car dependency.

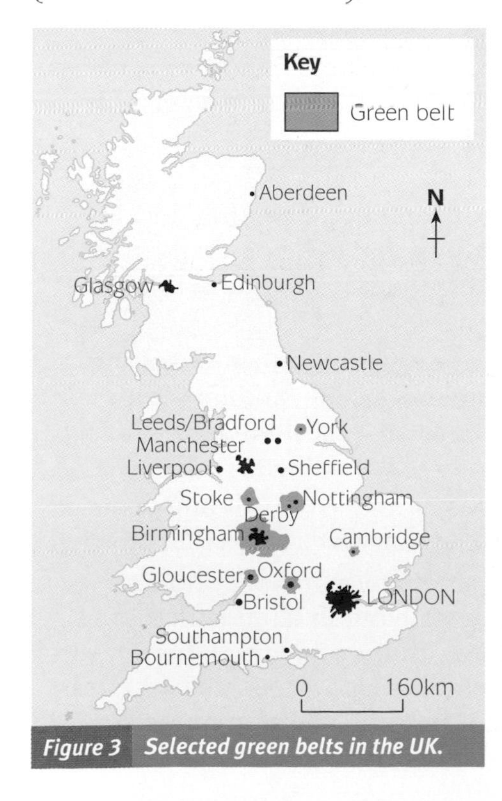

Figure 3 *Selected green belts in the UK.*

ACTIVITIES

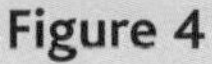

Figure 4

1 Study **Figure 4**, which shows green belt land around Birmingham:
   (a) Describe the land use shown.
   (b) Why would it not be possible to build a new housing estate on this land?

2 Participants in a planning meeting regarding a proposed out-of-town retail outlet within the green belt would include: a developer, the local council, an environmentalist, a farmer, a builder and a local shopkeeper.
   (a) Explain which parties would be against the proposal.
   (b) Make a case on behalf of the developer as to why the retail outlet should be built.

3 (a) Describe how accessibility to the city affects the development of rural villages.
   (b) Suggest how teleworking can reduce the pressure on the rural–urban fringe.

The characteristics and factors significant to a village expanding in size

Are all expanding rural settlements successful? If so, according to whom?

Suburbanised villages

A suburbanised village has a residential population who sleep in the village but who travel to work in a nearby large urban area. Suburbanised villages have expanded as part of the larger process of **counter-urbanisation**, whereby people are moving from urban areas into the surrounding countryside in developed countries. Look at **Figure 1**, which shows a suburbanised or metropolitan village. It has an original village core, with areas of infilling of new houses built between existing houses. More houses are constructed along roads out of the village in **ribbon developments** and larger planned estates are located outside the village core. The actual form of a suburbanised village is influenced by the existing buildings, the site of the village and local planning policy.

Changes in the suburbanised village

Many of the original village characteristics will change as a growing village becomes suburbanised:

- **Population** – More professionals, commuters and wealthy people with families move into the village. There are more newcomers than locals.
- **Housing** – The housing stock shows more diversity as many new planned estates, detached and semi-detached houses are added to the village core. Old detached, stone-built houses and cottages with slates/thatched roofs are repaired as newcomers invest in renovations. The increased demand puts house prices beyond the reach of young locals, many of whom rent accommodation or leave the village, unless affordable housing is supplied by the local council.
- **Employment** – New light industry is developed on the edge of the village. Many commuters work in the surrounding urban area, while labouring, manual and service jobs persist in the village.

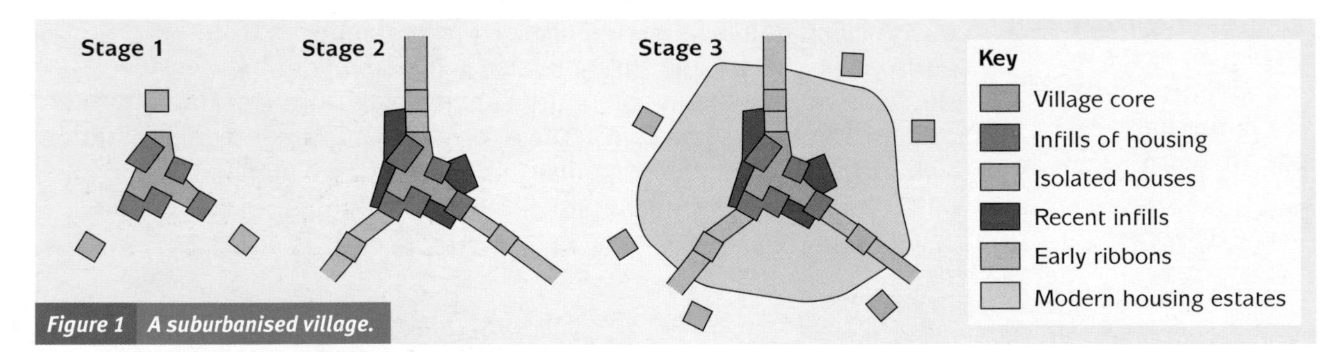

Figure 1 A suburbanised village.

- **Services** – Closure of convenience shops and local post offices. New specialist shops, enlarged or amalgamated schools and modern public houses provide services for the population.
- **Social** – The local community is swamped, resulting in divisions between the local people and newcomers. Some villages may be deserted ('ghost villages') during the day.
- **Transport** – Most families have one or two cars, especially if they commute to work on improved roads into the city.
- **Environment** – Increase in noise and pollution, especially from traffic, goes hand in hand with loss of farmland and open space. Traffic and congestion in rural areas and green belts will increase as people commute to work in urban areas. This puts pressure on narrow rural roads.

EXAM PREPARATION

The advantages and disadvantages of rural living	
Advantages of living in rural areas and villages	**Disadvantages of living in rural areas and villages**
Good range of community activities based in the village	Remote from urban services such as secondary schools and entertainment
Less congestion and pollution	Infrequent public transport for teenagers and elderly
More open space and freedom	Few activities for teenagers
Sense of community and belonging	Village life is perceived to be insular, with lots of nosy neighbours
Perceived to be safer, with low crime rates	Shop prices are higher and the range of products narrower
Better quality of life	Housing can be very expensive

The growth of commuting and commuter villages

Many villages in rural areas grew rapidly after the addition of railway and motorway links to the nearest city or conurbation. Known as dormitory or **commuter villages**, the function of these settlements has changed dramatically; they often lack schools and shops but retain pubs and restaurants.

EXAMPLE: Alvechurch, Worcestershire

Figure 2 OS map of Alvechurch at a scale of 1:25 000 (4cm = 2km).

Sited within the Worcestershire district of Bromsgrove, Alvechurch is a small parish with a population of just over 5000 people. The village borders both the city of Birmingham and Redditch new-town and is therefore a vital part of the West Midlands Green Belt that serves as a buffer between urban areas (**Figure 2**).

The form of the suburbanised village reflects the history of its growth (**Figure 3**). As far back as the eighth century, houses were built in the village core and along the Birmingham Road (B4120) as a ribbon development. Infilling and new estates have been built between the village core and the road and railway routes. The housing stock is mixed, with 10 per cent consisting of local authority rentals.

Work patterns have changed since the late 1800s when much of the local employment was agricultural. Among the current economically active residents, the average distance travelled to work is 14 kilometres.

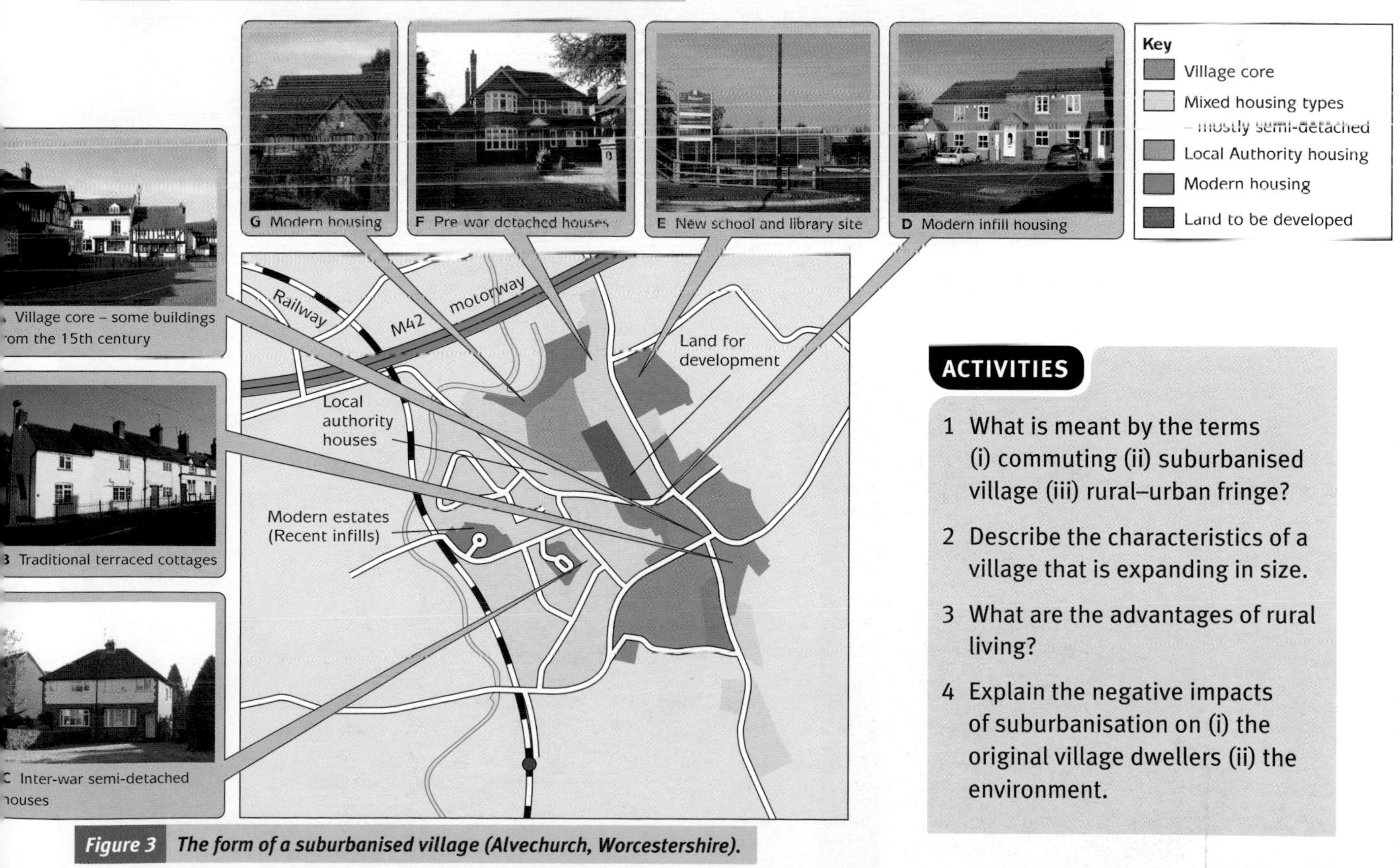

Figure 3 The form of a suburbanised village (Alvechurch, Worcestershire).

ACTIVITIES

1. What is meant by the terms (i) commuting (ii) suburbanised village (iii) rural–urban fringe?
2. Describe the characteristics of a village that is expanding in size.
3. What are the advantages of rural living?
4. Explain the negative impacts of suburbanisation on (i) the original village dwellers (ii) the environment.

Cornwall – the social and economic changes in rural areas

Rural depopulation in Cornwall

Cornwall is in a remote location in the south-west, far from the core region of south-east England. Why are remote rural areas in Cornwall in decline? Mainly, it is because of the reduction in traditional employment because of the declining labour requirements on mechanised farms, increased competition from abroad and exhaustion of natural resources, such as copper, tin and china clay. Rural depopulation (or the loss of people to other regions for job opportunities and better wages) occurred between 1861 and 1951. From the 1960s onwards, people began to move back into the countryside. The Cornish rural population is continuing to increase because of high in-migration of older people, attracted by the mild climate and beautiful scenery. A quarter of the half million population is over state-pensionable age, as compared with 19 per cent across the UK. Today, the number of births is below 4500 and the number of deaths exceeds 6000 a year, giving a natural population decrease. Out-migration of the younger, economically active population in search of education, jobs and affordable homes continues. Rural depopulation is now confined to only the most isolated rural areas. Such areas suffer from many problems, including an ageing population, a decline in services, and rural poverty and isolation.

Figure 1 ***Abandoned tin mines near Truthwall.***

The characteristics of a declining Cornish village – Truthwall

Truthwall, located on the Penwith Moors, on the extreme west coast of Cornwall, is now a much smaller settlement than during the peak period of tin mining in the 1820s–40s. Farming has continued in the area since medieval times, but once the mines shut, the village continued to shrink in size (**Figure 1**). Many of the village characteristics have changed as rural depopulation has occurred. These include:

- **Housing** – Poor housing lacks modernisation; some houses are left derelict; some converted into second and holiday homes.
- **Population** – Mainly elderly and retired people; some residents were born in the village and have stayed while others have moved away.
- **Employment** – Unemployment is high and triggers a vicious cycle of rural depopulation seen in **Figure 2**; few farm workers are left; low-paid seasonal jobs in the tourism industry are available.
- **Services and transport** – Shops, post offices and schools close; more pubs and restaurants open; infrequent public transport available.
- **Social** – A small isolated community with a few retired newcomers.
- **Environment** – Generally quiet, except at weekends when second-home owners fill the villages.

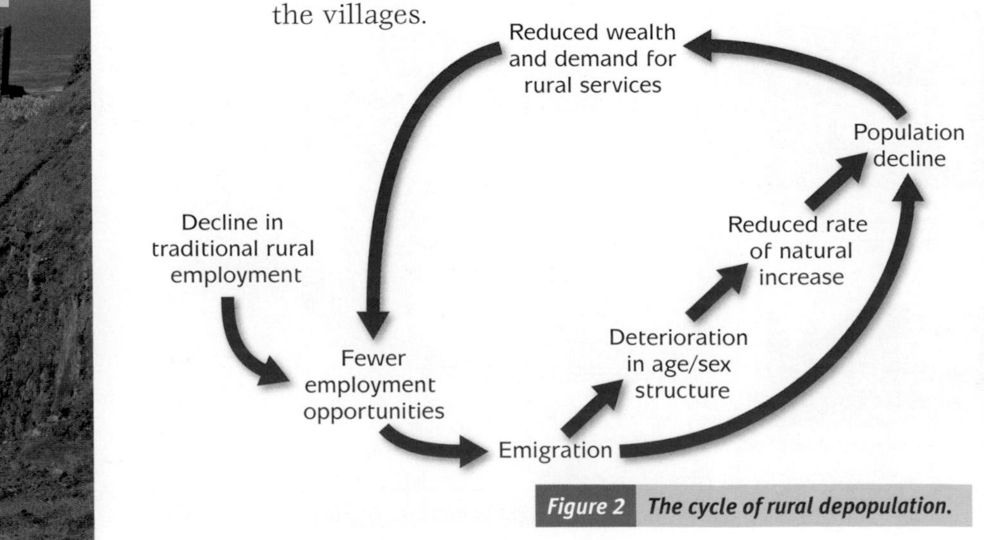

Figure 2 ***The cycle of rural depopulation.***

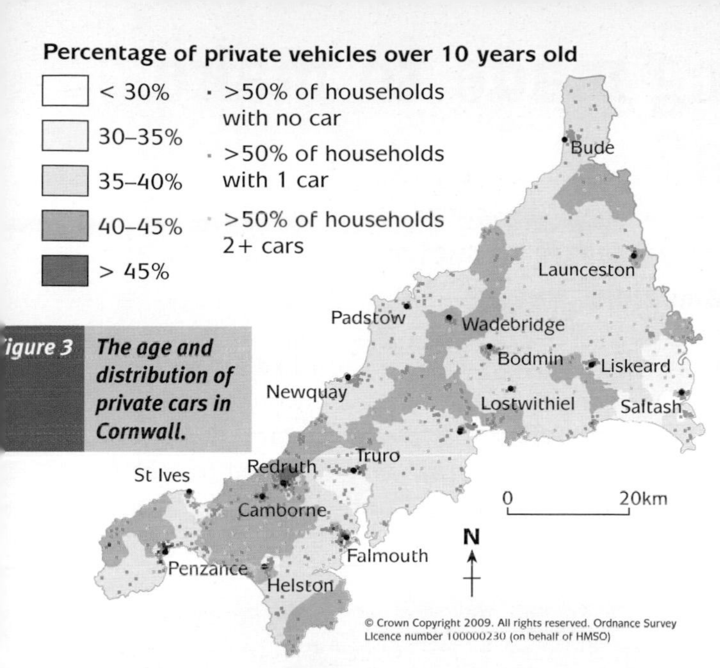

Figure 3 *The age and distribution of private cars in Cornwall.*

Today many of Cornwall's farm holdings fail to generate enough income to support even one person, and tourism in the English county with the longest coastline has become the dominant industry. Tourism contributes £1.5 billion to the local economy every year. The increasing demand for holiday homes, visitor services, catering, transport and retailing has led to more people being employed in seasonal, part-time and poorly paid jobs. Earnings in Cornwall are low, about 75 per cent of the national average.

The increase in second homes

One in ten Cornish properties is a **second home** or holiday let, bought in an idyllic setting, often by people who have benefited from well-paid city jobs or those planning to retire to the country in the future. These houses are not their owner's main place of residence. In popular resorts, such as Rock, Portscatho and Port Isaac, second-home ownership is between 50 and 80 per cent. Second-home owners have pushed the cost of housing beyond the reach of local people. A first-time home in Cornwall is now more unaffordable than in London. This problem was made worse by the 'right to buy' legislation of the 1980s, under which rented social houses were bought by their tenants.

Rural services

The availability of services in the almost 10 000 rural parishes in England is in decline, according to the 2008 'State of the Countryside Report'. This survey showed that the proportion of rural parishes without key services remained high. For nearly all service types, availability has fallen in rural areas, especially banks, petrol stations, dentists, post offices and Job Centres; but supermarkets, cashpoints, pubs and restaurants have seen an increase. St Mawes is a large community but the combination of retirees and the holiday market means that it is struggling to maintain a two-classroom school. Owing to the decline in rural provision and services, residents are now obliged to own a car or feel the effects of rural isolation (90 per cent of households in Cornwall have a private car). However, the average age of cars in Cornwall, shown in **Figure 3**, is among the highest in England, with a third being over ten years of age.

The consequences of rural depopulation

As rural areas like St Ives become swamped by second-home owners, those living near the **poverty line** (the minimum level of income deemed necessary to achieve an adequate standard of living) become hidden in statistics. In the four poorest wards located in west Cornwall, more than a quarter of households are living in poverty. Penwith has the lowest rate of economic activity in south-west England. Out-migration of young workers is still prevalent, even though the prosperous village of St Ives lies within its borders (**Figure 4**). By 2026 it is projected that one in three of the population will be retired; this will stretch health and recreational services and shrink services for the young.

ACTIVITIES

1. Explain why some rural services are thriving while others are under threat of closure.
2. Describe the causes and consequences of rural depopulation.
3. Using **Figure 4**, suggest why the shape of the population pyramid for Cornwall differs from that for the UK average.

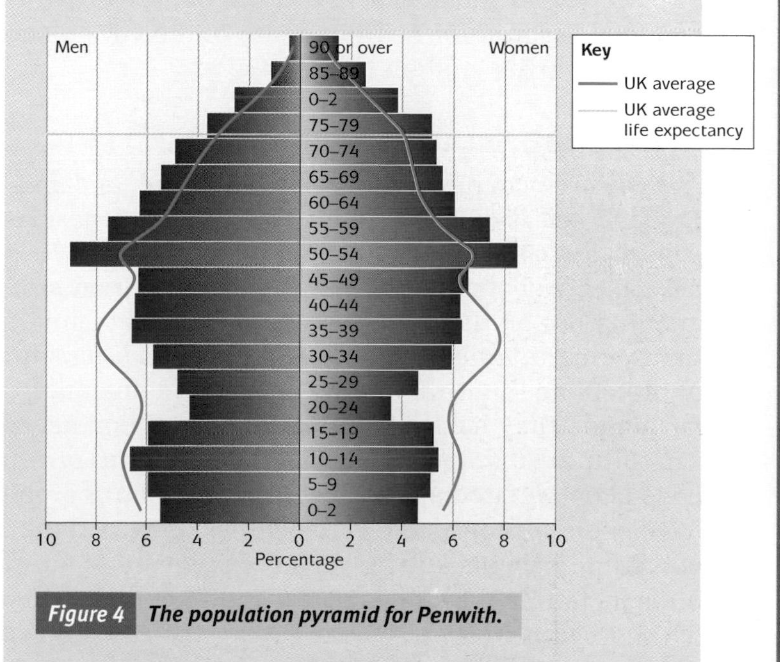

Figure 4 *The population pyramid for Penwith.*

What attempts are being made to make rural living sustainable?

Rural areas are disappearing under concrete at a rate of 54 square kilometres a year. Can rural areas be protected and sustainably developed at the same time?

Supporting the needs of the rural population and the rural economy

The Rural Delivery Review, published in 2003 by the Department for the Environment, Farming and Rural Affairs (Defra) recognised the problems of the countryside, including the lack of local control over economic and social issues, the loss of rural services and the complex and confusing system of rural funding. The government established Natural England, an organisation that aims to conserve the natural environment and manage tourism. Its three objectives are as follows:

1. **Conserving natural resources** – Deliver more competitive and sustainable farming and food industries.
2. **Protection of the environment** – Protect the environment and preserve biological diversity by increasing the number of farmland birds. Make the countryside attractive and enjoyable for all.
3. **Sustainable rural living** – Improve the productivity of all rural areas and accessibility to services for rural people. Services must be **sustainable** and work in harmony with the environment rather than against it.

Farmers in the UK currently receive nearly £3 billion in subsidies under the Common Agricultural Policy (CAP). Farming remains important. Agriculture employs over half a million people and is worth some £8 billion a year. Farming and food account for nearly a tenth of the UK's income and employ nearly 4 million people.

Conserving natural resources on the farm

Agriculture occupies over two-thirds of the land area in England and is essential to maintaining the landscape and conserving natural resources and wildlife. Since 2005, farmers have received one payment that is based not on production quantity but on the size of the farm, subject to environmental management. *Environmental Stewardship Schemes* are a way of protecting the rural landscape while maintaining profitable farmland. They have so far protected 5 million hectares of agricultural land. More than 17 000 kilometres of hedgerows have been restored and nearly 5000 kilometres of footpaths have been improved. Stewardship has increased the number of wildlife habitats and assisted the recovery of farmland birds including the stone curlew and bittern. An example of the *English Woodland Grant Scheme*, run by the Forestry Commission, can be seen in **Figure 1**, the aim of which is to continue replanting trees. Additional payments are available for environmental management of farms in upland England under the *Hill Farm Allowance Scheme*.

Figure 1 *Trees planted by the Forestry Commission in an upland area in England.*

Protection of the environment

Where the countryside is under threat from development the government must ensure that the planning process limits the type of development occurring on the rural–urban fringe by establishing green belt (**Figure 3**, page 175). This can be achieved by building more houses in cities and key market towns, and designating protected areas where no development is allowed. National Parks (**Figure 2**) protect areas of beautiful, spectacular and often dramatic countryside. There are 14 in the UK, with the South Downs designation currently under review. A total of 49 Areas of Outstanding Natural Beauty (AONBs) have the same protection as National Parks without their own Authority to protect them; 210 National

Nature Reserves (NNRs) are smaller sites with important wildlife habitats; and Sites of Special Scientific Interest (SSSIs) are sites that contain important wildlife species, habitats or geological features. There are 43 Heritage Coasts with special scenic and environmental value in England and Wales, most owned by the National Trust. Under the EU, 'Natura 2000 Network' sites have been designated as Special Protection Areas (SPAs) with a focus on bird habitats, and Special Areas of Conservation (SACs) have been put in place to prevent any further decline in vulnerable species.

Sustainable rural living

Sustainable rural living requires financial support and this is obtained from many different sources. These include:

- *The England Rural Development Programme (2007–13)*: £3.9 billion has been made available to invest in diversification or non-farming activities including tourism and recreation.
- Cornwall and the Isles of Scilly have gained an additional £55 million from the *European Union's 'Convergence' Objective*. These funds are invested pound for pound in small businesses.
- Funding, business advice and information are offered by the *Business Links Network*, to help small businesses in rural areas.
- *Regional Development Agencies (RDAs)* designate areas in need of assistance. Grants are available to assist individual enterprises and support rural development.
- The *Vocational Training Scheme (VTS)* aims to improve the occupational skills of farmers and foresters, offering grants to cover up to 75 per cent of training costs.
- In 2002 the government allocated just £8.7 million from a budget of £250 million to rural districts for its *'Home Starter Initiative'*. There is a desperate shortage of affordable homes in both urban and rural areas. Villages need affordable homes to help to maintain sustainability, services and village life. Full council tax should be paid on second homes to increase local council funds.
- *Broadband Internet* has been set up to reduce the effects of rural isolation. In Cornwall it has reached 13 Cornish telephone exchanges and 10 000 businesses, farms and voluntary groups.

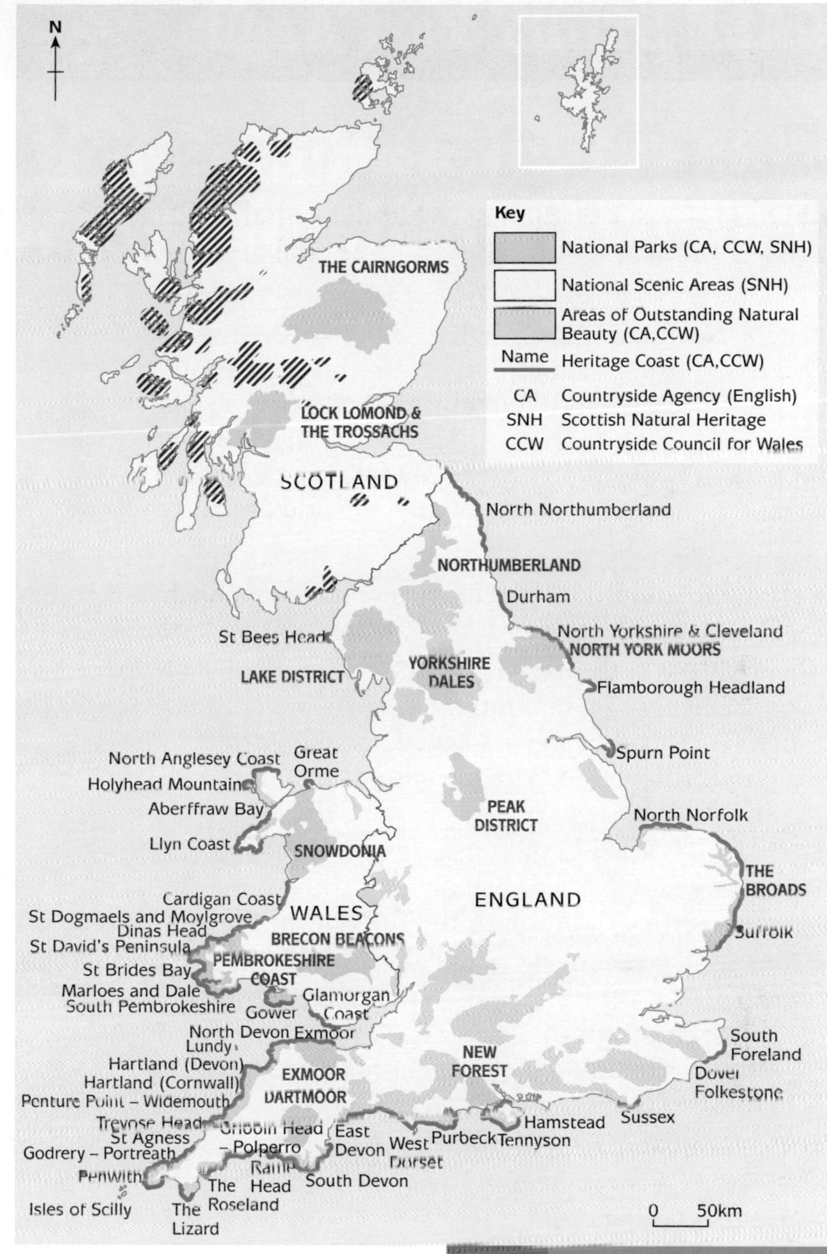

Figure 2 Protected areas of Britain.

GradeStudio

1. What do the following designations mean? (i) AONB (ii) SSSI (iii) NNR (3 marks)
2. What is environmental stewardship? (3 marks)
3. Describe two of the benefits of the Environmental Stewardship Schemes (4 marks)
4. With reference to **Figure 2**, describe the distribution of protected areas in the UK. (4 marks)
5. Name and explain government initiatives to support the rural economy. (6 marks)

East Anglia: Commercial farming in favoured agricultural areas

Arable farming in East Anglia is both **intensive** (because farmers attempt to obtain the maximum output from the land) and **commercial** (as the **cash crops** are sold for a profit). The following **physical factors** in East Anglia, in the arable core of eastern England, place few limits on farming and are ideal for growing crops:

- The flat land is ideal for the use of large machinery, such as combined harvesters, tractors and trailers.
- The fertile soils, mainly boulder clays, deposited by ice sheets as they retreated at the end of the last ice age 12 000 years ago. This makes the region suitable for growing cereals including wheat and barley.
- The sand, loam (a mix of sand and clay) and chalk soils are suitable for vegetables, fruit and root crops as they drain more freely than the clays.
- The long growing season allows for maximum crop growth from April to September. Summer temperatures around 17 °C and the long hours of summer sunshine allow fast growth and ripening of crops.
- The rainfall is sufficient for crop growth but not so heavy as to make the soil waterlogged so that the roots rot and die.

There are several **human factors** that also favour farming in East Anglia:

- The location is close to large urban areas, including London, which provides a large and wealthy market for the produce.
- The flat relief has enabled the East Coast railway, the M11 motorway and the A1(M) to develop, all of which allow the rapid transport of the produce to market.
- Almost 50 000 people work directly in farming and many more jobs depend upon it (in livestock feed production, transport, the veterinary profession and agricultural research and development). Educational establishments, such as Writtle College in Essex and Easton College in Norfolk, have developed in the area.

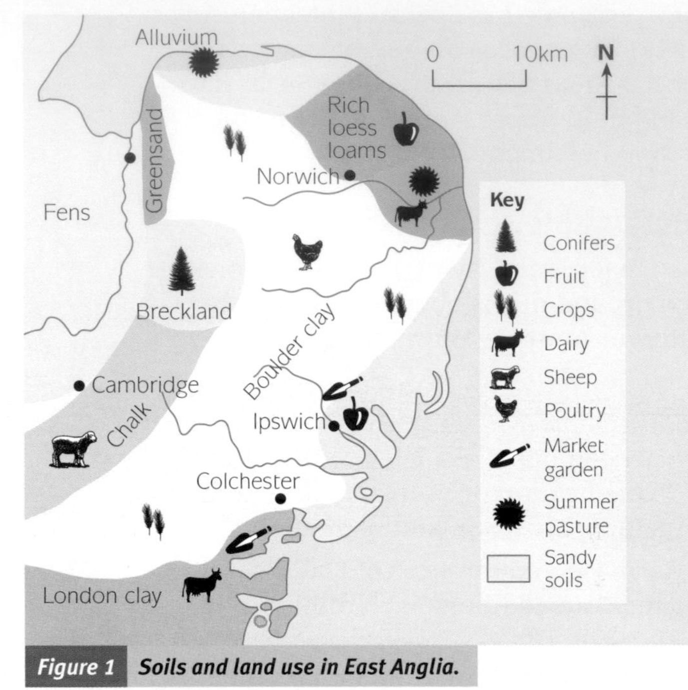

Figure 1 *Soils and land use in East Anglia.*

Around three-quarters of land in the eastern region is used for agriculture. The region of East Anglia covers Hertfordshire, Bedfordshire, Cambridgeshire, Norfolk and Suffolk. It is best known for its cereal crops, with farmers here growing more than a quarter of England's wheat and barley and over half the country's entire sugar beet crop. Over a million pigs are reared and 2 million eggs are laid every day in East Anglia. **Figure 1** locates the soil type and farm outputs across the eastern region.

The development of agri-businesses

Agri-business is a type of farming that is run as a large-scale, highly efficient big business – more likely a large food production company with a farm manager, cutting-edge machinery and a high-tech office. **Figure 2** illustrates how farms have changed between 1950 and 2000. Farms increased in size as they joined together (amalgamated) and intensified production. Farmers put all available land into production, by grubbing out hedgerows, cutting down trees and infilling ponds in order to increase areas of cultivation and maximise profits. This type

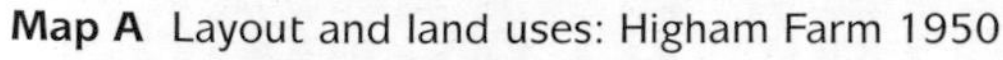
Map A Layout and land uses: Higham Farm 1950

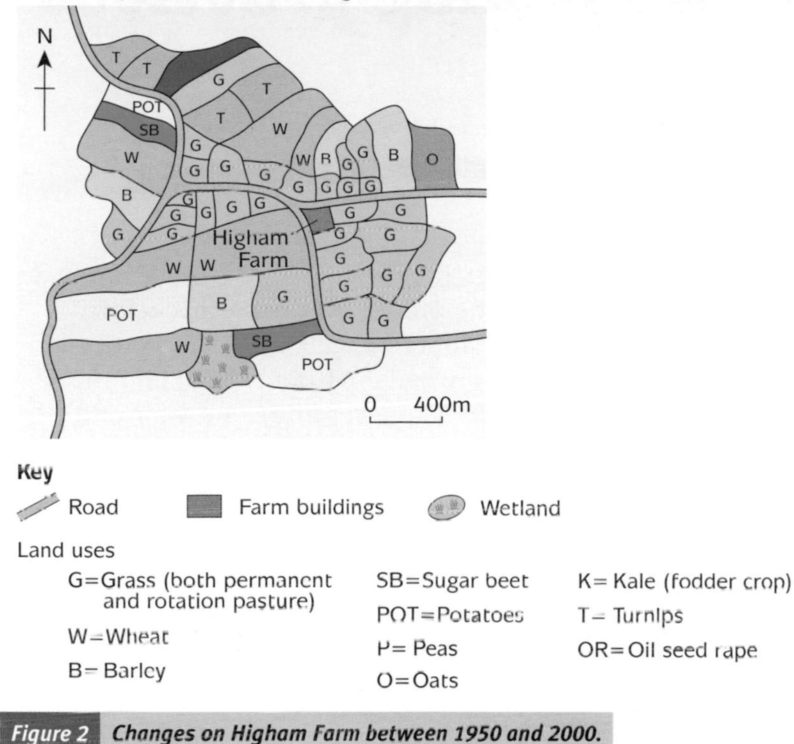

Map B Layout and land uses: Higham Farm 2000

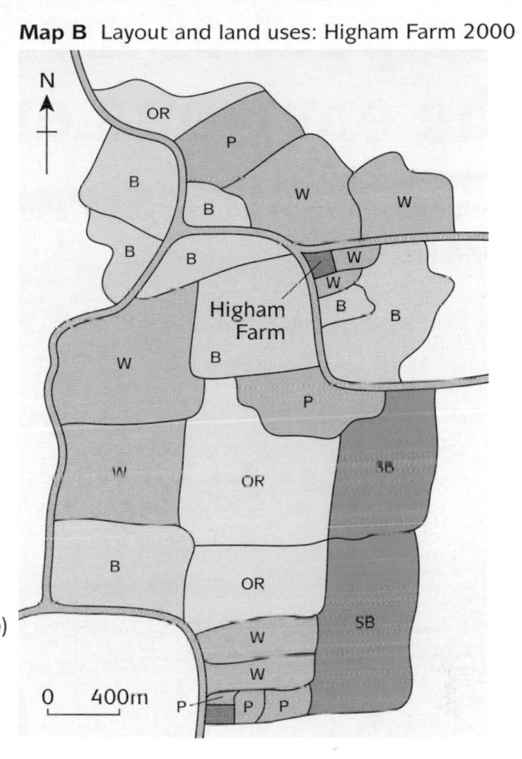

Figure 2 ***Changes on Higham Farm between 1950 and 2000.***

of farming includes a chain of suppliers of inputs (seeds, chemicals), food producers (farmers) and food processing businesses, such as British Sugar shown in **Figure 3** and Birds Eye who have high quality standards for cultivation.

Demands from the market

In the past the **Common Agricultural Policy (CAP)** guaranteed farmers a minimum price for their crop; the more the farm produced, the bigger the subsidy they received from the EU. Subsidies created huge surpluses and unfair global trade. The biggest fifth of farms received three-quarters of all agricultural subsidies. These subsidies are to be phased out by 2013. The huge purchasing power of the supermarkets gives them control over the choice of crops and cultivation methods used on the farm, driving the price paid for produce below the cost of production. The 'big four' supermarkets control a majority of the market. This has transformed farming into an intensely competitive industry, where animal welfare and environmental issues are given lower priority in the pursuit of cost reductions. Farmers were once guaranteed a price for farm produce; now UK farmers have to compete for business with overseas farmers, whose labour costs are cheaper. Sugar imports were effectively halted until the price guarantee was stopped in 2006. Today production is in decline in the UK and sugar imports are increasing.

Figure 3 ***Agri-business in East Anglia.***

ACTIVITIES

1 (a) Identify the human factors that favour farming in East Anglia.
  (b) Explain why the physical factors in East Anglia make it the most successful location for cereal growing in the country.

2 Study **Figure 2.**
  (a) Describe the main changes to farm size and land use on Higham Farm between 1950 and 2000.
  (b) Explain why the fields were increased in size on Higham Farm between 1950 and 2000.
  (c) Explain (i) the economic advantages and (ii) the environmental disadvantages of the changes.
  (d) Suggest what further changes might have been made on this farm since 2000.

FURTHER RESEARCH

Find out more about farming in the UK on the weblink www.contentextra.com/aqagcsegeog.

The impact of modern farming practices on the environment

Can government policies and consumer choice reduce the impact modern farming has on the environment?

The impact of modern farming practices on the environment

When farmers were paid to increase food output, wetlands, hedgerows, marshes and moorland were ploughed up. This allowed valuable agricultural land to be put into food production and larger machinery to be used. Between 1945 and 1995, 40 per cent of hedgerows were removed, making the landscape appear barren, rather like the North American Prairies (**Figure 1**). However, hedgerows act as a natural windbreak between fields, without which wind speeds increase. The roots of the hedgerow bind the soil together; once removed, the rate of soil erosion by wind and water increases. When the soil is left bare, wind and rain can easily pick up and transport soil particles away from the farm. Poor farming techniques can accelerate soil erosion. Harvesting in East Anglia coincides with autumn, when rainfall and storms increase in intensity. Water erosion can remove topsoil and reduce the productivity of the land. Moreover, hedgerows are important wildlife habitats and corridors, allowing the movement of animals, without which populations decline in size. Removing hedgerows caused the loss of many species. In Devon, this resulted in local extinctions of birds including the yellow wagtail.

Intensification of farming had other adverse environmental effects. The overuse of farm chemicals contaminated soils and local rivers as excess spray ran off the fields. Most farming relies heavily on artificial chemical fertilisers and pesticides. Fertilisers, including nitrates, phosphates, and slurry (farmyard effluent), are added to fields to improve plant growth. Chemical fertilisers replace the nutrients which have been used up in the soil. When they reach rivers, these nutrients cause a rapid growth of algae (algal bloom), which uses up oxygen, resulting in the suffocation of fish – the process of **eutrophication**. Pesticides and herbicides are chemicals applied to crops to control pests, diseases and weeds. Without these chemicals, productivity can decline by a quarter after one year and 45 per cent after three. Many of these target all life; they are not selective. Pesticides are blamed for the decrease in bee, butterfly and insect populations in England. The groundwater stores of East Anglia are being increasingly contaminated by chemical run-off.

Figure 1 *Since the removal of hedgerows, the East Anglian landscape now resembles the North American Prairies.*

Government policies aimed at reducing the environmental effects of high impact farming

The European Union (EU) spends £34 billion on the Common Agriculture Policy (CAP) annually. The UK government Department for Environment, Food and Rural Affairs (DEFRA) pays out £3 billion a year to UK farmers from the budget. Over the last couple of decades, farmers have been paid less for what they grow and more for the land that is not being cultivated. The policy of set-aside was the first such strategy, introduced to reduce the supply of farm food, with the outcome of improving the number of wildlife habitats on the farm. Under this scheme, fields were left empty of crops (fallow) for at least five years; farmers could also leave edges around their fields where crops were not sown (headlands).

The most recent reforms to the CAP break the link between subsidies and production. CAP funds are now being used to improve animal welfare and environmental standards across the EU. Under the *Environmental Stewardship Scheme*, farmers receive payment regardless of the type or amount of crop on the farm. Introduced in 2005, it offers incentives to conserve wildlife, maintain and enhance our landscape quality and protect our natural resources. Arable farmers can also grow biofuel crops under the *Energy Crops Scheme*. The world's largest straw-fired power plant at Sutton near Ely, Cambridgeshire, generates enough energy annually to power 21 000 homes. Under the *Environmental Stewardship Scheme*, farmers are able to take marginal land that is poor for farming out of production and still be paid under the CAP. This has allowed new hedgerows to be seeded and headlands of grass to be left around fields and watercourses where previously it had not been economically viable to do so.

Development of organic farming

Sustainable organic farming practices do not use synthetic chemicals such as pesticides and insecticides. This is better for the health of farm workers, wildlife and the environment. There is a growing consumer demand for our food to be produced organically. Sales of organic food in the UK have increased over the last 20 years and are now worth over £2 billion a year. Two-thirds of the baby-food market is now organic. People want to be informed about where their food comes from (provenance) and the journey it takes (food miles). More consumers are bypassing the supermarkets and obtaining their organic fruit and vegetables from box schemes seen in **Figure 2**, mail-order or straight from the farm. Sales direct from the farm were worth £150 million in 2006. As demand increases, greater numbers of farmers are switching to organic production, a slow process in which it can take three years for all traces of chemicals to be removed from the food chain. Organic holdings now represent 1.6 per cent of all farms in the UK and the market is continuing to expand.

Figure 2 ***Vegetable box deliveries.***

ACTIVITIES

1 Draw up a table to show the advantages and disadvantages of hedgerows.

2 Describe the cause and effect of the following farming issues: (i) hedgerow removal (ii) soil erosion (iii) artificial chemical use on the farm.

3 Why do you think the CAP has moved away from the use of subsidies?

4 Assess the benefits of organic farming – for (i) the consumer (ii) the farmer (iii) the environment.

5 Identify who benefits from a vegetable box scheme, compared to purchasing vegetables from a supermarket. Is this a sustainable form of farming?

FURTHER RESEARCH

Follow web-based farm trails on the weblink www.contentextra.com/aqagcsegeog.

Subsistence food production in tropical rural areas

Traditional farming is under threat. With 15 000 square kilometres of tropical rainforest disappearing every year, are subsistence farmers, like the Amerindians of Brazil, sustainable?

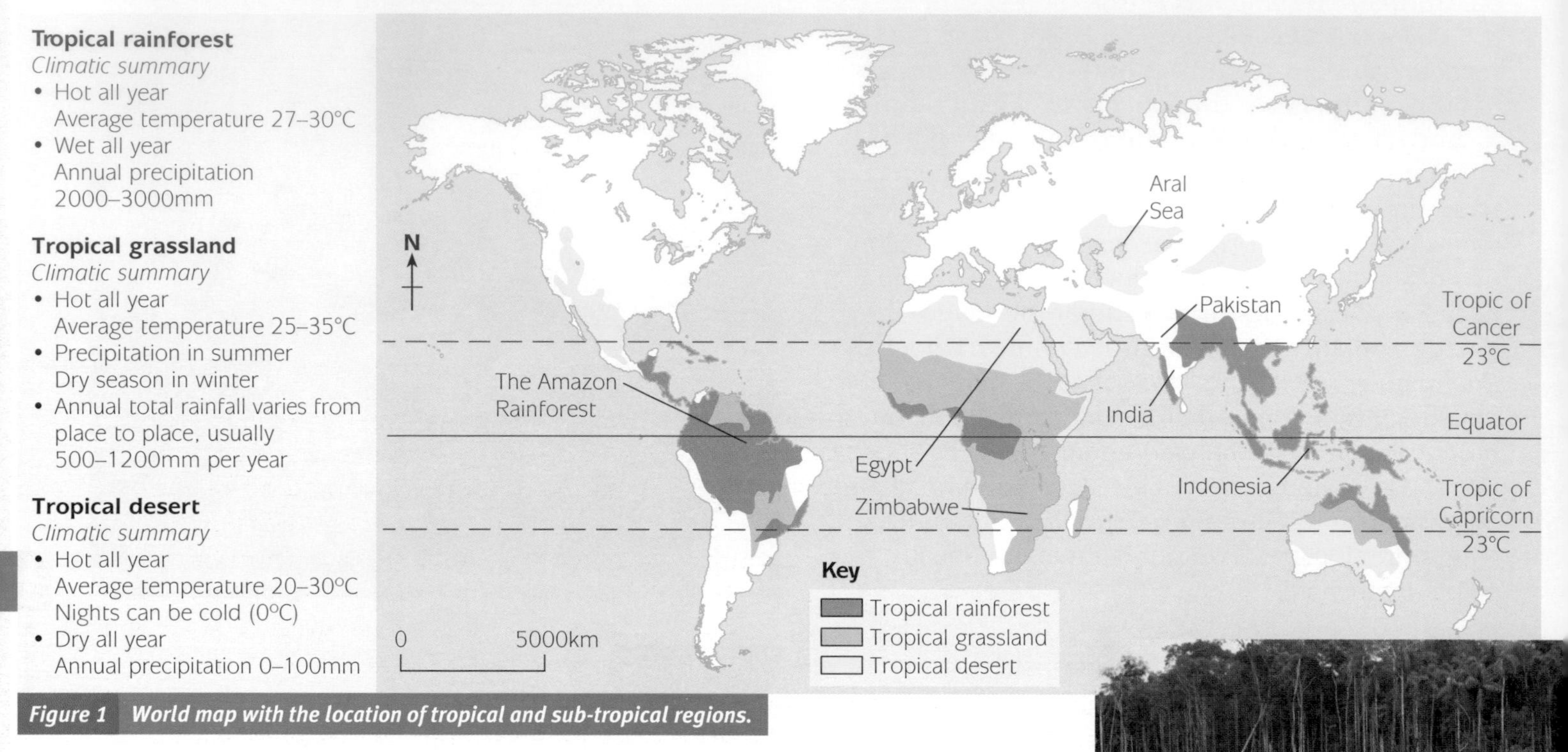

Figure 1 *World map with the location of tropical and sub-tropical regions.*

The physical conditions found in tropical and sub-tropical rural areas have traditionally resulted in low population densities. The people who live here are **subsistence** farmers who produce just enough food to feed their families. **Shifting cultivation** is an example of subsistence food production found in remote and inaccessible parts of the tropical rainforests, such as the Amazon shown in **Figure 1**, where the forest's resources have not yet been exploited.

Figure 2 *Slash and burn in the Amazon Basin.*

Subsistence food production

The Amazonian tribes clear a small area of about 1 hectare in the forest, leaving a few large trees for protection. After the vegetation and felled trees dry out they are burnt and the ash is used to fertilise the soil, as shown in **Figure 2**. This type of farming is known as shifting cultivation or 'slash and burn'. The main crop, manioc, is planted, along with yams, beans and pumpkins in the clearings called *chagras*. The Amerindian diet is supplemented by hunting, fishing and collecting fruit from the forest. This type of farming is *sustainable*, as it works in harmony with nature and does not permanently damage the ecosystem.

The impact of soil erosion

The productivity of the rainforest is dependent upon an unbroken nutrient cycle. Once the forest is cut down, then the cycle is broken. Without the natural vegetation cover, the source of humus for the soil has gone. The trees, cut down in the clearing, can no longer protect the bare soil from the heavy tropical rainfall, as there is nothing to intercept the rain. More water reaches the ground, which increases surface run-off and the formation of gullies across the landscape. Rivers receive more water and flood more frequently. More loose topsoil and sediment are washed away. This

Deforestation increases surface run-off, significant in tropical areas with intense rainfall.

Mining removes protective vegetation and topsoil. This is transferred into local rivers.

Plantations, often of one crop (e.g. coffee, bananas) require the natural vegetation to be removed.

Overcultivation removes organic matter and weakens the soil making it prone to erosion.

Cultivation on slopes makes gully erosion more likely if fields are left bare in winter.

Irrigation without soil drainage can cause salinisation and waterlogging.

Ranching can cause overgrazing, with too many animals. This removes protective vegetation and makes the land prone to erosion by wind and water.

Heavy machinery compacts the soil and stops water from draining freely.

Overgrazing and overcultivation in developing countries tend to occur as land is in constant use for food production, often for export.

Winter crops can reduce soil erosion.

Figure 3 *The causes and impact of soil erosion.*

clouds the water and reduces the channel's ability to transport water downstream, increasing the severity of flooding. There is less organic material to store water in the soil so the rate of leaching of minerals increases, further reducing the mineral content of the soil. The soil quickly loses its fertility and within four or five years crop yields decline and the clearing has to be abandoned. The community then shifts to another part of the forest and can only revisit the site when nutrients have built up again. This happens through re-growth of the forest, in which new soil formation replaces what has been lost, but the process often takes 25 years or more.

The impact of forestry and mining

The 5000 miles of the Trans Amazonia Highway built in the 1960s has brought development at a cost to the Amazon Basin. As much as 80 per cent of deforestation has occurred within 30 kilometres of official roads. Clear-felling, or the cutting of all trees, has supplied many jobs and income for the people of Brazil. Newcomers have introduced ranching and soya production, which now dominates much of the landscape. Known as **cash cultivation** or **commercial farming**, the produce is grown and sold for profit, often to overseas markets. The indigenous tribes, like the Matis and Kyopo of Amazonia, are therefore squeezed into smaller areas of forest or reservations, where patches of land are cleared before they have fully recovered.

Land in all tropical regions is being cleared for mining operations and oil exploration. Open pit mining for precious metals requires trees to be felled and vegetation to be removed. The forest soils are then pumped with water across sieves. The heavy metals remain on the sieves as the water and sediment are washed away. Mercury added to amalgamate gold then enters the water and pollutes the once pristine river systems. Silt added to the rivers from the mining operations causes the death of fish and other species as light cannot penetrate through the water. The traditional way of life in the tropics is becoming *unsustainable* due to the commercial pressures on the natural resources and the exodus of the young tribal members away from the forest. Some subsistence farmers join other migrants in nearby large urban areas or get work exploiting the forest. Many scientists believe that the Amazon is nearing a tipping point; if much more forest is destroyed the effects on the ecosystem will be irreversible.

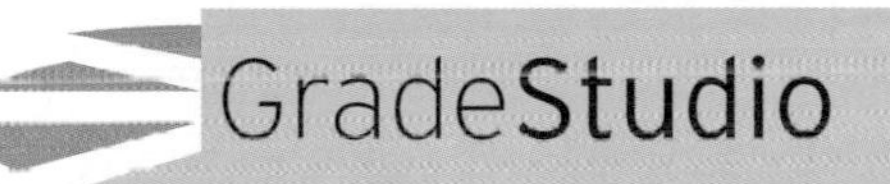

1 Use **Figure 1**.
   a Describe the global pattern of tropical rainforests around the world. (3 marks)
   b Compare the climate found in a tropical rainforest with that of a tropical desert. (4 marks)
   c Why do tropical rainforests only support low population densities? (4 marks)

2 State the difference between commercial farming and subsistence farming. (2 marks)

3 a Assess the effects of forestry and mining on the Indian tribes and their economy. (6 marks)
   b Explain the causes and effects of soil erosion in the tropics. (6 marks)

Exam tip

Make sure that you know definitions for the key terms in each topic.

Change in sub-tropical farming

Is irrigation a blessing or a curse? Can appropriate technology improve the lives of rural people in the sub-tropics?

The main form of agriculture in the sub-tropical region, where dry seasons result in water scarcity, is nomadism. Nomads have to move in search of grazing for their herds of goats, cattle or sheep. The vast areas wandered by nomads have reduced in size, and some have adopted a more sedentary lifestyle and now live and work in rural villages and towns, often as a result of irrigation schemes. Other nomads have settled in areas where there is access to rivers flowing through the desert such as the Nile in Egypt.

The impact of irrigation

Irrigation is the artificial watering of the land and is used by arable farmers to top up water levels in a dry climate. Water can be transferred to the field by simple *gravity-fed canals* dug away from rivers with earthen banks and sluice gates that open and allow river water and nutrients to flood fields and saturate the soil when the river rises. This also replenishes groundwater supplies. *Wells* or holes dug down to the water table provide water for individual farms.

- Assured water supply throughout Egypt
- Desert reclaimed for farmland
- Cultivated area doubled from 4% to 8%
- 2 or 3 crops per year instead of 1
- No longer any risk of summer floods
- Electricity supply for the whole country

Figure 1 *Advantages of the Aswan High Dam.*

In Egypt, water management of the Nile allows rice, sugar cane, barley, wheat, beans and cotton to be grown on the fertile flood plains. These crops have fed the growing population and have been exported for profit. The large-scale Aswan High Dam engineering project has controlled flooding on the Nile and allowed water for agriculture to be stored in Lake Nasser. Other advantages can be seen in **Figure 1**. Advanced sprinkler systems and miles of pipes have been constructed to transport water directly to crops and have reduced the amount of water evaporated. Crops can now be grown all year round and food security for the country has been improved.

Human mismanagement of land in the Indus valley in Pakistan on the edge of the Thar desert has converted marginal land used for agriculture into desert through the process of *desertification*. The region is densely populated, which has increased the pressure on land under cultivation. Soil erosion is a natural process, which has been enhanced by overgrazing and overcultivation of crops such as cotton, tobacco and sugar cane. The loss of plant cover has exposed the soil to erosion (see pages 186–7) and water mismanagement has caused **salinisation** and **waterlogging**.

Salinisation occurs when high temperatures draw water and salts up through the soil via capillary action, causing the soil to become increasingly saline. Salts form a hard crust on the surface and this slows root growth and the infiltration of water into the soil; it also causes waterlogging above the salty layer. The water table will also rise without proper drainage, drawing more salts to the surface (**Figure 2**). In the Nile delta, over-extraction of well water has lowered the water table, causing saline sea water to seep into the groundwater stores.

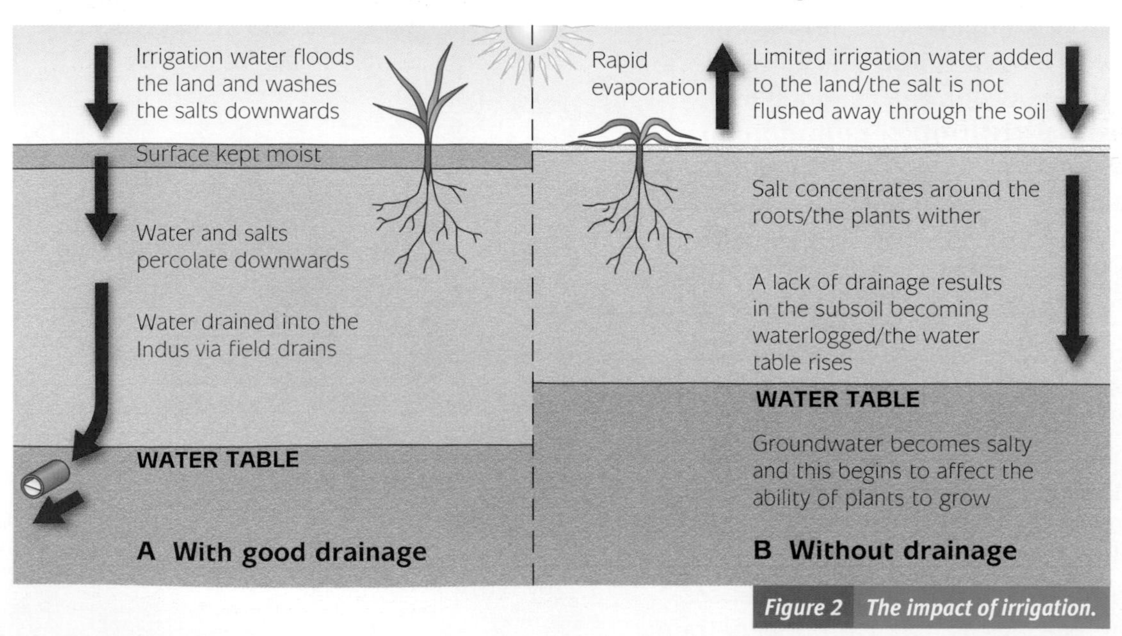

Irrigation water floods the land and washes the salts downwards
Surface kept moist
Water and salts percolate downwards
Water drained into the Indus via field drains
WATER TABLE
A With good drainage
Rapid evaporation
Limited irrigation water added to the land/the salt is not flushed away through the soil
Salt concentrates around the roots/the plants wither
A lack of drainage results in the subsoil becoming waterlogged/the water table rises
WATER TABLE
Groundwater becomes salty and this begins to affect the ability of plants to grow
B Without drainage

Figure 2 *The impact of irrigation.*

In Central Asia, poor management of the Aral Sea has been attributed to over-extraction of irrigation water for cotton crops. A total of 40 000 jobs in the fishing industry have disappeared, along with 90 per cent of the water. The plight of the Aral Sea is frequently described as an environmental catastrophe and has caused economic hardship for the local people. Environmental damage has reduced the amount of land that can be farmed.

Figure 3 *The result of salinisation in the Indus valley in Pakistan.*

Appropriate technology developments

Behind the concept of *sustainability* for the future is the idea that food production today should not be damaging the environment or reducing future generations' ability to provide for themselves. Many believe that **appropriate technology** (a level of technology in terms of size and complexity that makes it suitable for use by local people) is the key to sustainable food production. Projects should be small-scale, affordable, suited to the local environment and improve their lives for today and generations to come. Some successful projects include:

a) *Rainwater conservation* – Trapping water behind stones and planting trees on bare slopes in India. The improvements can be seen in **Figure 4**. Local farmers and village populations need only relatively small amounts of water, which is why this small-scale project is an appropriate solution.

Thunthi Kankasiya village	Before 1991	By 2000
Perennial drinking-water wells	0	23
River dams	0	1
Months of water availability	4	12
Land under cultivation (hectares)	85	135
Number of crops per year	0–1	2–3
Agricultural production (quintal/hectare)	900	4000
Migration rate (15–40 year olds)	78%	5%
Average period of migration (months)	10	2
Income per household (rupees per year)	8600	35 600

Figure 4 *Micro-watershed development in a village in Gujarat, India.*

b) *Soil conservation* – Terracing on slopes holds soil in place on the flatter land on the tops of the terraces so that steep slopes can be used for farming and erosion is reduced.

c) *Contour ploughing* – Farming around a slope reduces the movement of water and soil downslope. If trees or bushes are also added between crops, they give shelter after harvest.

FURTHER RESEARCH

Other examples of appropriate technology by the British charity **The Intermediate Technology Development Group** (ITDG) can be found on the weblink www.contentextra.com/aqagcsegeog.

Rural–urban migration

As countries develop, people move in greater numbers from rural to urban areas in search of better opportunities. The drive to intensify farming by irrigation and mechanisation has impacted on the environment. More prime arable land is being turned over for cash cultivation and urbanisation; thus marginal land such as forests and deserts are being converted into pasture for cattle or arable land for food production. Food shortages and riots have been seen in countries including Haiti, India and Senegal. Global dialogue between governments, encouraged by the International Food and Agricultural Trade Policy Council (IPC), must aim for a sustainable solution to feed the world's population of 6.9 billion and protect our rural areas for future generations.

ACTIVITIES

1 Describe the different methods of irrigation. Which methods are sustainable?

2 Highlight the causes of salinisation.

3 Explain how salinisation can lead to farmland having to be abandoned.

4 What is meant by 'appropriate technology'?

5 Using **Figure 4**, identify the significant changes which occurred in Thunthi Kankasiya between 1991 and 2000.

6 Research two forms of appropriate technology. For each one, describe and state where it would be used. Explain the impact it could have on the environment.

GradeStudio

Practice GCSE Question

See a Foundation Tier Practice GCSE Question on the weblink www.contentextra.com/aqagcsegeog.

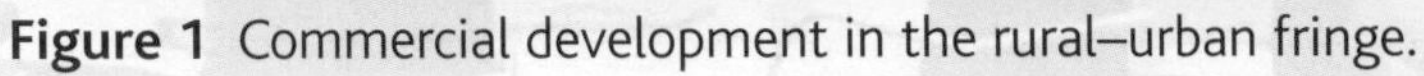

Figure 1 Commercial development in the rural–urban fringe.

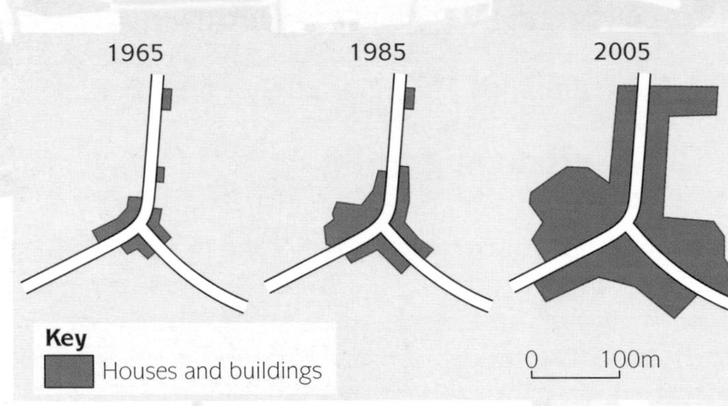

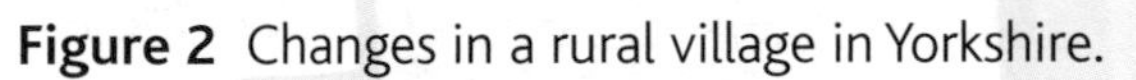

Figure 2 Changes in a rural village in Yorkshire.

Weekenders? Ooh, they come and tell us how to run our village and then disappear in the morning.

Londoners? Don't know who they are, they're too posh for us, too busy drinking wine they are.

It is not just about them that have come, it's about jobs and services. Even if local people could afford the homes, how would they earn a living?

Figure 3 Comments from long-time residents in a Wiltshire village, which has grown in recent years.

3 (a) Study **Figure 1**, which shows commercial development in the rural–urban fringe.

(i) State where the rural–urban fringe is located. **(1 mark)**

(ii) Name the two land uses shown in **Figure 1**. **(2 marks)**

(iii) Explain why rural–urban fringe locations in the UK are attractive places for this and other types of commercial development. **(4 marks)**

(iv) State two objections often made by local people when plans are put forward to increase the scale of developments in the rural–urban fringe, like the one shown in **Figure 1**. **(2 marks)**

(b) Study **Figure 2**, which shows how a village in rural England changed between 1965 and 2005.

(i) Describe how the village changed between 1965 and 2005. **(3 marks)**

(ii) What is the geographical name for villages that have changed like this? **(1 mark)**

(c) Read the comments in **Figure 3** from residents in a Wiltshire village.

(i) From these comments, identify and explain one social and one economic problem associated with growing villages in rural areas of England. **(4 marks)**

(ii) Suggest two reasons why, in the same area, one village grows while others do not. **(2 marks)**

(d) Choose one rural area in the UK that has been affected by rural depopulation and economic decline. Describe the reasons for this and the consequences. **(6 marks)**

Total: 25 marks

Exam tip

The question for this topic will be Question 3 in Paper 2.

Improve your GCSE answers

Describing and explaining change

The main theme of many exam questions will be **change** in the UK settlements. Be prepared with your answers.

Figure 1 Office on a business park in the rural–urban fringe. What are the attractions of this location compared with one in the middle of the city?

Change in rural–urban fringes

Know how they are changing

1. Spreading urbanisation – an increasing percentage of the area with urban instead of rural land uses. As urban areas sprawl, they bring urban land uses (housing, shops, offices and roads) with them (**Figure 1**).
2. Growing suburbanisation of villages – an increase in the number of houses and size of the built-up area. This brings with it changes in function from rural (mainly farming) to dormitory (residential for commuters) (**Figure 2**).

Figure 2 New housing estate increasing the size of the built-up area and changing the character of the area.

Know why they are changing

1. Greenfield sites are cheaper and easier to develop than brownfield sites. The land costs less to buy and new sites are cheaper to clear than urban derelict land.
2. Attractive for development because the area has so many possible uses. These include retail, business offices, leisure and transport.

Figure 3 Bucknell, Shropshire. The closed village shop has been converted into a house, with only a small part left to function as a post office. How long will it survive?

Change in rural areas further away from cities

Know how they are changing

1. Declining villages – shown by depopulation and buildings in poor repair or derelict.
2. Expanding villages – dormitory settlements for long-distance commuters and growth of second homes (for weekend and holiday use).
3. Loss of village services (whether declining or expanding) – including pubs, post offices, shops and primary schools.

Know why they are changing

1. Most rural settlements have only one function, often farming, compared with many in urban areas.
2. Decline in farm profits and increased mechanisation mean less rural employment.
3. Village location is a strong factor for decline or growth – does its location allow a change in function to dormitory or tourist?
4. Services go in both cases because of the private car and the great mobility it gives people for travelling and using services in urban areas.

Figure 4 Bucknell, Shropshire. New housing on farmland between different parts of the village, called infilling.

ExamCafé

REVISION

Key terms from the specification

Agri-business – type of farming that is run as a big business (no longer a way of life)
Appropriate technology – level in terms of size and complexity that makes it suitable for local people to use
Commercial farming – type of agriculture based on growing crops or rearing livestock for sale
Commuter – person who travels to work in another place every day by car or public transport
Organic farming – type of agriculture that does not use chemicals and artificial growth stimulants; farming in a natural and sustainable way
Rural depopulation – decline in numbers living in country areas, often due to out-migration
Second home – house (often in rural areas) that is not the owner's main place of residence
Soil erosion – loss of fertile topsoil by action of wind and water
Suburbanised village – small settlement in the countryside that has grown with new housing and now is less like the old rural settlement it used to be
Sustainable living – people working with the environment for a long future for their economic activities

Checklist

	Yes	If no – refer to
Do you know what is meant by rural–urban fringe?		pages 174–5
Can you state the main characteristics of villages that are growing into suburbanised villages?		pages 176–7
Can you state the characteristics of villages in decline?		page 178
Can you describe three ways of making rural living sustainable for the future?		pages 180–1
Name three ways in which intensive farming in the UK damages the environment.		page 184
Can you name and explain EU and UK government policies to reduce environmental damage from intensive farming?		pages 184–5
State the causes and impacts of soil erosion in tropical and sub-tropical areas.		pages 186–7
Do you know what is meant by appropriate technology and how it can help poor people in tropical areas?		page 189

Case study summaries

Rural area in the UK	Commercial farming area in the UK
Reasons for rural depopulation	Favourable physical and human factors for farming
Consequences of this	Development of agri-businesses
Characteristics of declining villages	Farming impacts on the environment
Growth of second homes	Changes due to market demands and government policy

Chapter 11

The Development Gap

Bhutan is the only country in the world that has Gross National Happiness as its aim for development. Instead of trying to become rich, Bhutan's rulers want the people to be happy, to look after the environment, and protect their heritage and culture.

QUESTIONS

- **What do we mean by development, and how can we measure it? Is wealth or quality of life more important?**
- **What are the physical and human blocks to development, which help make the world such an unequal place?**
- **How can people and countries make progress in development, and work together to close the development gap?**

An unequal world

What do we mean by developed and developing countries? Where are they? How can we measure development?

In spite of the progress humans have made, our world is still a very unequal one. The world has become richer in the past 30 years, but the gap between the richest third of countries and the poorest two-thirds has actually grown larger in the same period.

The gap shows clearly on **Figure 1**, where the area of each country is drawn according to the size of its **Gross National Product (GNP)**. GNP is the value of everything a country produces, measured in US dollars. North America, Europe and Japan have 75 per cent of the world's GNP. China's share is growing fast: in 2006 its GNP was the fourth biggest in the world, standing at a total of US$2 600 billion.

However a country's GNP figure on its own is not that accurate an indicator of its people's wealth. For example, in 2006 the UK's GNP was close to China's, but with a population of over one billion, on average China's people are actually much worse-off.

A better way is to divide each country's income by its population. **Gross National Income per capita** (per person) is the average income of people in a country measured in US dollars. GNI per capita gives us a better idea of how well-off people are, but **Figure 2** still shows great inequalities between different parts of the world.

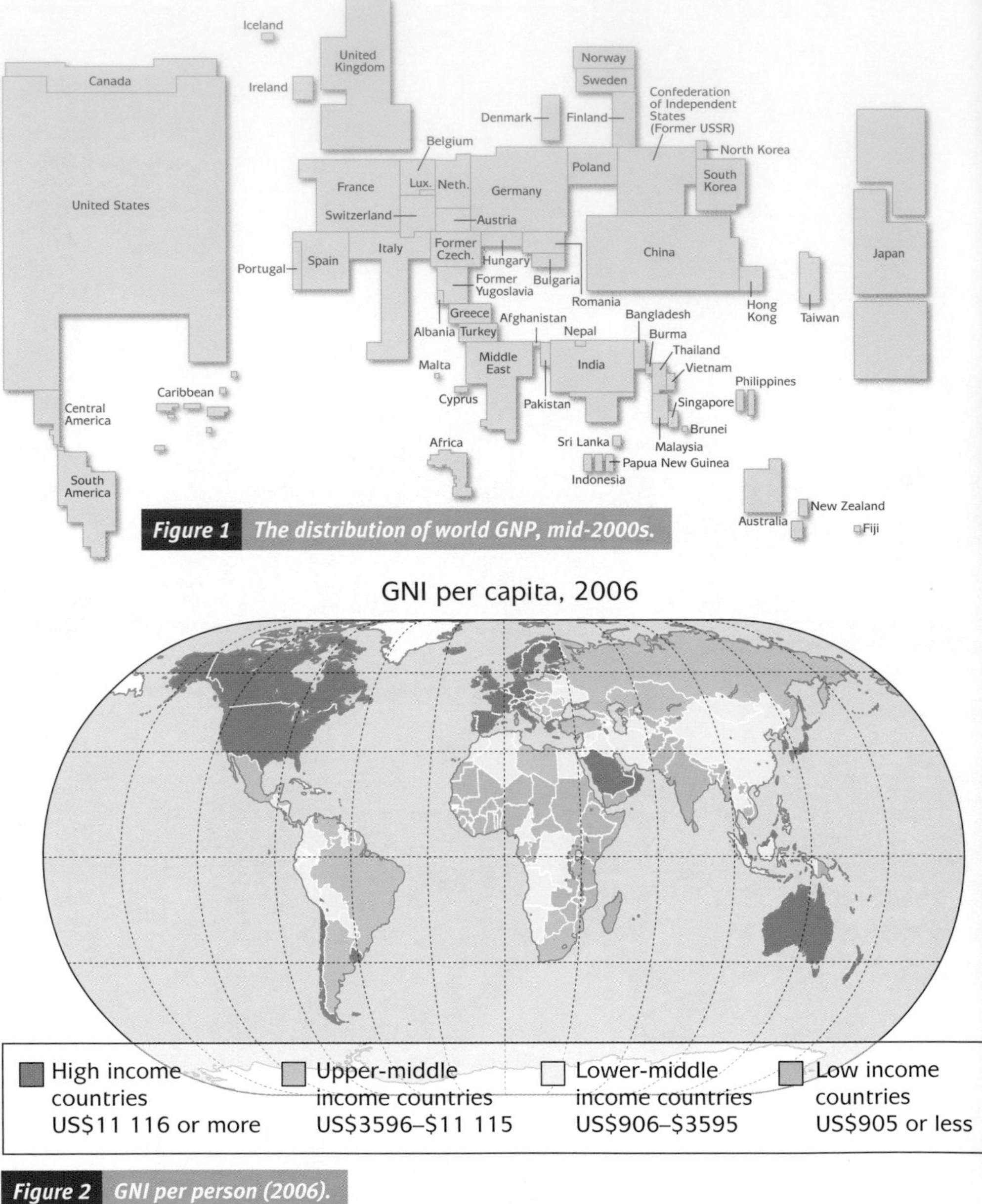

Figure 1 *The distribution of world GNP, mid-2000s.*

Figure 2 *GNI per person (2006).*

How can we classify different parts of the world?

In the 1970s geographers used to divide the world into the wealthy countries of the North, and poorer countries of the South. Even then, some countries did not fit the pattern. Today, some countries in the South have developed so rapidly that many people's standard of living is more like that of Europeans, rather than the poorest in the world. South Korea is a good example, compared with Bangladesh (**Figure 3**). Today the **World Bank** classifies countries like this:

- High-income countries with average incomes of US$11 116 or more (in 2006), are classified as **developed**.
- Low-income countries with average incomes of US$905 or less, and middle-income countries in between (US$906–US$11 115). Together these are classified as **developing**.

Figure 2 shows the developed countries, where most people are fairly wealthy, as well as the poorest, least developed parts of the world. In between are many middle-income countries. Many of them, like China and Brazil, are developing very fast.

2006	GNP or GNI $US billion	GNI per capita $US	Percentage employed in		
			farming	industry	services
USA	13 387	44 710	2	21	78
Japan	4935	39 630	4	28	66
UK	2456	40 560	1	22	76
South Korea	857	17 690	8	27	65
Brazil	893	4710	21	21	58
China	2621	2000	43	25	32
Kenya	21	580	19	20	62
Bangladesh	71	450	52	14	35

Figure 3 Developed or developing? Economic development in eight countries.

Wealthiest country: Luxembourg, US$70 330.
Poorest country: Burundi US$100 (2006 GNI per capita)

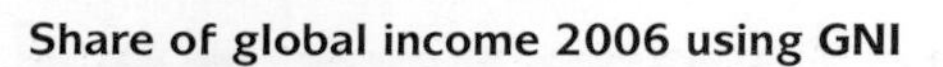

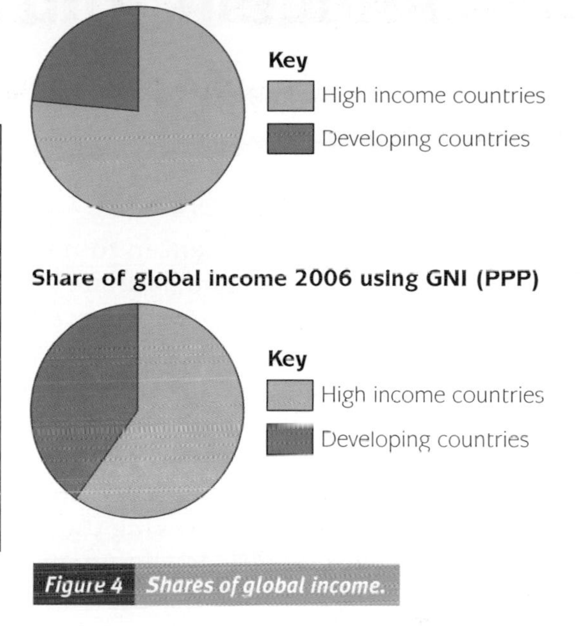

Figure 4 Shares of global income.

Problems with using wealth to measure development

Developed countries are often those with advanced industries and services that create jobs and wealth, compared with the poorest countries, which often have a higher percentage of people working in agriculture (**Figure 3**). People in wealthy countries usually have quite a good quality of life, because the wealth can be used for things like education, healthcare and other services. Therefore, looking at a country's wealth can be a good way to measure development, but it does have some drawbacks:

- It can be easier to collect data in wealthy countries so the figures may be more accurate than for poor ones.
- The data only measure products that are bought and sold. Food grown by farmers to feed their families is an important part of production in many developing countries, but this will not show up in the figures. Developing countries also commonly have large numbers of people working in the cash economy: their work is not recorded (and the government finds it difficult to collect taxes from them).
- The figures are only an average for each country, and do not tell us about inequality there. A wealthy Indian or Chinese may be far better-off than a poor American.

Finally, US$1 goes much further in some countries than others. To give a fairer picture, we use a measure called Purchasing Power Parity to adjust national income. Using this measure, developing countries have a bigger share of the world's wealth (**Figure 4**). It is also a reminder that there is a difference between income or wealth and standard of living. For example, someone living on $10 a day would be living in poverty in New York, but would be quite comfortably-off in some parts of the world.

ACTIVITIES

1. (a) Classify the countries listed in **Figure 3** into high-, medium- and low-income countries, using the World Bank's system.
   (b) Then classify the countries into developed and developing.
   (c) Compare the pattern of employment in the low-income and high-income countries. How does this help explain differences in wealth?
2. (a) Study **Figures 1** and **2.** Make a list of the continents in order of wealth, from high to low.
   (b) Which continent/s were difficult to classify, and why?
   (c) Explain any differences you can see between the maps of GNP and GNI per capita.
3. What is the difference between 'income' and 'standard of living'?
4. Write a short summary of what you have learned. Include sections on:
   - how the world's wealth is distributed, and the best way to measure this
   - differences between developing countries
   - some of the advantages and disadvantages of using wealth to measure development.

Development and quality of life

What do we mean by quality of life, and how can it be measured? How can people improve their own quality of life?

Development is about more than dollars

If you ask people what they want in future, many will say they want to earn enough money to live comfortably. But, often, they want more out of life (**Figure 1**). It is the same with countries. Although the two are not always linked, as countries become wealthier they usually become more developed, and people also become healthier, better educated and live longer. As well as making people better off, development is also about improving the quality of people's lives. To measure development, just looking at a country's income would not give us the whole picture.

Figure 1 Some ingredients of a good quality of life.

The Human Development Index

The United Nations **Human Development Index (HDI)** is one measure of people's quality of life. It gives each country a score based on its people's average life expectancy, education and standard of living. It is a broader measure of development than just using GNI per capita because it includes social and economic measures. Like maps of national income, **Figure 2** shows great differences in quality of life in different parts of the world, although the pattern is not the same as those shown in the maps on page 194. The UN classifies countries like this:

- High human development (HDI 0.80 and over)
- Medium human development (HDI 0.500–0.799)
- Low human development (under HDI 0.500).

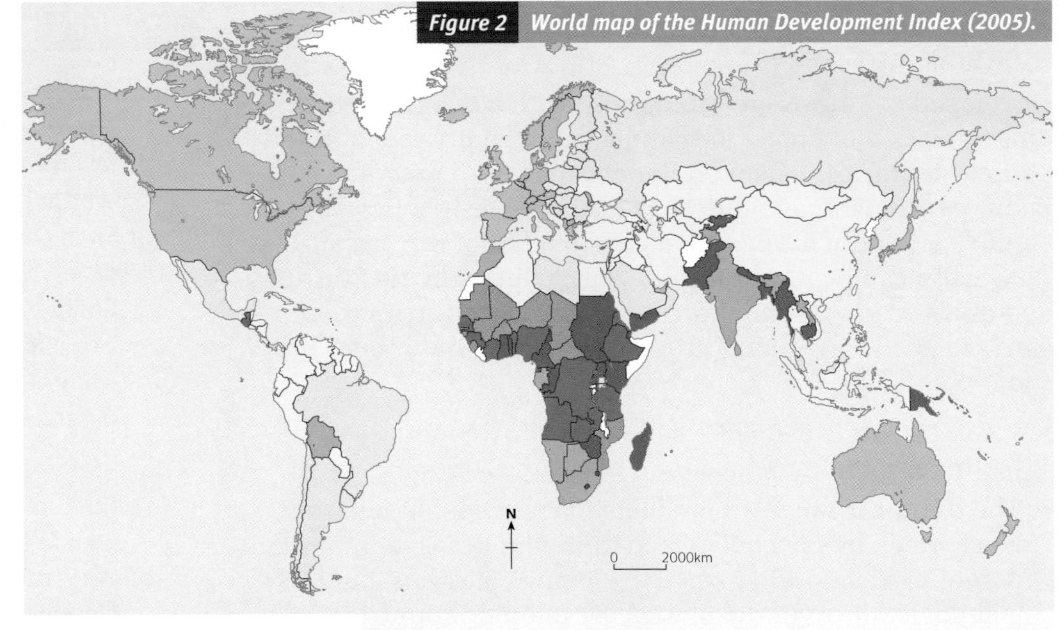

Figure 2 World map of the Human Development Index (2005).

Key

High human development	Medium human development	Low human development
Over 0.90	0.7–0.799	0.4–0.499
0.80–0.89	0.6–0.699	Under 0.40
	0.50–0.599	N/A

It is easy to think that lack of development is a real problem for the world. It certainly is for those one billion people living in abject poverty on less than US$1 a day. But development has been a real success story too, lifting billions out of poverty and helping them live longer, healthier lives. HDI is a useful way to show this progress in development over the years (**Figure 3**).

HDI	1960	1980	2005
Developed countries	0.799	0.889	0.916
All developing countries	0.260	0.428	0.691
Least developed countries	0.165	0.251	0.488
World	0.392	0.519	0.743

Figure 3 Changes in quality of life.

	HDI	Social measures of development					
		Birth rate per 1000	Death rate per 1000	Infant mortality per 1000 births	Life expectancy (years)	People per doctor	Adult literacy rate (%)
USA	0.951	14	8	6	78	435	99
Japan	0.953	9	9	3	82	476	99
UK	0.946	12	10	5	79	455	99
South Korea	0.921	9	5	5	78	625	99
Brazil	0.800	19	6	19	72	833	89
China	0.777	12	7	20	72	667	91
Kenya	0.521	39	12	79	52	10 000	74
Bangladesh	0.547	25	8	52	64	3333	47

Figure 4 *HDI and social measures of development in eight countries (2005–6).*

Social measures of development

We looked at wealth and economic measures of development on pages 194–5. **Figure 4** shows some social measures of development, in relation to population, health and education. These are often linked together. For example, lack of clean water and medical care causes illness; as a result, people die younger (life expectancy). Young children (infant mortality) and older people are most vulnerable and so the death rate is higher. Many development programmes focus on education because it helps improve people's health, their ability to earn money and to take part in decision-making. Educating women has particular benefits for their children. Educated women often choose to have fewer children and are able to care better for those they do have.

EXAMPLE: The Grameen Bank

The most effective development usually starts at home, when people improve their own lives. An example is the Grameen Bank, started by Nobel-prizewinner Muhammad Yunus in Bangladesh. A total of 41 per cent of Bangladeshis live on less than US$1 a day. The Grameen Bank specialises in lending small sums of money to the poorest people, whom other banks would turn away. The bank has now lent US$1 billion to over 7 million people, nearly all to women in the country's 78 000 villages.

The loans give women the capital to start their own small businesses, such as selling crafts or snacks, or renting out mobile phones. Once they are in profit, they can repay the loan, or take out a new one. The project is successful because it puts women in charge – they can make their own decisions, rather than depending on aid. Women are very reliable too; 99 per cent of loans are repaid. The income they make goes directly to their families and helps lift them out of poverty and improve their quality of life.

Grameen money reaches 16 million families – 80 per cent of the country's poor. As well as tackling poverty, it has helped reduce **fertility** and family size. It has been copied in **micro-credit** schemes around the world – including the USA.

GradeStudio

Look back at pages 194–7.

1 a Write definitions for standard of living and for quality of life. (2 marks)
  b Identify one way to measure (i) standard of living (ii) quality of life. (2 marks)

2 a Give an example of a social measure of development. (1 mark)
  b State one advantage and one disadvantage of using GNI per capita to measure development. (2 marks)

3

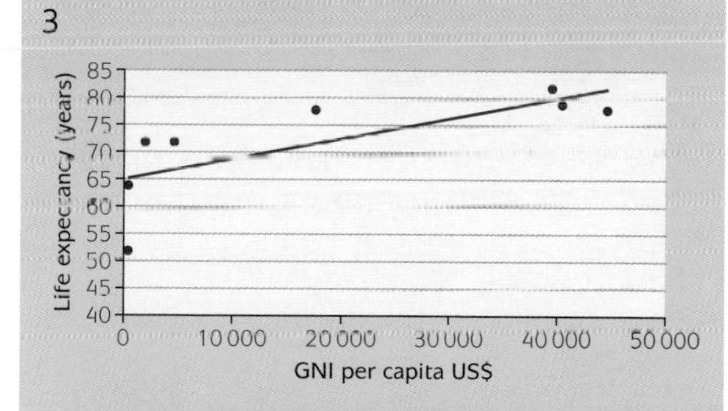

The scattergraph above shows the correlation between income and life expectancy.
  a Describe what happens to life expectancy when a country's GNI per capita increases. (1 mark)
  b Explain why life expectancy and income are linked like this. (1 mark)

4 From an example you have studied, describe a project that supports people in improving the quality of their lives. Include details of how people benefit from the project. (6 marks)

Obstacles to development

Why have some countries made good progress in development, but not others?

Many parts of the world have made progress in development (page 197). Others face great obstacles and only make slow progress. Some, especially countries like Zimbabwe in sub-Saharan Africa, have even gone backwards since 1990. The reasons can include natural hazards, political conflict, unfair trade, economic collapse and the impact of HIV/Aids (sometimes all at once).

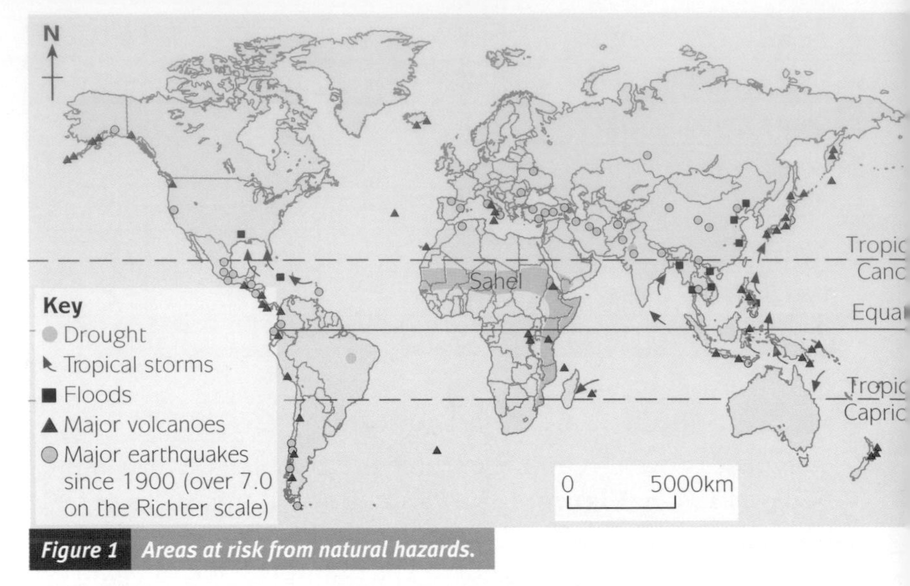

Figure 1 *Areas at risk from natural hazards.*

Natural hazards, such as earthquakes, hurricanes, floods and droughts, can slow or reverse development. They affect many parts of the world but have a particular impact on poor countries and people, because they are more vulnerable and lack the money to recover quickly (**Figure 1**). The Boxing Day tsunami of 2005 (page 21 and **Figure 2**) set back development in the worst affected countries for years. Climate change and growing populations will make many more people vulnerable to natural hazards in future. Reducing the risk of natural hazards, and helping people adapt to them, is a priority for development in many countries.

Drought affects food and water supplies and vegetation. It can be an inconvenience in wealthy countries like the UK (page 51), but in the worst affected parts of the world like the Sahel, drought can be a matter of life or death.

Floods may cause severe property damage in developed countries but most lives are lost in the densely populated lowlands of developing countries (pages 90–91). The poorest often live in the places most at risk.

Hurricanes and cyclones affect tropical coasts. Island states, for example in the Caribbean, and delta regions like Bangladesh are most vulnerable. Like other hazards, in developed countries they often cause fewer deaths but more property damage (pages 56–7).

Earthquakes and volcanoes have less of an impact in developed countries because their buildings are built to withstand shocks, and they have better early warning systems (Chapter 1). For example, in December 2003 two earthquakes, both 6.5 on the Richter scale, killed three people in California but 30 000 in Iran.

Political conflicts can also have devastating impacts on development. Like natural hazards, they often result in:

- widespread deaths and injuries
- large numbers of displaced people and refugees
- destruction of **infrastructure** (roads, power supplies, schools, etc), factories and farms.

Many recent political conflicts are located in developing countries, and are often caused by disputes over control of land and resources. Some are hangovers from the colonial period, when Europeans drew up boundaries that divided peoples, and threw local people off their land.

Political mismanagement and corruption can slow or reverse development in any part of the world. An example is Zimbabwe, once one of the most developed African countries. Since the 1990s it has suffered some of the worst setbacks in welfare and human rights, as a result of HIV/Aids and poor government. It is an extreme example of a government that has failed its people.

Zimbabwe	1990	2005–6
HDI	0.654	0.513
Life expectancy (years)	61	43
Infant mortality rate (per 1000)	52	68

EXAMPLE: Adapting to climate change in Bangladesh

Most of Bangladesh is flat and less than 10 metres above sea level. In a normal year about 25 per cent of the country is flooded in the **monsoon season** (pages 90–91). People there are used to living with normal floods, but global warming is a new hazard. Summer rainfall is increasing and sea levels are rising, causing more flooding and erosion in the delta and islands. People here risk losing their land, animals, homes and lives. By 2100, sea level will rise by up to 90 centimetres, and cyclones will become stronger and more frequent. Already, hundreds of thousands have fled the countryside to live in the slums of Dhaka.

Oxfam's South Asia River Basin Programme aims to reduce deaths caused by flooding and the destruction of people's homes and livelihoods, and to give poor people security to plan for the future. In Bangladesh, Oxfam works with communities through eleven local development partners to:

- build banks of earth 1.5–2 metres high to raise homes and villages above flood level
- build emergency flood shelters, which double up as community halls and schools, with solar panels for electricity
- build 200 toilets and 55 tube wells a year
- provide health education and training to improve farming techniques.

Figure 3 **Flood platforms in Bangladesh.**

Kodvanu, from Fulhara village, said: 'Because these homes were raised, they weren't damaged by the floods. We can grow vegetables on the roof, and fruit trees such as mango that can be eaten during the floods'.

Indian Ocean tsunami	224 000 dead, 2.4 million displaced
Hurricane Katrina (USA)	1300 dead, 1 million displaced
Pakistan/Kashmir earthquake	74 500 dead, 4 million displaced or bereaved
Darfur, Sudan	5000 die each month, 2 million displaced by conflict
Hurricanes Stan and Wilma (Caribbean, Central America, USA)	2000 dead, 2 million displaced
Colombia	3000–4000 die each year, 2 million displaced by conflict
Niger famine	73 000 deaths, 3.5 million hungry
Congo	31 000 dead in one month, 2.8 million displaced by conflict
Southern Africa famine	12 million hungry
Northern Uganda	4000 die each month, 2 million displaced by conflict

Figure 2 *2005: a year of disasters.*

FURTHER RESEARCH

Research two recent natural disasters, one in a developed and one in a developing country. Compare their impact on people, property and infrastructure, and how each country was able to respond.

ACTIVITIES

1 Classify the events of 2005 (**Figure 2**) into natural and political disasters. Then locate them on a world outline map, using graphics to show the impact of each. Were the worst impacts in the developed or developing worlds?

2 (a) Draw pie or divided bar graphs to show this data:
  - Distribution of natural hazards: Europe 14 per cent, Africa 21 per cent, Americas 20 per cent, Asia 42 per cent and Oceania 3 per cent.
  - Distribution of people killed: 2 per cent high human development, 32 per cent medium and 66 per cent low human development

  (b) Giving examples, explain why natural hazards have a different impact on developed and developing countries.

3 Put together a case study of the development project in Bangladesh.
  - Draw a sketch map to show the location of Bangladesh and its delta (pages 90–1).
  - Write notes using these headings: What the problem is; How the project aims to help; Who is involved; How it makes a practical difference.
  - Extend your work by investigating the links to projects in Bangladesh on the weblink www.contentextra.com/aqagcsegeog.

Water: the world's most precious resource

Why are there differences in people's access to water? Why is lack of clean water an obstacle to development?

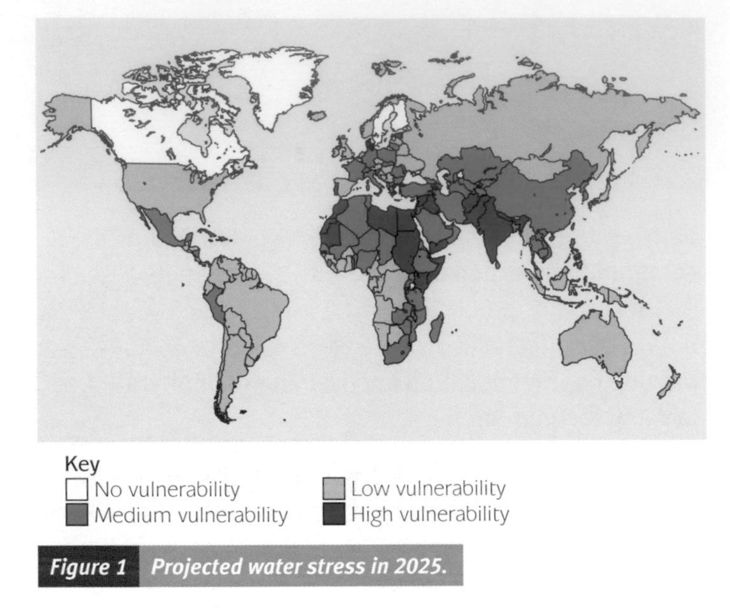

Key
No vulnerability
Low vulnerability
Medium vulnerability
High vulnerability

Figure 1 *Projected water stress in 2025.*

A vital (but uneven) resource

Water is a vital natural resource for all life on earth. Only 3 per cent of the Earth's water is fresh, and most of that is locked up in ice sheets. But that still leaves 10 million cubic kilometres of usable water. It is constantly circulating as precipitation refills rivers, lakes and underground water stores. This should be more than enough for the world's 6.5 billion people, but water supplies are unevenly distributed between countries and within them:

- Most water-rich countries are in the tropics and temperate regions where rainfall is high.
- Most water-poor countries are in the desert regions of the Middle East and North Africa. Many of these countries rely on underground **aquifers** or dams, such as the Aswan High Dam on the River Nile in Egypt (page 228).

	USA	China	Kenya	Bangladesh
Agriculture	41	68	64	96
Industry	46	26	6	1
Domestic	13	7	30	3

Figure 2 *How water is used around the world (percentages).*

The UN estimates that one in three of the world's population lives in a country already suffering from **water stress**, where supplies are limited compared with the population. The world's thirst for water will be a pressing resource issue in the twenty-first century. By 2025 global water use is expected to rise by 40 per cent, when the UN expects two-thirds of the world's population will live in countries with moderate to high water stress (**Figure 1**). Water supplies are under increasing pressure for a number of reasons:

- Most water (90 per cent) is used for agriculture, especially in developing countries (**Figure 2**).
- As populations grow, **fresh water availability** is reduced.
- Global water consumption is rising at twice the rate of population growth as people's standard of living rises.

Much of the new demand will come from developing countries, due to population growth, rising demand for food, and expanding industries and cities, especially in China and India. This situation will be worsened by changing weather patterns brought by climate change (pages 54–5).

Unequal access to clean water

Wealthy countries can invest huge amounts of money to ensure a plentiful water supply and reliable sewage treatment, so that most people take 100 per cent access to clean drinking water and **sanitation** for granted.

In developing countries, 2.4 billion more people have had access to an **improved water supply** in the last twenty years, and 600 million to sanitation. But this leaves about 1.1 billion people without access to improved water (**Figure 4**), and 2.5 billion without safe toilets. Around 20 per cent of the population of developing countries are still forced to get water from potentially dangerous sources like rivers and waterholes (**Figure 5**). In rural areas, access to clean water and sanitation is almost always worse than in cities.

- **135 litres a day:** average UK water use (cost: £1.22 per 1000 litres)
- **10 litres a day:** average developing world water use
- **6 kilometres:** average distance women in developing countries walk to collect water (weight 20 kilograms)
- **£360:** cost of 1000 litres from a water vendor in Kibera slum, Nairobi

Figure 3 *Water in figures.*

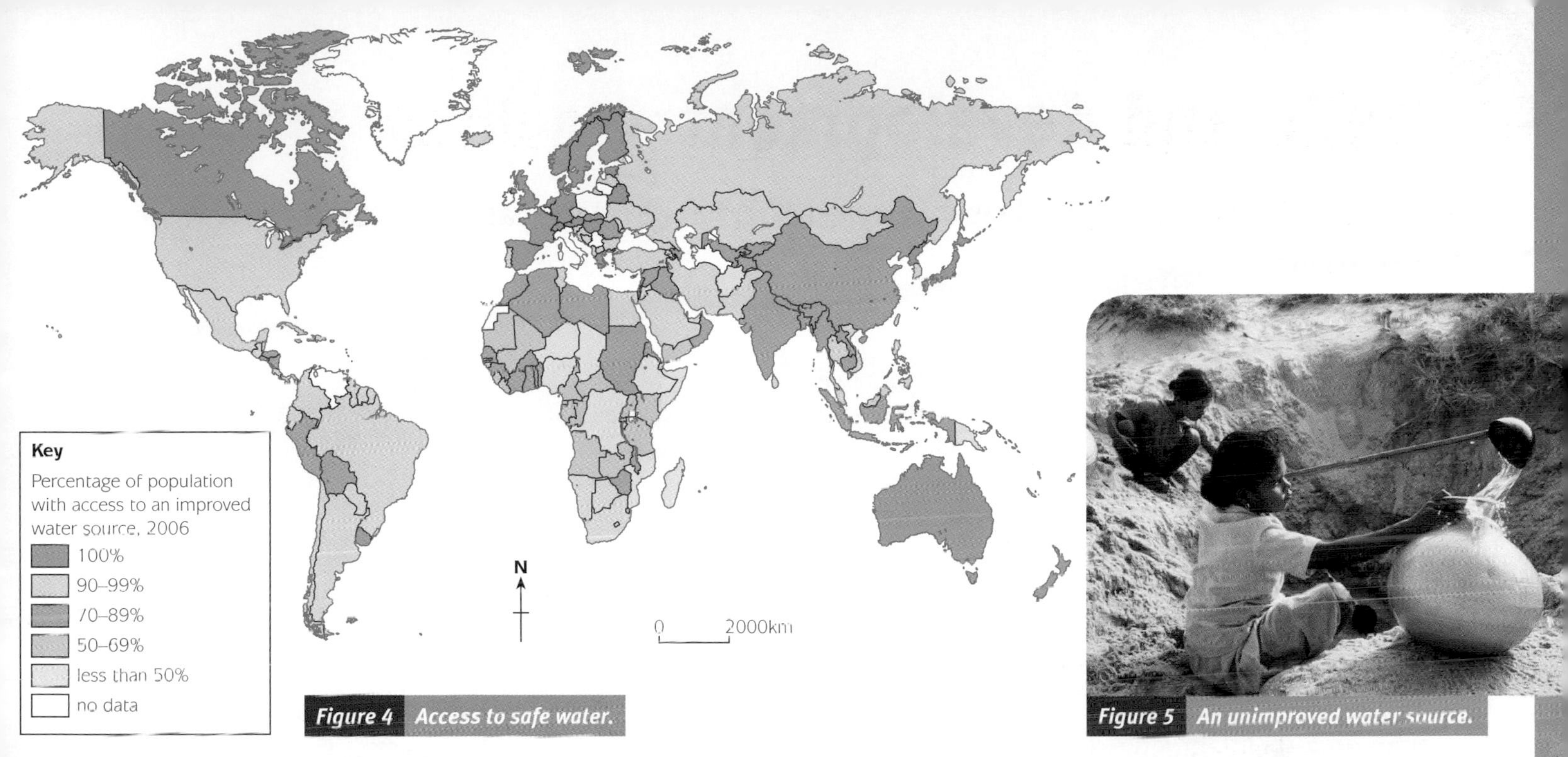

Figure 4 Access to safe water.

Figure 5 An unimproved water source.

Worldwide, 94 per cent of city people have access to clean water, compared with 72 per cent in rural areas. Even within urban areas, there are often great differences in access for wealthy people and the urban poor living in shantytowns.

Poor access to clean water is still a major cause of poverty and an obstacle to development, especially for the poorest. One reason is that (usually) women and girls waste hours every day collecting water when they could be in school, or working to improve their lives.

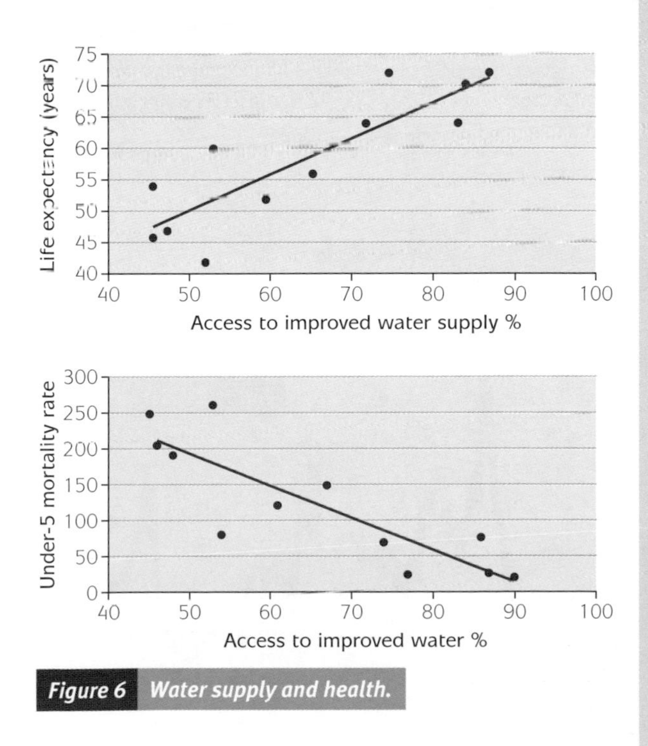

Figure 6 Water supply and health.

Another reason is that contaminated water can have severe effects on people's health. Diseases spread by dirty water can make people too weak to work or go to school, and kill the most vulnerable, including young children and the elderly. As much as 80 per cent of all developing world disease is water-related, claiming over 2 million lives a year – a death every 15 seconds.

ACTIVITIES

1 (a) Study **Figure 1**, then describe which parts of the world will have **no vulnerability** and **high vulnerability** to water stress by 2025. Explain the patterns you have described, using maps of rainfall and population distribution to help you.
(b) Using **Figure 1** and **4**, which parts of the world will suffer from future water stress and poor access to safe water?

2 (a) Study the scattergraphs in **Figure 6**. Describe and explain the **correlation** each shows.
(b) Draw a table like the one below and use the figures for safe water to identify the four countries on the graphs. Then add the figures for life expectancy and under-five mortality.

	Brazil	China	Bangladesh	Kenya
Access to improved water supply	90%	77%	74%	61%
Life expectancy				
Under-five mortality rate				

3 Draw a spider diagram with **'Unequal access to clean water'** at the centre, and three legs: developed/developing, urban/rural, women/men. Use the figures and text from these pages to add details and examples.

Trade and development

What is trade, and how can it help development? What obstacles can trade put in the way of development?

International trade

International trade happens when one country sells or **exports** its goods or services to another. Trade in goods is called **visible trade**, and trade in services like tourism is **invisible trade**. Exports earn money for the exporting country, so they are usually good for development. One country's exports become another's **imports** – the importing country has to pay for these. The difference between exports and imports is called the **balance of trade**, and it tells us something about a country's economy and prosperity:

- If a country's exports are greater than its imports, it has a **trade surplus** – and, therefore, gets wealthier.
- If its imports are greater than its exports, the country has a **trade deficit** – it may build up debts and, therefore, get poorer.

International trade is dominated by the developed countries (**Figure 1**). This is no accident – they have grown wealthy through trade, and because they are wealthy they are also powerful. Some of the wealthiest countries club together to form **trade blocs** like the European Union (page 208). Some newly industrialised countries, like China, India and Brazil, are quickly increasing their share of world trade. But for many other developing countries, instead of helping, trade is another obstacle to development. There are two big problems: **the pattern of trade**, and **unfair trade** (page 204).

The pattern of trade is a problem for those developing countries that depend mainly on primary products, often since colonial times (**Figure 2**):

- Some primary products like iron ore are valuable, but more value is added by manufacturing – turning them into something useful (like steel). Most manufacturing is done in developed countries, so they benefit more. The finished goods are more expensive – they may even be sold back to the country where the resource originally came from. A country that mainly exports primary products, but has to import manufactured products, will find it difficult to become wealthy.
- Primary products are traded in the financial centres of places like London and New York, far away from the producers. Prices go up and down on world markets, and may go so low that producers do not cover their costs. It is difficult for producers to plan ahead; for example coffee bushes take several years to grow, and farmers cannot suddenly change their crops if coffee prices fall (**Figure 3**).

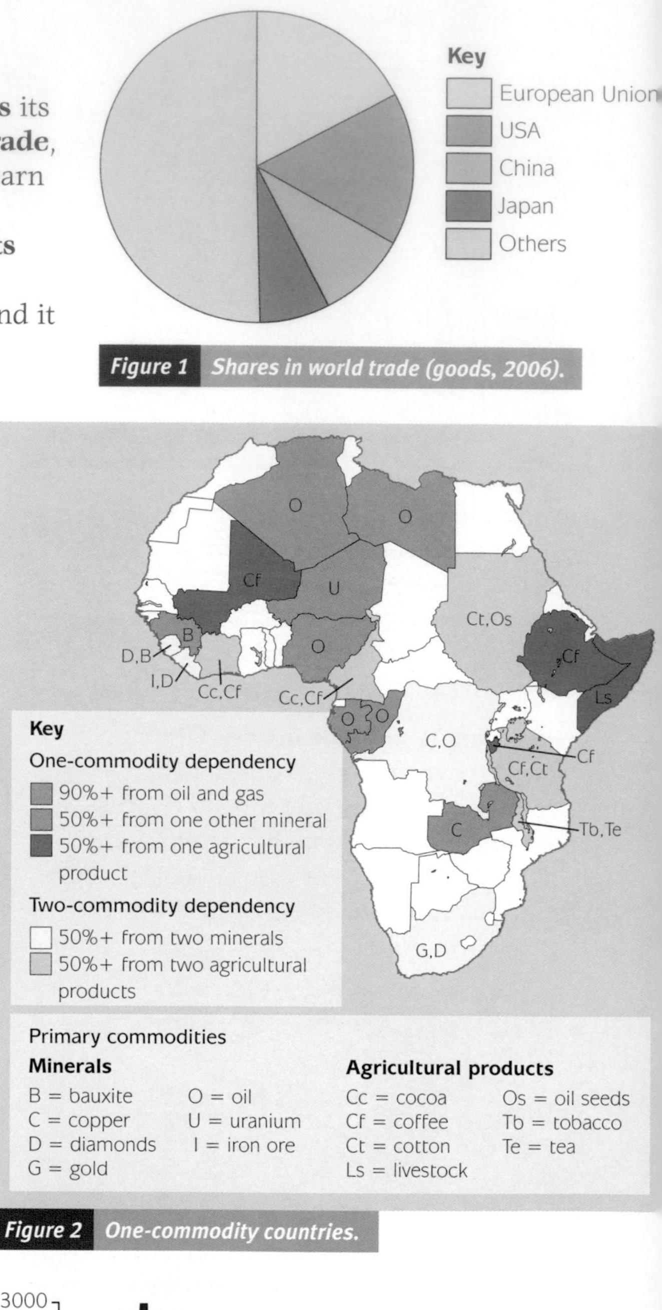

Figure 1 Shares in world trade (goods, 2006).

Figure 2 One-commodity countries.

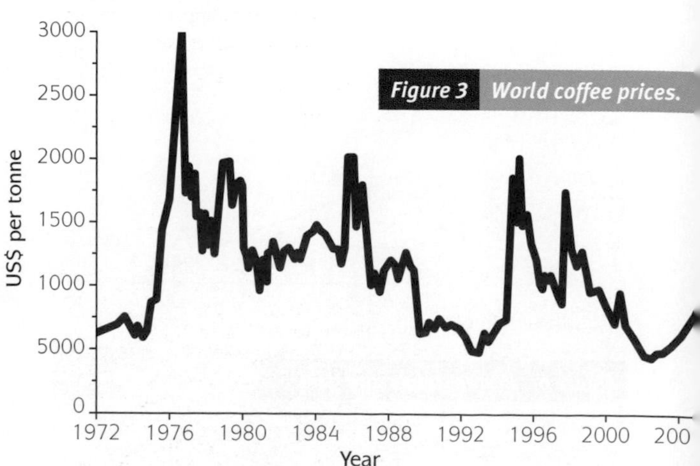

Figure 3 World coffee prices.

Trade stories: Korea and Bangladesh

In the 1950s South Korea's economy was mainly based on primary products, especially farming, and manufacturing exports were almost zero. In the 1960s South Korea built up its manufacturing industries, helped by massive investment from the USA. It protected its industries from competition by an import ban. Korea began a period of **export-led growth** – by the 1990s it had become one of the world's biggest electronics suppliers and the twelfth largest trading country.

By contrast Bangladesh's economy and exports depend more on primary products (**Figure 4**). Bangladesh is building up its industry and services through foreign investment, and the exports from its big clothing industry can be found on every UK high street. An important source of money for development comes from Bangladeshis living abroad who send money home. These **remittances** are greater than official aid to the country.

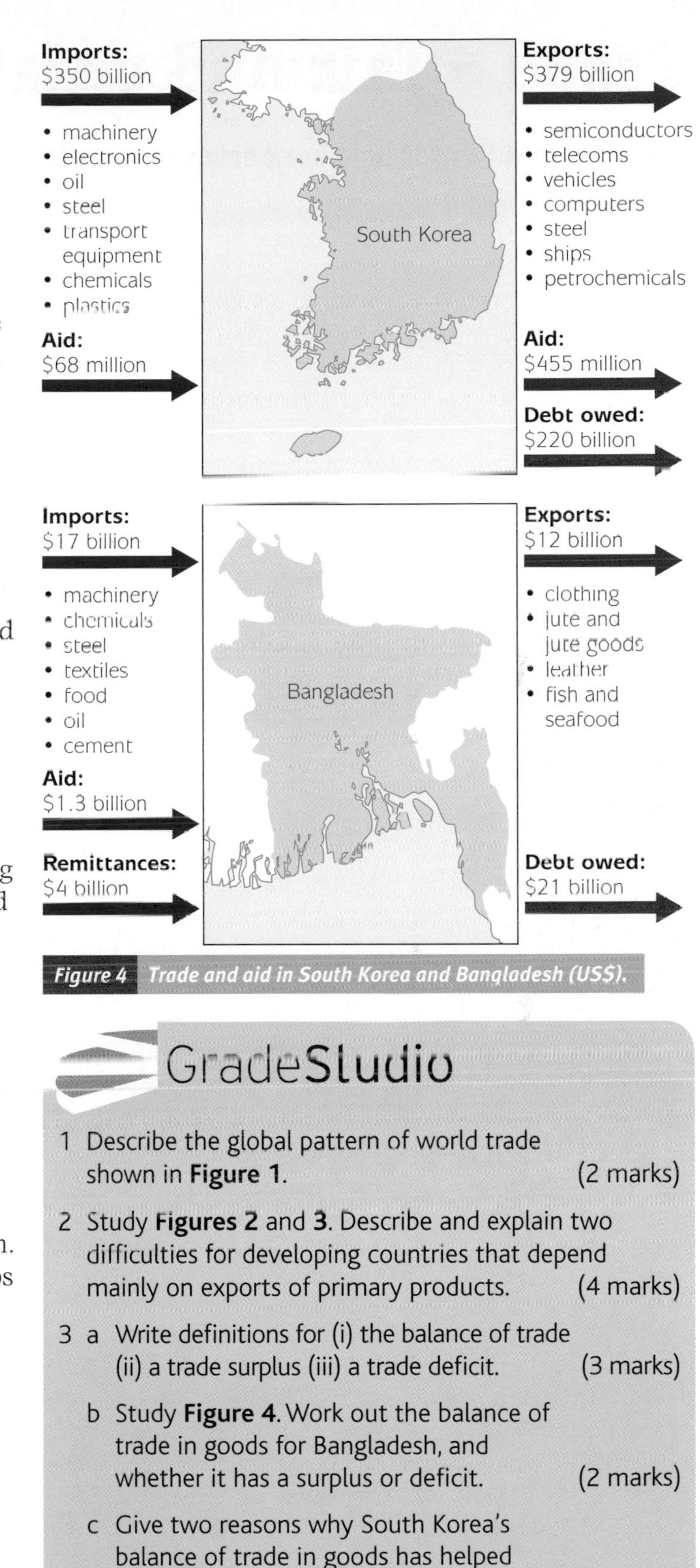

Figure 4 *Trade and aid in South Korea and Bangladesh (US$).*

Debt for conservation swaps

In the 1980s and 1990s many developing countries built up big debts through trade deficits and loans for big development projects. Some ended up paying out more in interest on their debts than they earned from exports or received from aid. One solution is **debt abolition** – but lenders do not always agree to this. Some tropical countries have suggested an inventive solution: swapping their debt for nature.

Costa Rica has large areas of untouched tropical rainforest (**Figure 5**), some of the most **biodiverse** environments in the world. In 2007 its debt to the USA was $90 million. Costa Rica has agreed to spend $26 million to increase the area of protected forest, support local communities and encourage ecotourism. In return the US government and conservation groups will cut Costa Rica's debts by the same sum.

Deforestation contributes 17 per cent of global CO_2 emissions, and forests are a natural **carbon sink**, as well as being rich in **biodiversity**. Even though halting deforestation would cost US$100 billion a year, forests are worth far more to humankind as living forests than as logs.

Figure 5 *Costa Rica rainforest areas.*

GradeStudio

1 Describe the global pattern of world trade shown in **Figure 1**. (2 marks)

2 Study **Figures 2** and **3**. Describe and explain two difficulties for developing countries that depend mainly on exports of primary products. (4 marks)

3 a Write definitions for (i) the balance of trade (ii) a trade surplus (iii) a trade deficit. (3 marks)

b Study **Figure 4**. Work out the balance of trade in goods for Bangladesh, and whether it has a surplus or deficit. (2 marks)

c Give two reasons why South Korea's balance of trade in goods has helped make the country wealthier. (2 marks)

4 Using an example you have studied, explain how 'debt for nature' swaps can help developing countries with tropical forests. (5 marks)

Exam tip

For questions like 4 above, it's sometimes a good idea to draw a rough spider diagram to help you remember all the angles before you start writing.

Trade – fair and unfair

How can unfair trade hold back development? How does fair trade work, and how can it help?

Unfair trade

International trade has grown massively in the last 50 years but it has mainly benefited the developed world. In theory, free trade should make everyone wealthier, but developed countries often seem to have an unfair advantage. For example, farmers in the EU and USA have a powerful political voice, and they are well-supported in a number of ways:

- Many are **subsidised** to produce food, making it cheaper to export (sometimes to developing countries).
- They are protected by **quotas**, which restrict food imports, and **tariffs**, which make imports from developing countries more expensive.

The World Trade Organisation (WTO) sets the rules of trade, and it has reduced some subsidies, quotas and tariffs. But developing countries complain that it is really dominated by the most powerful countries, and their corporations. If developing countries' products reach developed countries, the farmers often see little of the final sales price (**Figure 1**).

Fair trade

One way for small farmers in developing countries to get a better deal is to join a fair trade scheme or **trading group**. Fair trade gives them a fairer price and protects them against very low prices. Fair trade agreements also aim to protect the environment and strengthen the local community, so they are also an example of **sustainable development**.

Around the world, people spent £1.6 billion on fair trade products in 2007. Seven million farmers, workers and their families benefited in 58 developing counties. Fair trade products are often more expensive to buy, and retailers still take a big share of the price. But sales in the UK have grown every year – in 2007 they were worth nearly £500 million (**Figure 4**).

Most fair trade products are foods like coffee and bananas, but the fair trade idea has spread to other goods. For example, fair trade clothes have led to improved working conditions for garment workers in places like Bangladesh.

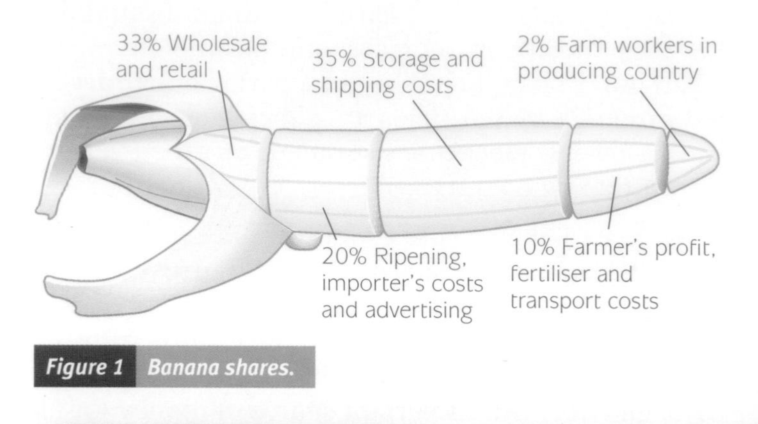

Figure 1 Banana shares.

Figure 2 Mechanised rice farmers in the USA.

Free trade?

- Rice farmers in the USA are supported by government subsidies, which help keep their prices down ($530 million in 2006). They grow more rice than the USA needs – the **surplus** is exported to countries like Honduras. The imported rice is cheaper than the rice local farmers grow, putting them out of business and increasing the country's import bill (**Figure 2**).
- In 2006 EU sugar farmers got a guaranteed price for their sugar of €670 a tonne, three times the world market price. They exported the surplus sugar, costing 12 000 sugar workers in Swaziland their jobs. Even though Mozambique's sugar costs €280 a tonne, tariffs stopped its sugar exports to the EU, losing €100 million a year. Meanwhile EU consumers paid an extra €800 million for their sugar.

EXAMPLE: Fair trade bananas in Dominica

Regina Joseph's 1-hectare (2.5 acre) farm is a short walk from her home on Dominica, a small island in the Caribbean's Windward Islands. She grows a variety of cash crops, some of which are sold in local markets, and bananas for export, which provide 70 per cent of her family's income. She grows bananas without using chemicals and respects the environment.

Every fortnight Regina and her farm workers harvest, sort, wash, pack and label bananas, ready for export to the UK. The farm produces 35 boxes of bananas a fortnight. Regina joined a fair trade group in 2000. The farmers in her fair trade group sell their bananas for a minimum of US$5.75 a box, $1 more than usual, plus a social premium of $1.75.

Regina's extra fair trade income has helped her make improvements to her farm, including better facilities for her workers in the packing area. The income has also helped put her children through school and university. The social premium has been invested in business improvements such as better farm transport, and in the community, for example paying for computers for local schools.

Demand for fair trade bananas in the UK has grown so much that three-quarters of Dominica's bananas are sold through fair trade buyers. Many small-scale Windward Islands growers have been hit hard by increased competition from big banana plantations in Central and South America. For them and the Windward Island economies, fair trade has been a lifeline.

Figure 3 A farmer tending bananas.

EXAM PREPARATION

Fair trade principles

- A minimum price guarantee, to cover costs even when world prices fall.
- Producers in control: they own the land and make democratic decisions.
- Decent working conditions and no child labour.
- A social premium: money to improve life in the community.
- Protection for the environment.

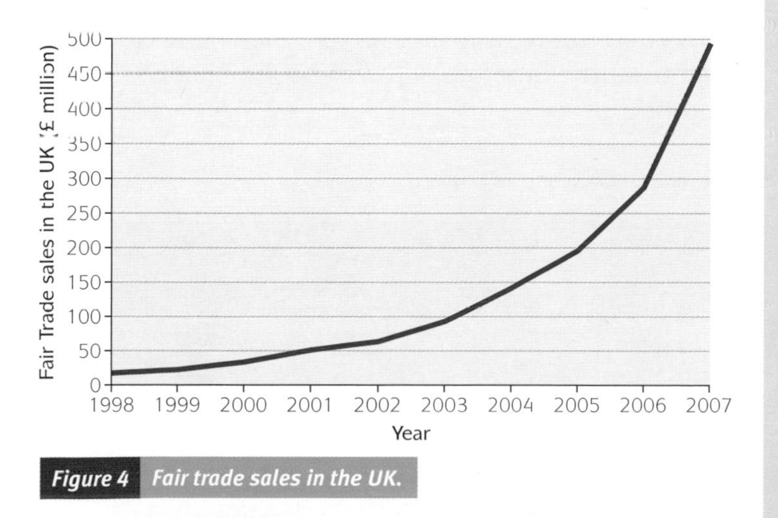

Figure 4 Fair trade sales in the UK.

FURTHER RESEARCH

Read more about fair trade on the weblink www.contentextra.com/aqagcsegeog.

ACTIVITIES

1 How much of the final price of bananas (**Figure 1**) ends up in the producer country?

2 (a) Use an atlas to find the countries mentioned in the information about sugar and rice. On a world outline map, use graphics to show how unfair trade affects sugar and rice farmers in these countries.
(b) Explain how subsidies and tariffs work against developing countries.
(c) The WTO wants to get rid of subsidies, tariffs and quotas. Work out how this would affect: sugar farmers in the EU, sugar farmers in developing countries, and consumers in the EU.

3 Read the case study about fair trade bananas.
a) Find Dominica and the Windward Islands in your atlas and draw a sketch map to show their location in the Caribbean.
b) Make a list of the fair trade principles, then use them as headings to make notes on fair trade bananas in Dominica
c) Finally describe the main benefits to farmers like Regina and to the Windward Islands.

4 Research a fair trade case study for a contrasting product and country. Present your work in the same way as you did in question **3**.

Aid

Can aid close the development gap?

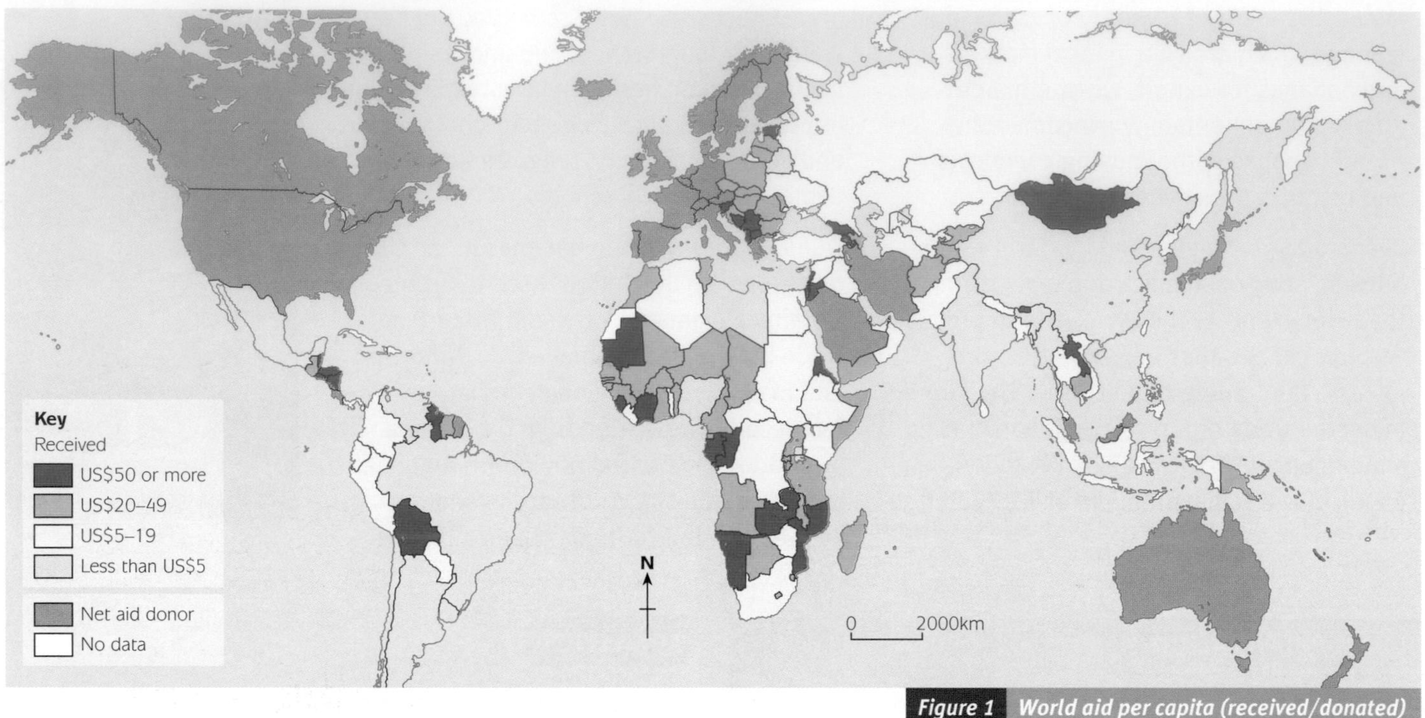

Figure 1 World aid per capita (received/donated)

Closing the gap between the wealthiest and poorest countries is one of the biggest challenges facing humankind. Aid programmes are one way to help people and communities break out of poverty and recover from disasters or conflicts. In 2006 the governments of wealthy countries gave US$104 billion in aid, and aid agencies $15 billion. Some developing countries, like Saudi Arabia and China, are increasing their aid programmes **Figure 1**.

Aid comes in different forms:

- goods such as machinery, food, medical supplies
- people with special skills such as engineers, teachers or doctors
- money to buy local goods, invest in development projects or reduce a country's debts
- loans for big projects (which have to be repaid).

Emergency or **humanitarian aid** helps people recover from natural hazards and conflicts in the **short term**. It is a lifeline for people in great need. People in low-lying Bangladesh are well-adapted to normal floods, but are vulnerable to severe river and coastal flooding and to tropical cyclones. Cyclone Sidr in November 2007 killed 500 people and left 5 million without food, water and shelter. Emergency aid helped people survive and get back on their feet again (**Figure 3**).

Development aid aims to improve people's standard of living and quality of life in the **long term**, for example in health, education, farming or the environment. Providing safe water supplies, like this water pump in Bangladesh (**Figure 4**), is an example of practical long-term aid. Most aid agencies prefer to invest in projects like this, working with local partners to identify what people really need. If people are involved in the development of their own community, it is more likely the changes will be **sustainable**; therefore, aid will not be needed in future.

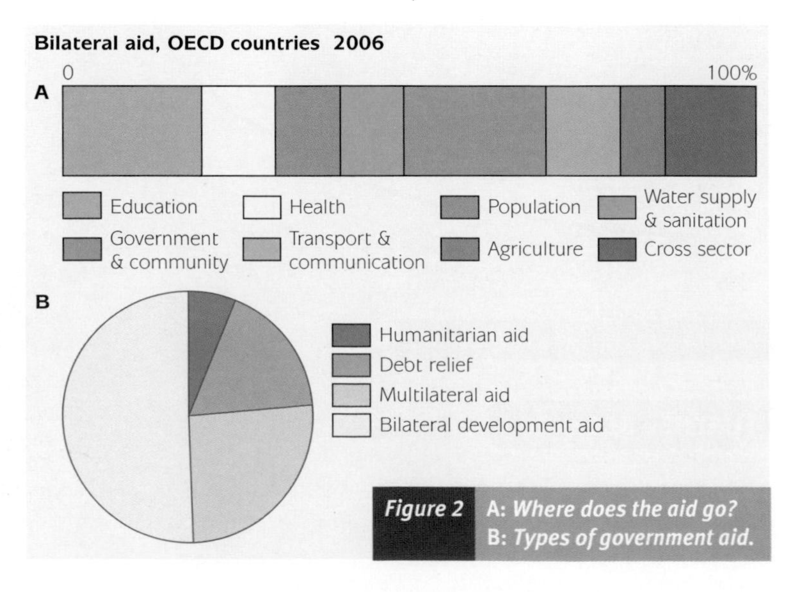

Figure 2 A: Where does the aid go? B: Types of government aid.

Figure 3 Emergency aid reaches the survivors of Cyclone Sidr in 2007.

Types of aid

Voluntary aid comes from charities like Oxfam and Action Aid. These are non-governmental organisations (NGOs), which raise money from the public, and are mainly based in developed countries. They usually work closely with communities and other NGOs in developing countries, so the aid is more effective because it is based on local needs. Because they are good at working on community projects, many charities also receive grants from governments. But many NGOs are tiny, so they face a huge challenge.

Bilateral (government) aid is given by the government of one country to another. Sometimes this type of aid is used for large-scale development projects such as dams, which do not always meet the real development needs of the country. Another criticism is that donor governments can place conditions on how the aid is used (**tied aid**). For example, they may sell goods or services as part of the deal, so that much of the aid money finds its way back home. Many governments also give bilateral aid to friendly countries for political reasons, rather than those in greatest need. But most governments are now much more careful to make sure their aid is used to meet the needs of the poor (**Figure 2**). The UN sets a target of 0.7 per cent GNI to be spent on aid – but few wealthy countries do so.

Multilateral aid is given by governments to international agencies such as the World Bank, which then fund development projects. The United Nations also has many agencies such as the World Health Organisation and UNESCO. These agencies are large organisations with lots of money to invest in development, but they are sometimes criticised for being out of touch with people's real needs.

Problems with aid

- The public are more likely to give to emergencies, like Cyclone Sidr, than to help with longer-term development projects – even though these are more successful.
- Sometimes food or goods sent as emergency aid puts locals out of business – making the problem worse in the long term.

Figure 4 A Bangladeshi woman draws safe, clean water from the village hand-pump at a time of severe flooding. The water pumps are an example of practical long-term aid, useful and appropriate to the local community.

- Aid can make weaker countries dependent on stronger ones; in the long term, countries need to develop their own trade and industries to be self-sufficient.
- Aid can help link the rich world and developing countries, but it may also make some givers feel superior.

ACTIVITIES

1 Study **Figure 1**, then describe which parts of the world give most aid, receive most, and neither give nor receive much. The maps of GNI and HDI (pages 194 and 196) will help you.

2 Look at the four forms of aid in the bulleted list on page 206. Copy the list, then for each one say whether it benefits the receiving country, the donor country or both.

3 Make a copy of the table below and add details and examples.

Type of aid	Main features	Advantages	Disadvantages
Bilateral Multilateral Voluntary Emergency Development			

4 In pairs investigate two or three different websites from aid and development agencies such as Oxfam, Action Aid or Islamic Relief. Describe, and give examples of:
- emergency aid
- how they work with local people for long-term development
- the countries they work in
- two different development projects, and explain how they are sustainable for people and the environment.

Closing the EU development gap

What contrasts in development are within the European Union? How can the EU close them?

Development in the EU

In 1957 Belgium, France, West Germany, Luxembourg, Italy and the Netherlands became the first six members of the European Economic Community (EEC). Only 12 years earlier, these countries had been at war. By 1995 the EEC had grown into the European Union (EU) with 15 members. On its fiftieth birthday in 2007, the EU had 27 member states with a population of 495 million people. Today the EU is the world's largest single market and the most powerful trade bloc (page 202).

The EU has been very successful in increasing trade between its member countries, improving the environment and spreading security and democracy. As a result, people in the EU have become wealthier and enjoy a better quality of life. One reason more countries have joined the EU is to trade with other EU countries: another is the hope that the EU will help them develop and improve the standard of living of their citizens.

There have always been contrasts in development between EU countries (**Figure 1**). The wealthiest part of Europe in the second half of the twentieth century included West Germany, France, Belgium, the Netherlands, northern Italy and parts of the UK. Geographers call this Europe's **core** region. It has the biggest and wealthiest population, the most advanced industry and services, and the best **communications**. Other countries like Spain, Portugal, Greece and Ireland, on the western and southern **periphery** (edge) of Europe, were poorer. But EU membership helped them to develop and prosper. The EU has also helped poorer regions within countries to develop, for example southern Italy and northern Ireland.

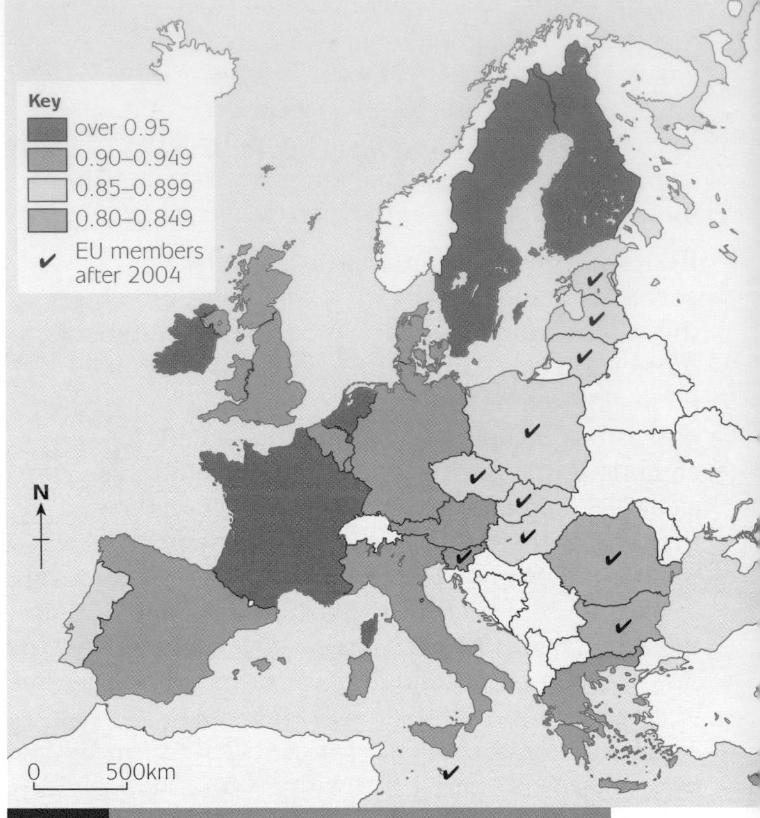

Figure 1 The Human Development Index in the EU (2005).

Eastern Europe

One of the biggest challenges for the EU is how to reduce the development gap between countries. The biggest gap now is between the wealthier countries that joined the EU up to 1995, and the ten countries from Eastern Europe that joined after 2004. These ten countries were all allies of the USSR until the end of communism in 1989. Then many of their industries collapsed and many people faced great hardship, with rising unemployment, falling living standards and low wages. As well as suffering from economic problems, parts of Eastern Europe are remote and have poor communications. They are in Europe's new Eastern periphery.

These differences in wealth and quality of life (**Figure 1**) help to explain why some people are migrating within the EU from east to west (pages 146–7). But the differences also make Eastern Europe attractive to companies looking for places to invest in new businesses. For example, in 2007 the multinational Peugeot-Citroën closed its Coventry car factory and relocated to Slovakia, where the wages of skilled workers were one-third of those in Western Europe.

FURTHER RESEARCH

Read more about the EU on the weblink www.contentextra.com/aqagcsegeog.

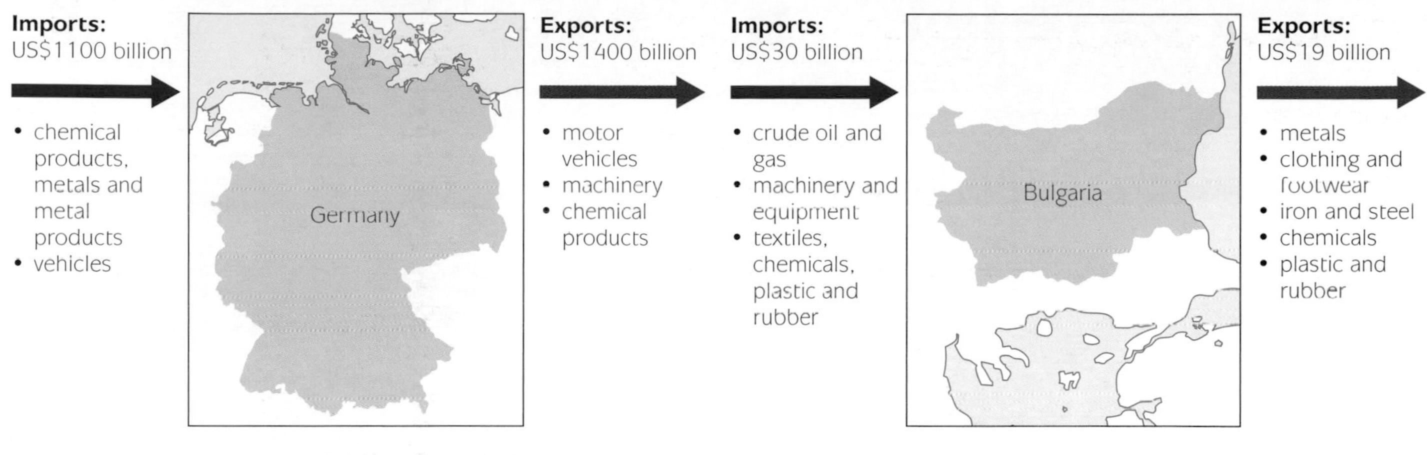

	Population (million)	GNI per capita (US$)	Jobs % in farming/industry/services	HDI	Infant mortality per 1000 births	Life expectancy (years)
Germany	82	38 860	2/30/68	0.935	4	79
Bulgaria	8	4590	9/34/57	0.824	12	73

Figure 2 *Germany and Bulgaria compared (2006–7).*

How the EU reduces inequalities

The EU member states contribute about 1 per cent of their total wealth to the EU budget, €130 billion in 2008. The wealthiest countries make the biggest contributions – over 40 per cent comes from Germany and France. Over a third of the EU budget is then spent on regional aid for less developed countries and regions. The idea is that regional aid will help them develop and benefit the people there, and then – as they become better-off – the wealthier parts of the EU will benefit too. There are three types of regional aid programme:

- the European Regional Development Fund, to improve investment and **infrastructure**
- the European Social Fund, to pay for education, training and job creation
- the Cohesion Fund, to improve the environment and transport and develop renewable energy.

Regional aid has helped countries like Ireland and Greece, and regions like southern Italy in the past. For example, Ireland's wealth was only two-thirds the EU average when it joined in 1973; by 2008 it was one of the EU's wealthiest countries. Regional aid will help the new Eastern European members to develop in future, especially by creating jobs and improving transport and the environment. But closing the development gap is likely to take some time, and these countries will get over half of EU regional aid between 2007 and 2013.

ACTIVITIES

1 Use an atlas and **Figure 1.** On an outline map of Europe, label:
   a) where the core region of Europe is, and the new Eastern EU countries
   b) add notes to show why there is a gap in development between them.

2 Study **Figure 2,** which shows contrasts in development between Germany and Bulgaria.
   a) Decide which data shows the biggest development gaps. Choose and plot the best type of graphs to show this data.
   b) Work out the balance of trade for Germany and Bulgaria and compare their imports and exports. What do these tell you about the two countries? (Pages 202–3 will help you).
   c) Write a short summary describing the most important development gaps between the two countries, then look back at the text on page 208 to help you explain them.

3 Give examples of how the EU is trying to reduce the development gap, especially between Western and Eastern Europe.

4 Comparing development in two other EU countries.
   a) Choose two other contrasting countries, one in Western Europe (not the UK). Compare the countries' economic and social development, and their quality of life. **Figure 2** and pages 194–7 will help you identify the data you need from the weblink www.contentextra.com/aqagcsegeog. You might share the countries with your group and produce a joint report.
   b) Summarise the patterns your work shows about the development gap between your countries, the reasons for it, and how it might be closed.

GradeStudio

Practice GCSE question

See a Foundation Tier Practice GCSE Question on the weblink www.contentextra.com/aqagcsegeog.

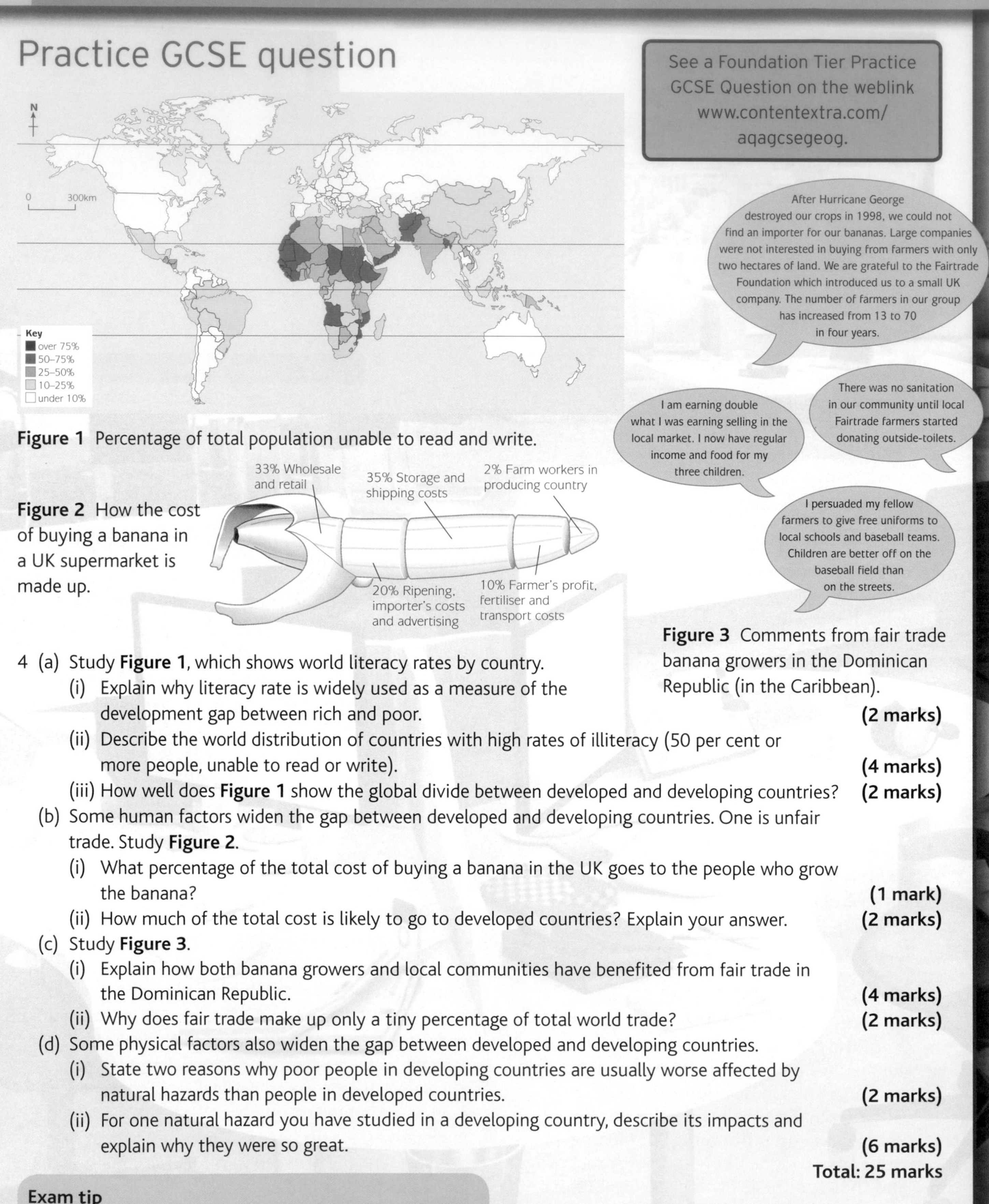

N
0
300km
Key
over 75%
50–75%
25–50%
10–25%
under 10%
33% Wholesale and retail
35% Storage and shipping costs
2% Farm workers in producing country
20% Ripening, importer's costs and advertising
10% Farmer's profit, fertiliser and transport costs
After Hurricane George destroyed our crops in 1998, we could not find an importer for our bananas. Large companies were not interested in buying from farmers with only two hectares of land. We are grateful to the Fairtrade Foundation which introduced us to a small UK company. The number of farmers in our group has increased from 13 to 70 in four years.
I am earning double what I was earning selling in the local market. I now have regular income and food for my three children.
There was no sanitation in our community until local Fairtrade farmers started donating outside-toilets.
I persuaded my fellow farmers to give free uniforms to local schools and baseball teams. Children are better off on the baseball field than on the streets.

Figure 1 Percentage of total population unable to read and write.

Figure 2 How the cost of buying a banana in a UK supermarket is made up.

Figure 3 Comments from fair trade banana growers in the Dominican Republic (in the Caribbean).

4 (a) Study **Figure 1**, which shows world literacy rates by country.

(i) Explain why literacy rate is widely used as a measure of the development gap between rich and poor. **(2 marks)**

(ii) Describe the world distribution of countries with high rates of illiteracy (50 per cent or more people, unable to read or write). **(4 marks)**

(iii) How well does **Figure 1** show the global divide between developed and developing countries? **(2 marks)**

(b) Some human factors widen the gap between developed and developing countries. One is unfair trade. Study **Figure 2**.

(i) What percentage of the total cost of buying a banana in the UK goes to the people who grow the banana? **(1 mark)**

(ii) How much of the total cost is likely to go to developed countries? Explain your answer. **(2 marks)**

(c) Study **Figure 3**.

(i) Explain how both banana growers and local communities have benefited from fair trade in the Dominican Republic. **(4 marks)**

(ii) Why does fair trade make up only a tiny percentage of total world trade? **(2 marks)**

(d) Some physical factors also widen the gap between developed and developing countries.

(i) State two reasons why poor people in developing countries are usually worse affected by natural hazards than people in developed countries. **(2 marks)**

(ii) For one natural hazard you have studied in a developing country, describe its impacts and explain why they were so great. **(6 marks)**

Total: 25 marks

Exam tip

The question for this topic will be Question 4 in Paper 2.

Improve your GCSE answers

How to answer map-based questions

- Many exam candidates are put off by having to study a world map full of information, like that in **Figure 1**.
- They are equally uncomfortable with question words such as *pattern* and *distribution*.
- When asked to **describe a distribution** from a map, you should concentrate on saying where things are found, but you can also mention where they are not found.
- Do the same when asked to **describe a pattern**; say where there are many (or values are high) and where there are few (or values are low).
- It is important to end with a comment summarising the main feature of the distribution or pattern.

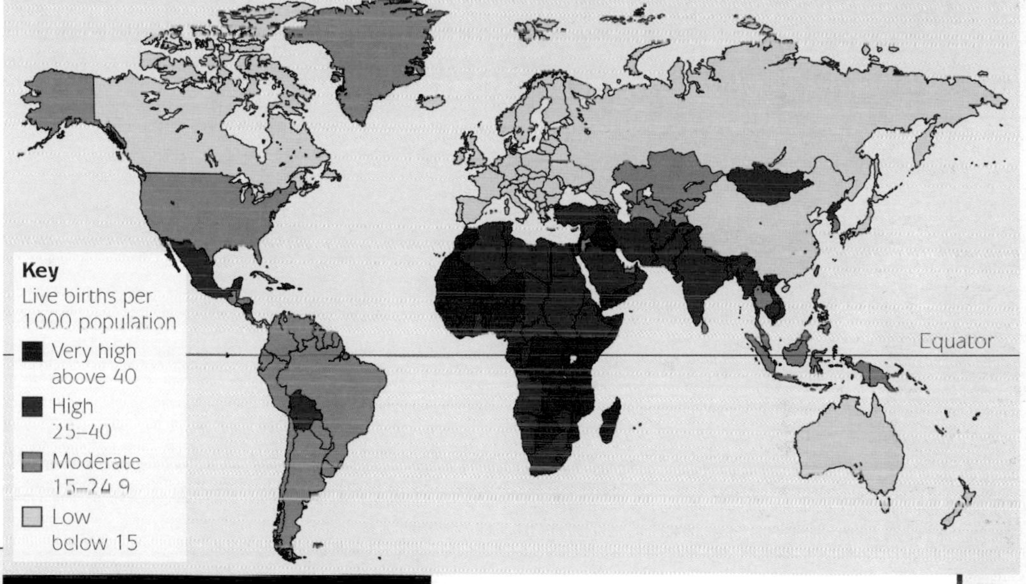

Figure 1 World distribution of birth rates.

Question: Study **Figure 1**. Describe the world distribution of high and low birth rates. (5 marks)

Guide for answering

1 Take note of the command word – 'Describe'. This is telling you to write about what the map shows.
2 Identify the question theme – 'high and low birth rates'. Study the key to see how they are shown on the map.
3 Study the map. It is essential that you know the names of the continents.

Possible answers for high birth rates

- Most countries with high birth rates are in Africa.
- All African countries are above 25 per 1000.
- About half of African countries are very high – above 40 per 1000.
- A band of high birth rates runs west to east from the Mediterranean through South Asia.

Possible answers for low birth rates

- These are found in most countries of Western Europe.
- Also found in other continents such as Australasia.
- And in other countries like Canada and Japan.
- All are countries in the developed world.

And the summary comment? It could go along the lines 'High birth rates are concentrated in developing countries in Africa and South Asia, whereas low birth rates are mainly found in developed countries in temperate latitudes.'

ExamCafé

REVISION

Key terms from the specification

Aid – money, goods and expertise given by one country to another, either free or at low cost
Development – level of economic growth and wealth of a country
Fair trade – farmers and producers in developing countries are given a fair deal by buyers in developed countries; prices paid are always higher than their costs of production
GNI (Gross National Income) per head – total income of the country, divided by the number of inhabitants, to give average income per person
GNP (Gross National Product) – total value of all the goods and services produced by people and companies in the country in one year
HDI (Human Development Index) – is a measure of people's quality of life using more than one measure of development, based on life expectancy, education and standard of living
Infant mortality rate – number of child deaths under one year old per 1000 people
Life expectancy – average number of years that a new-born child can expect to live
Measure of development – statistical way to show the size of differences in levels of economic growth and wealth between countries

Checklist

	Yes	If no – refer to
Can you name and explain three different ways of measuring the development gap between countries?		pages 194–7
Do you know what is meant by quality of life and how it can be improved?		pages 196–7
Can you give the causes and impact of a natural disaster in a developing country?		pages 198–9 (and see below)
Do you understand the problems caused by poor access to clean water supplies?		pages 200–01
Can you explain how the pattern of world trade favours developed countries and what can be done to make it fairer to developing countries?		pages 202 and 204–5
What is meant by development aid and how it differs from other types of aid?		pages 206–7
Can you describe the ways in which the EU is trying to reduce differences in levels of development between member countries?		pages 208–9

Case study summaries

Natural hazard – study options in the book		pages	Development project
Location	earthquake (India and Indonesia)	20–21	Location
Cause	volcano (Montserrat)	16–17	Project information
Impact	tropical storm (hurricane Mitch)	56–7	Changes made
Responses	flooding (Bangladesh)	90–91	Results

Chapter 12
Globalisation

In January 2007 people descended on Branscombe Beach (Devon) to scavenge bounty from the globalisation of manufacturing industry – all sorts of things, from £12 000 BMW motorbikes to carpets, beauty cream and wine casks, were washed ashore in containers from the shipwrecked *Napoli*.

QUESTIONS

- **What is globalisation and why is it happening?**
- **In what ways does globalisation affect all of us?**
- **Can future demands for energy be met in more sustainable ways than by using fossil fuels?**
- **What problems are caused by trying to meet increasing world food needs?**

Understanding globalisation and the rise of transnational corporations (TNCs)

What is globalisation? Why is everyone affected by it? Are the effects of TNCs good or bad?

Globalisation refers to the *increasing* importance of international operations for people, countries and governments. People in one country or region of the world are being affected *more and more* by economic events and decisions made in other parts of the world, often thousands of kilometres from where they live. Both manufacturing and service companies are operating in ways that are increasingly international. The result is that few countries, people or industries are untouched by what happens in other parts of the world. They share the gains in good times and the pain in bad times. When one country 'sneezes', others 'catch a cold'.

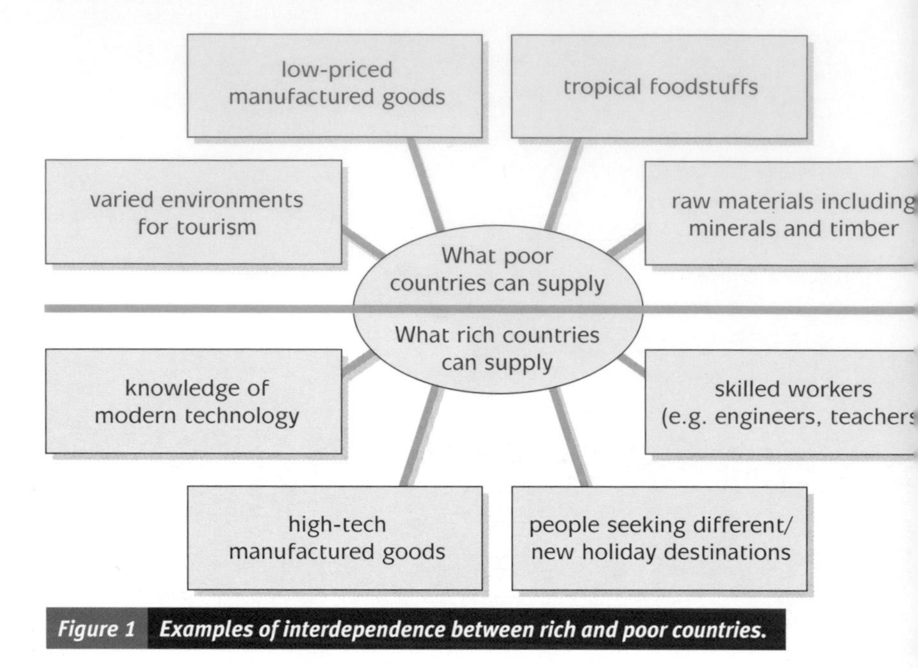

Figure 1 *Examples of interdependence between rich and poor countries.*

The consequence of globalisation is increasing **interdependence** between countries in all parts of the world. Interdependence describes countries' shared needs to exchange goods and services. One country has something that another one does not, or cannot make as cheaply; they exchange (or trade) and, in doing so, they become dependent on one another. There is plenty to exchange between rich developed countries (mainly in temperate lands) and poorer developing countries (mainly located in the tropics), as shown in **Figure 1**.

What made globalisation possible?

The simple answer is the revolution in global communications.

For people:

- Instant contact by e-mail, text, fax, phone, and audio/video links and conferencing, made possible by satellite communications and submarine fibre-optic cables.
- Jet aircraft allowing company executives, managers and skilled personnel to reach virtually any country in the world within 24 hours.

For goods:

- Containers for moving goods by sea, which are interchangeable between ships and road and rail transport on land (**Figure 2**).
- Cargo aircraft and cargo carried on scheduled passenger flights for goods that are too perishable or too valuable for transport by sea.

Global connection is faster, easier and cheaper than it has ever been.

Figure 2 *Container ship.*

- Ease of transport – interchangeable between land and sea transport.
- Cheapness – lifted from one type of transport to another by crane.
- Security – sealed in factory/warehouse and not opened until destination.

The growth of transnational corporations (TNCs)

TNCs are large multinational companies with business interests in many different countries. They include BP (based in the UK), Nestlé (Switzerland), Ford (USA) and Sony (Japan). They are truly global companies, making decisions globally rather than nationally. Some information about them is given in **Figure 3**. Most have their headquarters in rich, developed countries; this is where important decisions about global operations affecting other countries are made. They set up operations in rich and poor countries alike, wherever there is an opportunity to make a profit. They can take advantage of the fact that overall business operating costs are much lower in some countries than in others. Although more TNCs are based in Europe, the majority of the biggest ones are based in the USA. The business interests of the TNCs can be many and varied and include mining, plantation farming and manufacturing. Everyone knows the names of many of them.

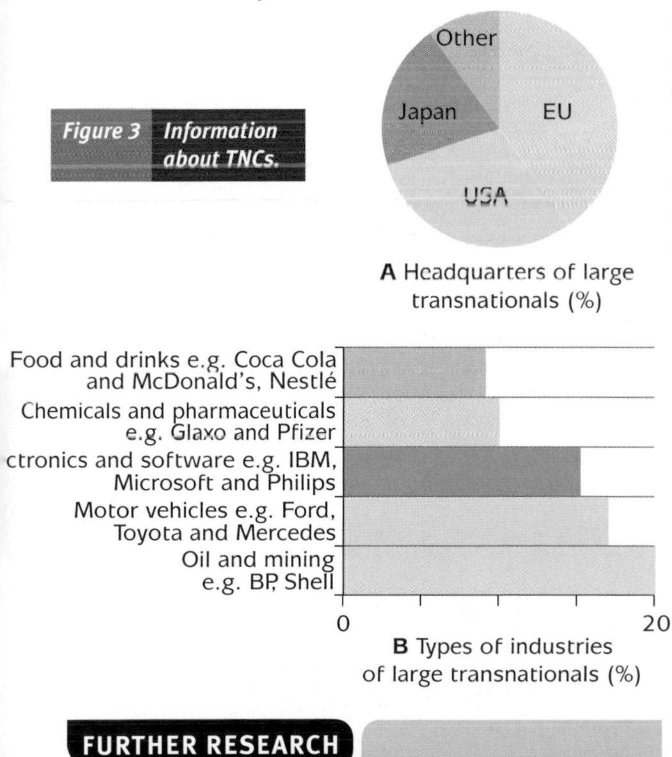

Figure 3 Information about TNCs.

FURTHER RESEARCH

Choose a TNC and visit its website to discover the location of its head office, where it operates and its main types of business. For an example, visit the weblink www.contentextra.com/aqagcsegeog.

Advantages and disadvantages of TNCs

Countries that have no attractions for TNCs, mainly in Africa and Asia, are among the world's poorest countries. Factors such as landlocked location, political instability, government corruption and wars deter these companies. Countries where TNCs set up manufacturing industries, such as Mexico, Brazil and South-East Asia (pages 218–21), have profited most.

Advantages	Disadvantages
These are mainly **economic** • They bring **capital**, **modern technology** and **skills** which the counrty does not have. • The country's **infrastructure** (e.g. transport and energy supply) is improved for them or by them. • They create **jobs** which increase exports; if these are manufactured goods, the dependence on low-value primary products is reduced. • There may be other benefits such as a **multiplier effect** for service-sector jobs and economic development.	These are a mixture of **economic**, **social** and **environmental**. • **Economic** disadvantages include low wages, tax avoidance and the fact that profits are taken out of the country and sent back to developed countries. Also, TNCs can leave a country as quickly as they came. • **Social** disadvantages include poor working conditions and lower safety standards than would be allowed in developed countries. • **Environmmental** disadvantages include water and air pollution, because local pollution controls are either weak or ignored.

Figure 4 Summary of the advantages and disadvantages of TNCs.

However, always remember that TNCs are driven more by profit than a desire to achieve economic development for the host country; they are the world's most competitive companies, and will leave a country as quickly as they came if there are no profits to be made.

The main arguments against TNCs are that they exploit and take advantage of poor people and poor countries, that they cream off large profits, and that they do not care about the environment.

ACTIVITIES

1 Name and describe two different ways of providing
   (a) quick communication between a company boss in London and a factory manager in Shanghai
   (b) easy transport of factory goods made in China to the UK.

2 (a) State three characteristics of TNCs.
   (b) Explain how McDonald's (or one of the other companies named in **Figure 3A**) fits the TNC label.

3 Do the economic advantages of TNCs in **Figure 4** outweigh the disadvantages? Give reasons for your answer.

4 Go to pages 243–5 in Chapter 13.
   (a) Describe the attractions of Kenya for tourists from the UK.
   (b) Name and explain one positive and one negative effect of globalisation for Kenya.

CASE STUDY Globalisation in services

ICT and call centre growth in India

One aspect of globalisation is the movement of services and support industries from rich to poor countries. The main advantages for companies are the same as for manufacturing: cheaper labour and lower operating costs. One of the countries benefiting from this switch is India. This huge country of over 1 billion people produces 2 million graduates a year, 80 per cent of whom are fluent in English. Although call centre jobs have received the greatest media attention, the range of work transferred is broader than this and includes computer programming and data processing. This is having a multiplier effect on service-sector employment, because each new phone and IT job is reckoned to support another job such as driving, catering or cleaning.

The Indian city that is benefiting the most is Bangalore (Karnataka), the 'IT capital' of India. Although it all began when Texas Instruments set up a successful design centre here in the mid-1980s, it only took off during the late 1990s after the Indian government adopted a more welcoming attitude to investment by overseas TNCs. This southern city, home to 6 million people, houses more than 250 high-tech companies, located in impressive new technology parks that have sprung up around the city. One of the largest and most popular with overseas companies is Electronics City (**Figure 1**). It was built for companies working in information technology, software development, telecommunications and financial services. Here are located a mixture of Indian companies (Infosys, Wipro) and international companies (Digital, Siemens, Motorola).

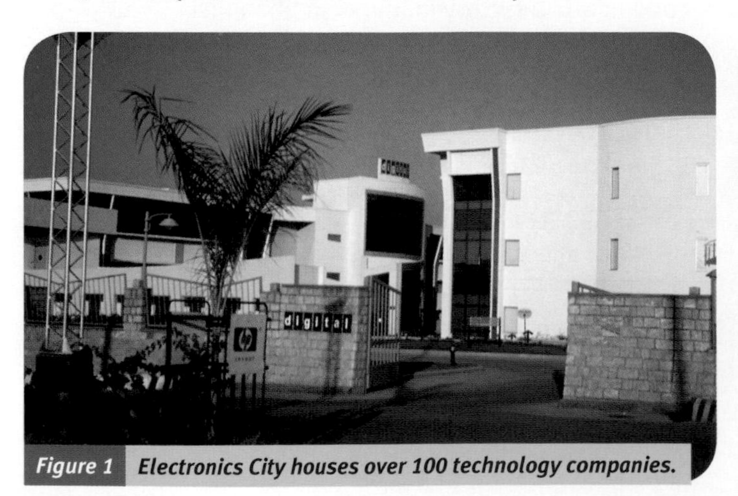

Figure 1 ***Electronics City houses over 100 technology companies.***

Nice work, if you happen to live in Bangalore

Fancy earning twice as much as a doctor and being driven to and from work every day? At the office there's a free canteen, a fully equipped gym and a pool table. Now here's the snag. You'll be earning £150 per month and may have to work until 2.30a.m. And you'll be living in India.

According to the Communication Workers Union, 33 large companies, including Barclays, British Airways, Lloyds TSB, Prudential and Reuters have together outsourced 52,000 jobs serving UK customers to India. Trade Union Amicus has predicted that 200,000 jobs could be lost overseas by 2010.

It is not hard to see why. It has been estimated that British companies usually save a minimum of £10m a year for every 1,000 jobs they move overseas. India produces more graduates per year than any other underdeveloped nation in the world. Call centre advisers are paid around £150 per month, while a junior doctor or teacher earns £60.

Norwich Union, Britain's biggest insurer, is one example. It has quickly built up a workforce of 3700 people in India and plans to have 7000 by 2007. According to the company, cheaper labour is not the only reason for turning to India. 'The work ethic is tremendous. It is partly cultural and reinforced by the education system. They have a high competitive attitude and status is important.'

One Indian customer manager from Bangalore said 'My parents (a civil servant and teacher) walk to work each day. My company send a car for me. I avoid the heavy traffic of old rickshaws, bicycles and dirty cars, plus the odd cow'.

Like all boom cities, Bangalore is not without its problems. Most of the problems, typical of big cities in developing countries, are present and in pressing need of attention, including shanty housing, clogged and fume-ridden roads, frequent power cuts and shortages of clean water supplies. People continue to flock into the city.

Source: *Daily Telegraph*, 24 January 2005

Figure 2 ***Newspaper report from January 2005.***

ACTIVITIES

1 Give evidence (including numbers and values) for each of the following statements:
   (a) Bangalore is a major IT centre.
   (b) UK companies save money by transferring call centre and IT operations to India.
   (c) Growth of call centres and IT work brings economic and social benefits to Indians.

2 Describe two possible disadvantages for Bangalore of rapid growth supported by investment from overseas.

CASE STUDY A TNC

Siemens of Germany

Based in Berlin and Munich, Siemens is the largest engineering group in Europe. It is the most international of transnational companies with a presence in 190 countries and regions (i.e. in almost every country in the world). Worldwide it employs over 400 000 people. The focus is upon innovation and technological development, to develop and support its wide product range (**Figure 3**). The company can be said to make everything from 'trains to chips' (of the computer kind, of course). Research and development are concentrated in Germany, where a long tradition of technology exists and a highly qualified and skilled workforce is available. However, much of the actual manufacturing has been moved to countries where production costs are lower (pages 218–21). **Figure 3** shows Siemens operations in Penang in Malaysia, on a large government-planned industrial estate near the international airport.

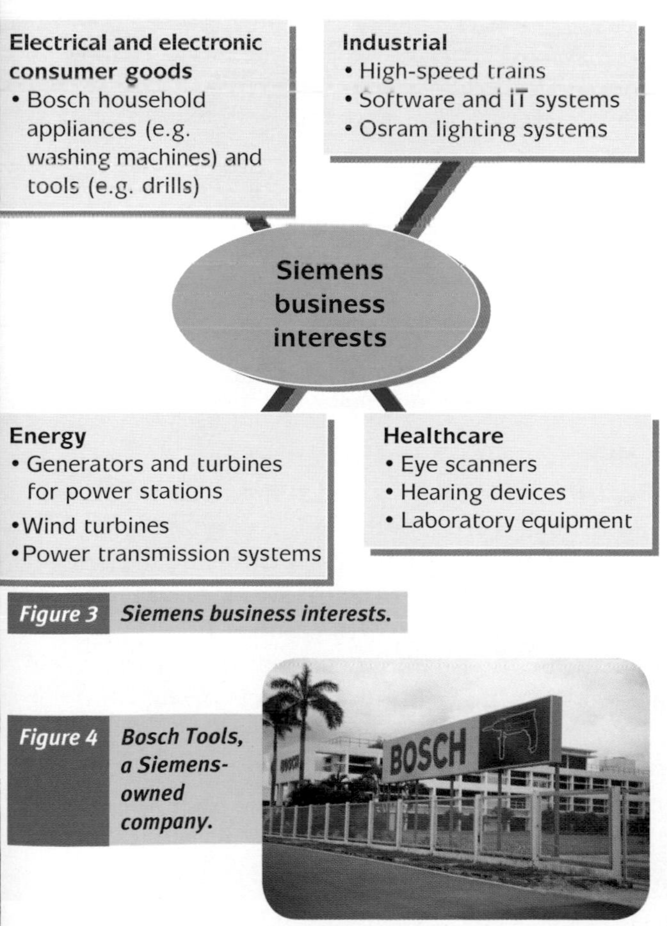

Figure 3 *Siemens business interests.*

Figure 4 *Bosch Tools, a Siemens-owned company.*

GradeStudio

1 Read the report from 1984.

Bhopal, Central India, December 1984

A leak of gas from a pesticide plant, owned by Union Carbide, an American transnational corporation, killed over 2500 people. It affected the health of up to a quarter of a million people, most of whom were slum dwellers. They had been attracted to the city by its rapid industrial growth and were unaware of the risk of living next to a factory producing toxic chemicals.

The cause of the accident was put down to design faults in the plant's safety systems, made worse by a decline in maintenance standards. Significantly, in the USA, plants producing these chemicals are required to be located 50 kilometres away from settlements.

a State three characteristics of transnational corporations like Union Carbide. (3 marks)

b How could Union Carbide have reduced the risk of a gas leak? (2 marks)

c Why was there a greater loss of life in India than there would have been from a similar incident in the USA? (3 marks)

2 Read the report from 2004.

Bhopal still bears the scars

Bhagwan Singh and his family live in the shadow of the most notorious factory in India. Their modest house lies only a stone's throw from the disused Bhopal chemical plant, where in 1984 a catastrophic toxic leak claimed the life of their baby and more than 3000 of their neighbours.

Today the factory site lies abandoned and closed to the public. But city investigators say that thousands of tonnes of toxic waste are still stored there. In the monsoon season, the rains wash the chemicals into the ground water, contaminating local wells.

Campaigners claim that almost 20,000 have since died from the effects of the disaster and that 150,000 continue to suffer from the symptoms of chemical poisoning. They say these include cancer, anaemia, infertility and birth defects.

For families such as the Singhs, finding uncontaminated water to drink is a pressing priority. Residents say that water piped in or delivered by tankers is inadequate. People are still drawing stinking water from condemned wells. The cycle of death threatens to engulf another generation.

Source: *Sunday Telegraph*, 21 November 2004

a Describe the long-term effects of the Bhopal disaster. (2 marks)

b Explain why these effects are still being felt 20 years later. (2 marks)

c Would this have been the case in a developed country? Explain your answer. (2 marks)

3 Suggest reasons why both the Indian government and people of Bhopal were in favour of the plant being built. (6 marks)

Exam tip

In newspaper reports, underline and highlight key points likely to be of use in your answers.

The changing global distribution of manufacturing industry

Where is manufacturing industry declining and why? Where is it growing and why?

The Industrial Revolution began around 1750 in the UK and gathered pace with the invention of the steam engine, using coal, and its use to drive machinery in factories. The UK was the first country in the world to become an industrial country; at its peak before the First World War, over half the UK workforce was in manufacturing (**Figure 1**). What were then thriving traditional heavy industries, such as steel, shipbuilding, metal smelting and engineering, along with cotton and woollen textiles, employed huge numbers in the coalfield regions of Clydeside, the north-east, South Wales, Lancashire, Yorkshire and the West Midlands.

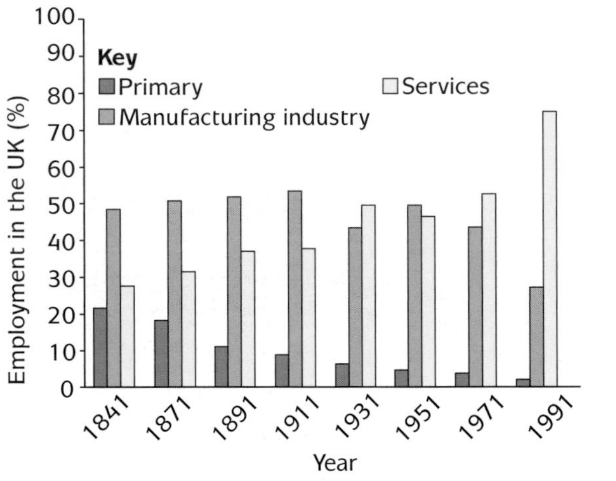

Figure 1 **Changes in employment in the UK (1841–1991).**

The UK was also the first country to experience **de-industrialisation** – a decline in the relative importance of manufacturing industry, which resulted in a sharp fall in industry's share of the total workforce to 18 per cent in 2001 and under 15 per cent by 2005. The decline began after 1951, and speeded up during the 1980s (**Figure 1**). The early effects of globalisation were already being felt; British steelworks, shipyards and textile mills were increasingly unable to compete on costs and delivery times with the **newly industrialising countries** (NICs) in the Far East. Closure of coal mines, partly through exhaustion of their coal, coincided with the decline in heavy industries and led to massive job losses in places like South Wales, the north of England and lowland Scotland where there were few alternative employment opportunities. This led to severe socio-economic problems, including high rates of unemployment, poverty and deprivation, made worse by falling property prices as whole streets of houses were boarded up. Abandoned houses, factories, mines and railway sidings further blighted the appearance of these areas, already suffering from environmental problems such as polluted rivers and canals, waste tips and derelict land, caused by many years of heavy industry and mining.

Reasons for rising industrial growth in NICs

For many years developed countries in Europe and North America, plus Japan from 1960, dominated world manufacturing. They still do, but not to quite the same extent. All are showing some of the symptoms of de-industrialisation as their heavy and mass market consumer goods industries face increasing competition from imports made in developing countries. **Figure 2** shows that the developing world's share of global manufacturing output has reached 30 per cent. It is still rising.

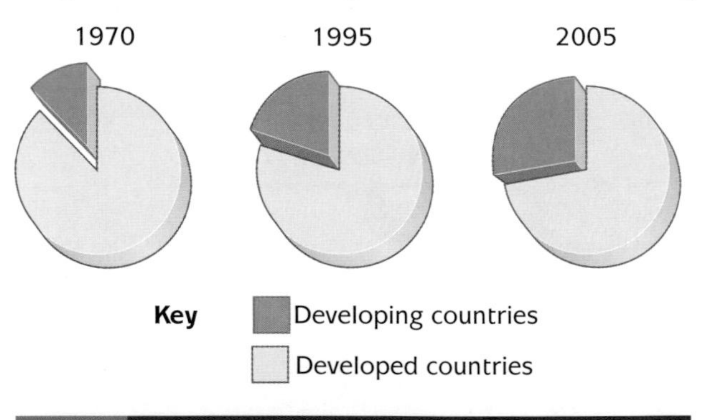

Figure 2 **Percentage share of world manufacturing output.**

Some developing countries experienced rapid industrial growth in the 1970s and 1980s. The most successful of these were labelled NICS. The majority were located in East Asia (page 220), but they also included Mexico and Brazil in Latin America. The trigger for manufacturing growth came from big American, European and Japanese manufacturing companies making the commercial decision to transfer some of their operations overseas. The strongest **pull factor** was lower labour costs; however, there were other advantages, which meant that total production costs were significantly lower than in their home countries (**Figure 3**).

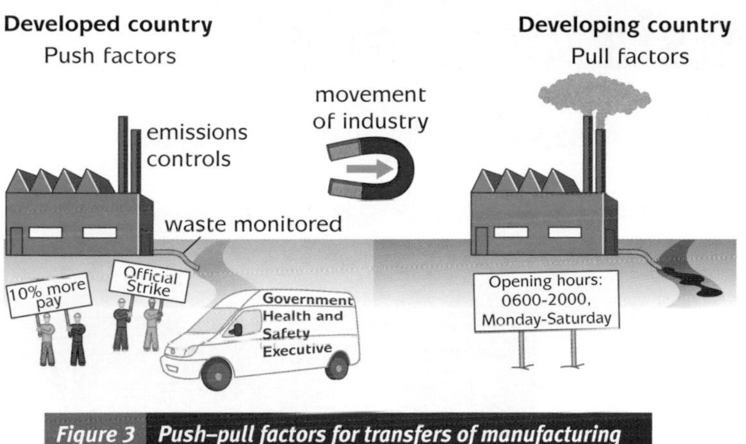

Figure 3 *Push–pull factors for transfers of manufacturing industry from developed to developing countries.*

In most developing countries trade unions, which fight for workers' rights, are either absent or weak. For most people, the prime concern is survival; plenty are willing to work exceedingly long hours to earn extra money. Health and safety regulations and laws, where they exist, are routinely ignored and weakly enforced. The name 'sweat shops' is often used for places with working conditions and practices that would be deemed unacceptable in Europe or the USA. An additional pull factor can be tax-free zones. Sometimes governments create these to allow goods to be moved in and out without payment of customs duties; often companies are given exemption from paying local taxes. Tax rates tend to be lower in developing than in developed countries, and companies may not be forced to pay national insurance or make healthcare contributions. The amount of government legislation and regulation in the European Union (EU) can be a strong **push factor** for TNCs.

Uneven distribution of manufacturing in the developing world

This is quite marked. Opportunities for industrialisation are not evenly spread within the developing world, as **Figure 4** shows. East Asian countries have been responsible for most of the growth away from the developed world since 1980 (page 220); in particular, the significant overall increase since 1995 was largely due to industrial growth in China (page 221). The comparative industrial backwardness of sub-Saharan Africa (that is, all of Africa except for the five countries with a border on the Mediterranean Sea) is very apparent. Many African countries cannot begin to offer what the former prime minister of Singapore considered essential for successful industrial growth (Exam Preparation Box). Instead Africa is a continent that is known for armed conflicts, tribal rivalries, corrupt leaders and politicians, high birth rates and a general economic environment that is hostile to outside investment.

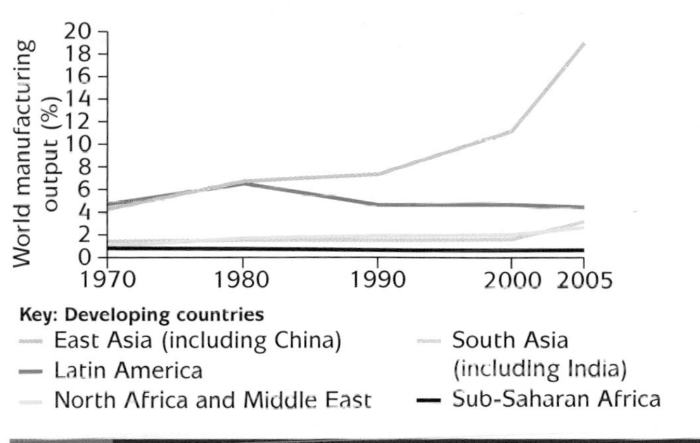

Figure 4 *Shares of world manufacturing output between different parts of the developing world 1970–2005.*

EXAM PREPARATION

Factors needed for successful economic growth:

- political stability
- non-corrupt government
- education for all
- easy access for foreign businesses
- family planning.

ACTIVITIES

1 Use **Figure 1** and employment information in the text.
   (a) Draw a line graph to show percentage employment in manufacturing in the UK 1841–2005.
   (b) Explain how the graph shows the effects of globalisation on the UK.

2 (a) What is meant by de-industrialisation?
   (b) Describe its (i) economic (ii) social (iii) environmental effects.
   (c) Why have its effects so far been felt most in Europe and North America?

3 Study **Figure 3.**
   (a) Identify the push factors for industry transfers from developed to developing countries.
   (b) Which are stronger – push or pull factors? Explain your choice.

4 Choose one country in sub-Saharan Africa to investigate.
   (a) Name its main sources of income.
   (b) Give reasons why it is not a major manufacturing country.

Industrial growth in Asia

What are you wearing and what do you have at home that was made in Asia? Why has China become the workshop of the world?

Hong Kong, Singapore, Taiwan and South Korea developed first and fastest in Asia; they were referred to as the 'East Asian tigers'. They are far enough along the road of economic development to have many of the characteristics of developed countries. Other NICs in the region are Thailand, Malaysia, Indonesia and the Philippines. More recently, Asia's two population giants, China and India, have begun to stir. Between them, they make the full range of manufactured goods from 'ships to chips', and everything in between (**Figure 1**).

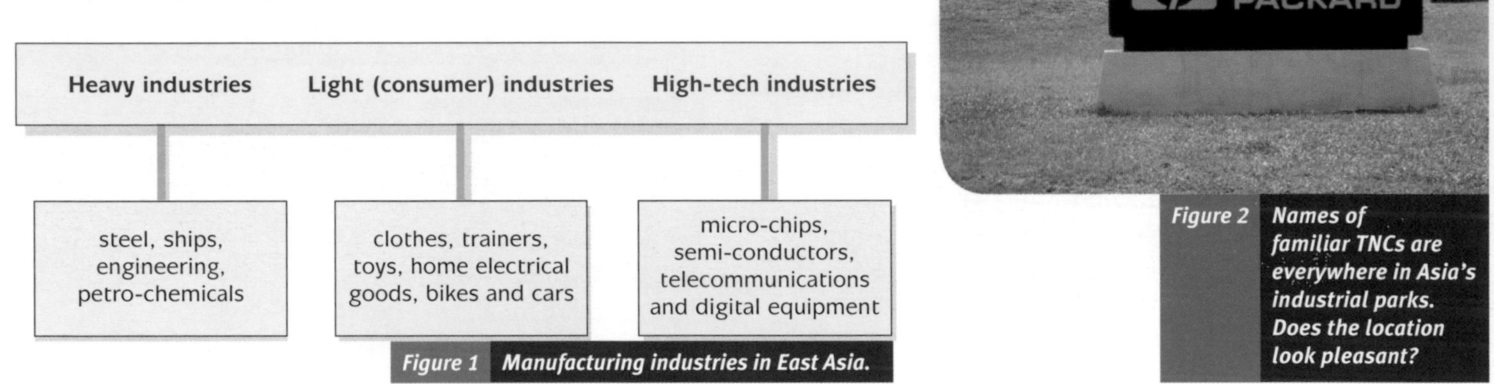

Figure 1 **Manufacturing industries in East Asia.**

Figure 2 **Names of familiar TNCs are everywhere in Asia's industrial parks. Does the location look pleasant?**

Advantages of East Asia for industrial growth

1 Undoubtedly the most important is cheap **labour** supply. Wages are low by world standards (**Figure 3**). The textile industry is used as an example because it is labour-intensive. Asian workers are reliable and work hard for long hours, often in factories that would not meet all the health and safety standards of those in the West.

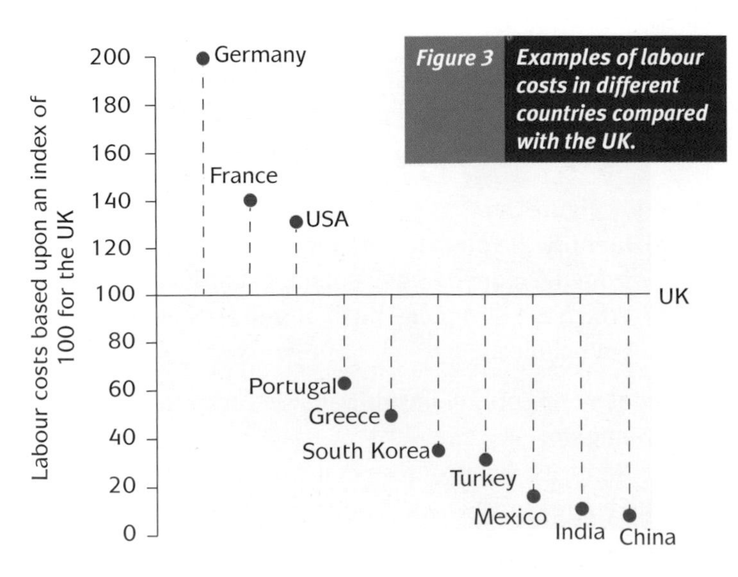

Figure 3 **Examples of labour costs in different countries compared with the UK.**

2 **Transport** is important. All countries in the region have access to the main shipping lanes. The use of containers (see **Figure 2** on page 214) has reduced the cost of transporting manufactured goods by sea, as well as making it much easier and more secure. Sea has always been the cheapest way to transport goods over long distances.

3 The great push towards industrialisation in NICs would not have been possible without **government backing** and a commitment to economic development. Governments in the region encouraged the import of overseas companies' capital and technology to establish factories and provide employment. They were also responsible for providing the political stability necessary for successful industry and trade.

4 **Markets** at home are increasingly important. Although many factories were set up to export all their products, home markets are expanding as people become more prosperous. Asia is the most populous continent; the future potential of Asian markets is enormous as economies grow and personal wealth increases.

China, the new industrial giant

Manufacturing growth since 1979 has been nothing short of spectacular (Information Box). After years of blocking trade with western countries, the communist government did an abrupt U-turn in 1979 and began to allow investment from overseas. Two of the first transnational companies to set up factories were VW cars (Germany) and Pepsi-Cola (USA). Today China is the world's workshop – is it possible to buy a pair of trainers without 'Made in China' stamped inside them?

Why the U-turn? The Chinese leaders accepted that they were never going to modernise the country's factories without the new machines, modern technology, technical skills and efficient working practices that only overseas companies could provide. China remains a communist state; the government still controls the banks, large heavy industries and public services. It is the production of consumer goods, many of them for export, that is in the hands of transnational companies and their Chinese partners.

Why are transnational companies rushing into China?

The simple answer is that most of the advantages 1–4 described on the previous page are greater in China than elsewhere. China has a huge pool of cheap labour, and so far it has undercut other Asian countries on costs (**Figure 3**). The market for consumer goods is already rising in the cities as the Chinese become better-off. The potential size of the home market is enormous. Can any big company afford not to be in a country that houses 20 per cent of total world population? Only for transport is China no better placed than its East Asian neighbours. However, all the big industrial centres are in the east (**Figure 4**). Shanghai is a major port and business centre and Hong Kong remains China's outlet to the world.

Figure 4 *Main industrial areas in China.*

INFORMATION

China and world manufacturing

- World's third largest manufacturing country in 2005.
- Produced 7 per cent of world manufacturing output in 2005.
- Dominates world production of ready-to-wear clothes, footwear and toys.
- Has over 25 per cent of world production of TV sets and washing machines.
- Makes half of all the world's cameras, mobile phones and photocopiers.

The 'China effect' on the rest of the world

China's apparently unstoppable growth is causing pain elsewhere. No one else can match its low production costs for making clothes, not even poor Asian countries like Bangladesh, nor major cotton-producing countries like Pakistan. In traditional industrial countries of Europe and North America, the China effect is speeding up the pace of industrial decline and change. This is even happening in the original NICs; Samsung of South Korea is moving its low-cost manufacturing to China while concentrating on the development of digital products at home.

ACTIVITIES

1 (a) Describe what **Figure 2** on page 218 shows about East Asia's share of world manufacturing output.
(b) State how it is different from that of developing countries in other regions.

2 (a) Describe what **Figure 3** shows about differences in labour costs between Europe and Asia.
(b) Explain fully why many clothes, sports goods and electrical appliances on sale in shops in the EU are manufactured (i) in East Asia in general (ii) in China in particular.

3 (a) Make two lists of the advantages and disadvantages of TNCs and globalisation for Asian countries.
(b) Explain why some Asian countries have benefited more than others.

Increasing global demand for energy

Why is global energy consumption increasing? How great are the environmental impacts of the world's continuing dependence on fossil fuels?

There are two underlying reasons for increasing global energy use. One is world population growth; the other is higher levels of economic development, bringing with it increased wealth and technological advances.

World population growth

World population has increased and the rate of increase has speeded up, particularly in the second half of the twentieth century and in the developing countries (see pages 134–5). It took almost 100 years for the world's population to double from 1 to 2 billion, but then took less than 50 years for it to double from 2 to 4 billion (**Figure 1**).

1830	1 billion	1974	4 billion
1927	2 billion	1984	5 billion
1960	3 billion	1999	6 billion

Figure 1 **Total world population.**

Higher levels of economic development

Modern economic development dates back to about 1750 in the UK and the start of the **Industrial Revolution**. The technological breakthrough came with the invention of the steam engine, which was capable of driving machinery. This led to the growth of factories and towns. **Coal** was the natural resource upon which the Industrial Revolution was based; the Industrial Revolution soon spread to other countries that had their own coal. Germany and the USA industrialised from 1870 onwards. During the last 100 years the number, range and diversity of manufactured goods have increased, leading to today's **consumer-orientated society** in developed countries. Economic growth since 1750 has been accompanied by a massive increase in consumption of natural resources of all types.

People everywhere are seeking an improved standard of living. Homes in developed countries are full of electrical and electronic goods; some do jobs that were formerly done by hand, such as washing and washing-up, while others are for leisure and entertainment, such as TVs, stereo systems, computers and iPods. Private cars use more energy per person than public transport, especially when occupied by only one person. People in developing countries are desperate to follow the example of those in developed countries: once a home is hooked up to an electricity supply, the minimum requirements are a TV and fridge. Factories, offices, homes and transport in those Asian countries that have experienced rapid economic development since the 1970s are consuming more and more energy. China (1.3 billion people) and India (1.1 billion), the world's population giants, are experiencing record rates of economic growth.

Together they explain why global energy consumption has reached the record highs shown in **Figure 2**. Note the continued dominance of **fossil fuels** – oil, coal and natural gas formed over millions of years from the remains of plants and animals.

Figure 2 **World commercial energy consumption, 1988–2007.**

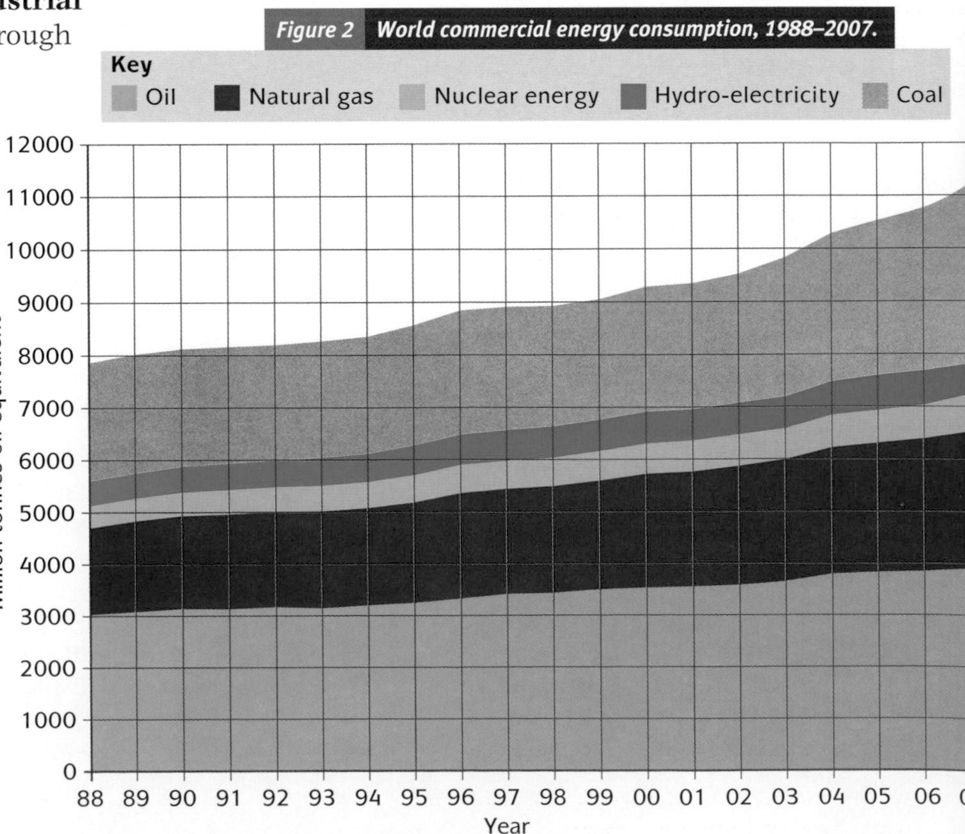

World primary energy consumption slowed in 2007, but growth of 2.4% was still above the 10-year average. Coal remained the fastest-growing fuel, but oil consumption grew slowly. Oil is still the world's leading fuel, but has lost global market share for six consecutive years, while coal has gained market share for six years.

Source: *BP Statistical Review of World Energy 2008*

Global variations in energy use

Energy consumption per head is regarded as a reliable indicator of a country's level of economic development (**Figure 3**). Well-developed manufacturing and service sectors, high levels of transport and many movements of goods and people (both for work and leisure) all have high energy demands. In contrast, the energy demand of an Andean farmer and his family living in a village without electricity is low (**Figure 4**). So too is their quality of life.

In Canada and the USA consumption per head is double that in Europe, and more than 800 times that of the average developing country. Climate provides a partial explanation; many places in North America experience marked seasonal variations between very cold winters (much heating needed) and very hot summers (air conditioning a necessity). However, waste of energy is also a relevant factor. The American love of big 'gas-guzzling' cars is waning but still continues. North Americans tend to overheat their buildings in winter, while in summer the air conditioning is quite severe.

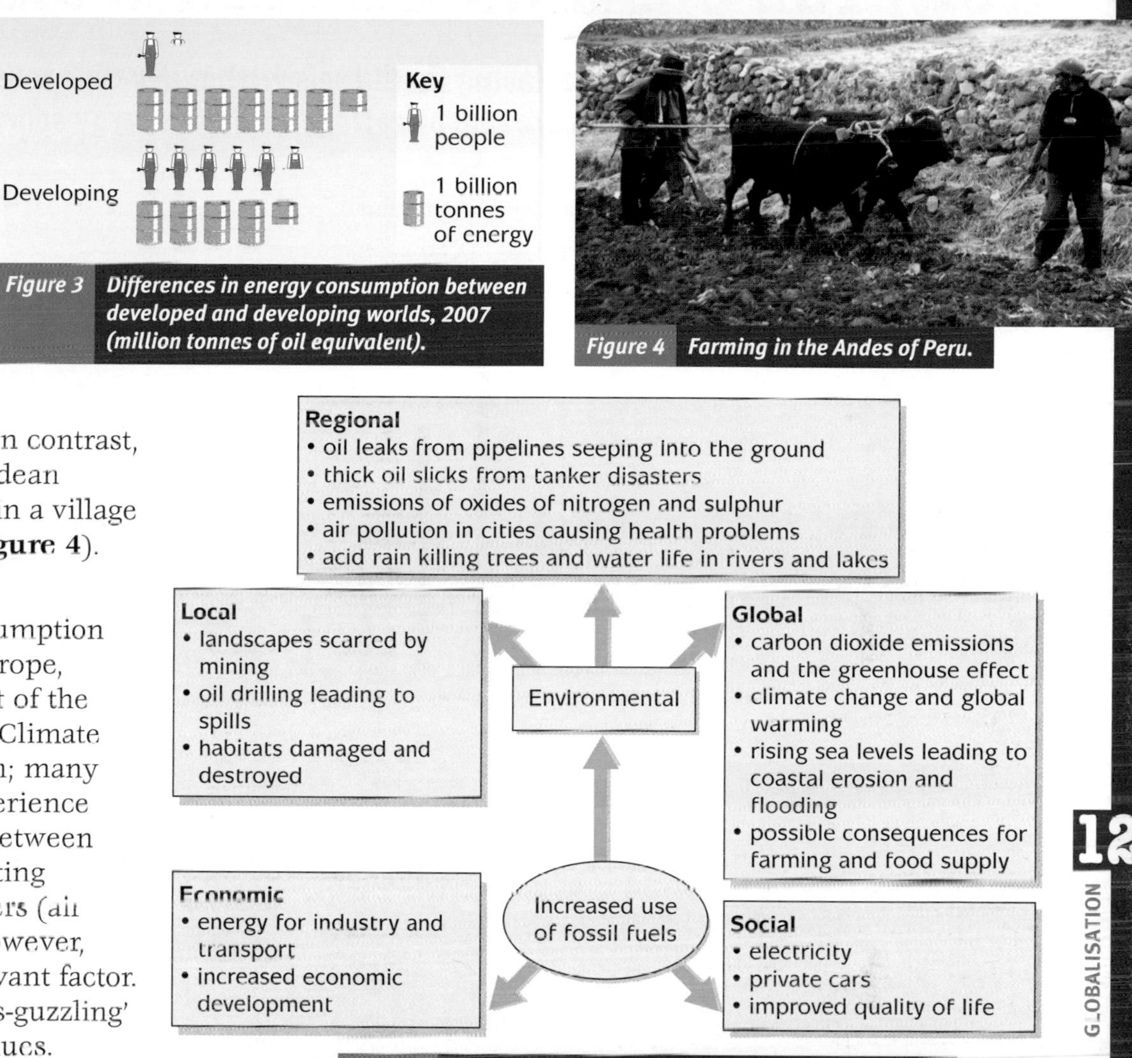
Developed
Developing
Key
1 billion people
1 billion tonnes of energy
Regional
• oil leaks from pipelines seeping into the ground
• thick oil slicks from tanker disasters
• emissions of oxides of nitrogen and sulphur
• air pollution in cities causing health problems
• acid rain killing trees and water life in rivers and lakes
Local
• landscapes scarred by mining
• oil drilling leading to spills
• habitats damaged and destroyed
Environmental
Global
• carbon dioxide emissions and the greenhouse effect
• climate change and global warming
• rising sea levels leading to coastal erosion and flooding
• possible consequences for farming and food supply
Economic
• energy for industry and transport
• increased economic development
Increased use of fossil fuels
Social
• electricity
• private cars
• improved quality of life

Figure 3 **Differences in energy consumption between developed and developing worlds, 2007 (million tonnes of oil equivalent).**

Figure 4 **Farming in the Andes of Peru.**

Figure 5 **Impacts of increased energy use from fossil fuels.**

Impacts of increased energy use

Economic and social impacts are mainly positive (**Figure 5**) – for economic development and raising people's quality of life. However, given that most of the world's energy comes from fossil fuels, environmental impacts are dire.

Controlling them is an **international** issue. Pollution, particularly air pollution, does not stop at country borders. The world is warming up (**Figure 1**, page 54); the effects of global warming are worldwide (**Figure 3**, page 55). The Kyoto climate change conference in 1997 was a pioneering attempt at an international solution. Page 55 gives further information.

ACTIVITIES

1 Using **Figure 2**, work out (i) the difference in world energy consumption between 1988 and 2007 (ii) the approximate total and percentage contributions of fossil fuels to world energy consumption in 2007.

2 (a) Describe what **Figure 3** shows about differences in energy consumption between developed and developing countries.
(b) Explain why developed countries use more energy than developing countries.
(c) How does the scene in **Figure 4** support your answer to part (b)?

3 Read pages 54–5.
(a) Describe how **Figure 1** shows that global warming already exists.
(b) Explain how burning fossil fuels contributes to the greenhouse effect.
(c) State the possible economic, social and environmental consequences of global warming.
(d) Answer questions 2(a) and (b) in the Activities on page 55.

Towards sustainable future energy supplies

What are the good reasons for replacing fossil fuels with renewable sources of energy? Why has it been slow to happen? Why are there still objections to clean energy sources like wind power?

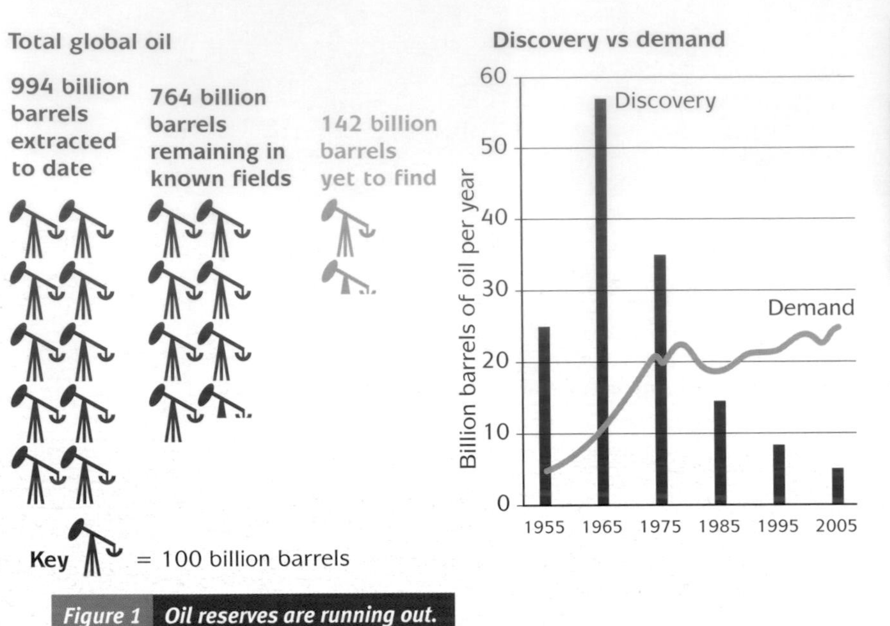

Figure 1 *Oil reserves are running out.*

Figure 2 *Examples of renewable sources of energy.*

The era of cheap, easy-to-extract oil (costing US$10–20 per barrel) is coming to an end. Annual consumption now exceeds new discoveries (**Figure 1**). Future oil supplies will come at a higher cost either from deep water offshore fields or oil sands (in the range US$70–90 per barrel). In contrast, sources of **renewable energy** such as sun, wind, running water, waves and tides, which use natural resources, will never run out. Only these can provide guaranteed **sustainable** sources of energy that will be available to people long after fossil fuels have been used up.

At present renewables contribute only a tiny percentage of world energy supplies, 7 per cent at most. Hydro-electric power (HEP) is the only one that is a major global energy source (**Figure 2**, page 222). Yet the advantages of using renewable sources of energy make impressive reading (**Figure 3**). No atmospheric pollution and complete sustainability are the two big advantages that they enjoy over fossil fuels. In developing countries, a small dam or a few wind turbines may be able to generate enough electricity for village needs and lead to a great improvement in quality of life.

Why is the contribution from renewables still tiny?

The necessary physical conditions are not present everywhere, such as waterfalls with large volumes of water for HEP, or regular strong winds/sunlight for wind turbines/solar panels to operate efficiently. Nor are they necessarily present all the time; there may be a dry season with low water, days without wind or with cloud cover. Thermal power stations burning coal, gas and oil will still be needed as back-ups. Despite the green credentials of their power, they can still arouse opposition from the public and environmental groups – due to habitat losses from dam and reservoir construction, loss of scenic beauty with wind farms, and use of toxic substances such as cadmium sulphide in making photovoltaic solar panels (PVs). But the main factor is simply cost. Even after the big rises in oil and gas prices in 2007–8, fossil fuels remain the cheapest (and easiest) way of producing electricity. Why gamble with investments developing new energy technologies? The incentives have never been great enough during the many years of low oil prices.

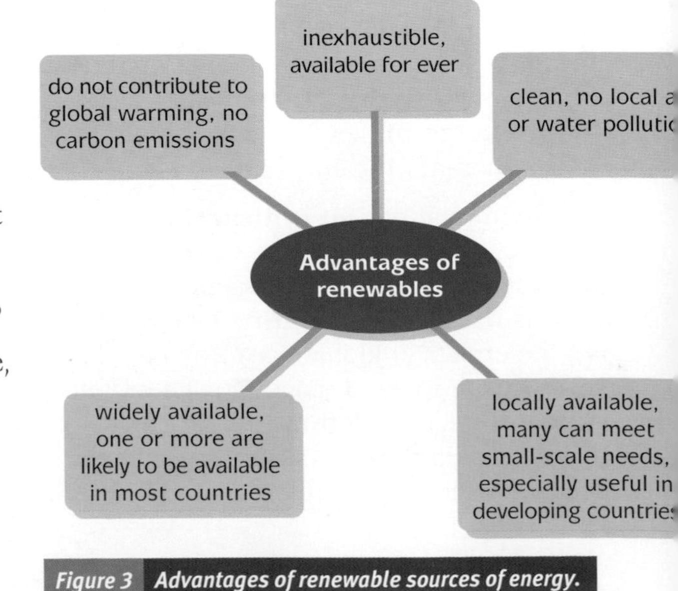

Figure 3 *Advantages of renewable sources of energy.*

Wind power in the UK

Research over two decades has produced a modern, electronically controlled wind turbine, which stands over 30 metres high and has fibreglass blades 35 metres or more across (**Figure 2**). The technology is now well tested. Wind turbines are located in open, exposed places, mainly on hilltops, but increasingly along the coast. The UK has better physical conditions for wind power than most other EU countries. Strong westerly winds are a feature of its maritime climate, and being an island country gives it a greater length of exposed coastline. The investment needed to set up a wind farm is high and has been financially possible only because of government support. If the government is to meet its climate-change targets, such as 10 per cent of electricity from renewables by 2010, the only possible way is 7.5 per cent from wind, which would mean 3000 more turbines on top of the 1000 already in place by 2004. This is why 2004 saw a great increase in wind farm activity (**Figure 4**). Wind power technology is well ahead of wave power, for which the UK also has great natural potential; a new wave machine that 'snakes' across the sea surface where it is rocked by waves to make electricity only began trial operations off Cornwall in 2007.

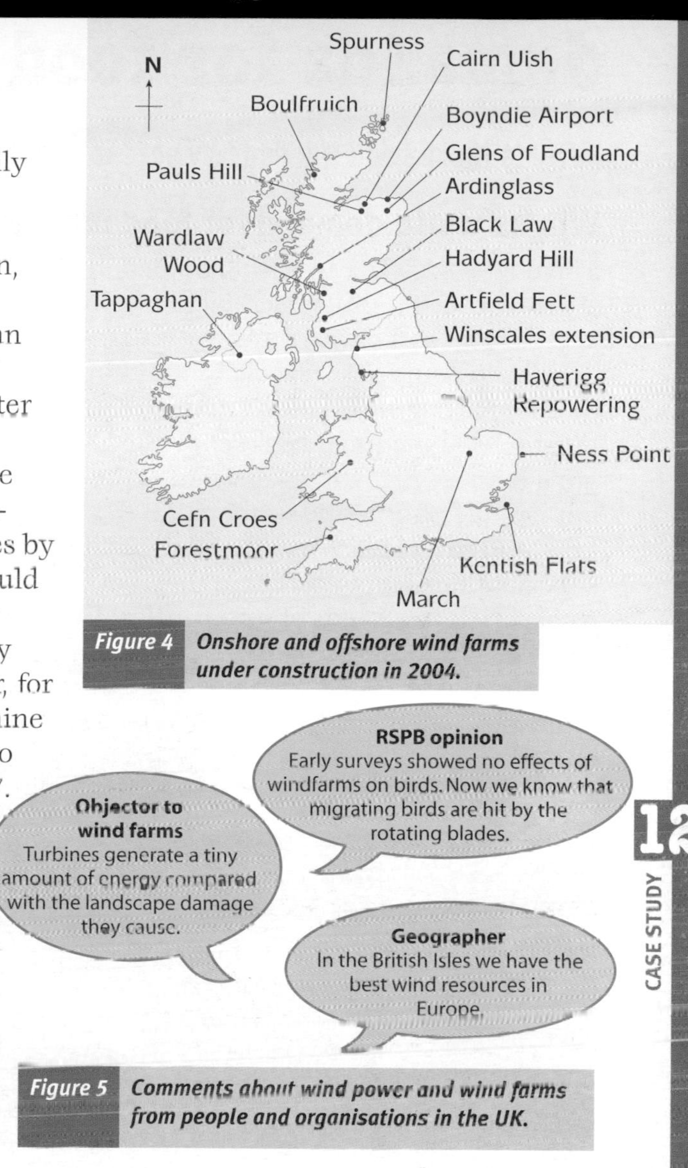
N
Spurness
Cairn Uish
Boulfruich
Boyndie Airport
Glens of Foudland
Pauls Hill
Ardinglass
Black Law
Wardlaw Wood
Hadyard Hill
Tappaghan
Artfield Fett
Winscales extension
Haverigg Repowering
Ness Point
Cefn Croes
Forestmoor
Kentish Flats
March
Objector to wind farms
Turbines generate a tiny amount of energy compared with the landscape damage they cause.
RSPB opinion
Early surveys showed no effects of windfarms on birds. Now we know that migrating birds are hit by the rotating blades.
Geographer
In the British Isles we have the best wind resources in Europe.

Figure 4 *Onshore and offshore wind farms under construction in 2004.*

Poll finding
80 per cent of UK people support wind power.

***Country Life* survey of what people hate most**
Wind farms.

Government agency
If we are serious about tackling climate change we have to look at renewable energy and that means wind power.

Figure 5 *Comments about wind power and wind farms from people and organisations in the UK.*

Wind turbines are pollution-free and the ground between them can still be used for farming. Local people are often less enthusiastic about them because they are noisy and can disrupt TV reception. Many turbines are needed to produce the same output as one thermal power station, and on calm days no power at all is produced. No two people agree on the way they affect the scenic beauty of landscapes; some think they enhance the view, others believe they destroy it.

FURTHER RESEARCH

Check on the progress in harnessing power from wind, wave and tidal energy in the UK at the weblink www.contentextra.com/aqagcsegeog.

GradeStudio

1 Cost of electricity generation for different technologies in the UK in 2004 (pence per kilowatt hour): coal 3.2p; gas 2.5p; nuclear 4.2p; onshore wind 4.2p; offshore wind 5.1p; wave and tidal 14p; photovoltaic solar 45p.
   a Draw a graph to show these values. (3 marks)
   b Describe what they show about wind power in the UK as a source of energy compared with
      (i) fossil fuels (2 marks)
      (ii) other renewables. (2 marks)

2 a Describe the distribution of wind farms under construction in 2004 from **Figure 4**. (3 marks)
   b Suggest reasons for the distribution shown. (3 marks)

3 Study the comments and views about wind energy in **Figure 5**.
   a State the arguments supporting the greater use of wind energy in the UK. (4 marks)
   b How strong are the arguments against use of wind energy in the UK? Explain your point of view. (3 marks)

Exam tip

Start answering straight away. Do not write out the question before answering.

Use the number of marks as a guide to length of answer needed.

Global search to satisfy increasing demands for food

Why is more food needed? Where is it coming from? How bad are the consequences?

Why is world demand for food increasing?

The obvious answer is world population growth (**Figure 1**, page 222), which means more mouths to feed. The less obvious answer is demand from developed countries, especially in Europe. In developing countries, most (if not all) the food that is eaten is home-produced. People in rural areas grow their own; town dwellers go to the market and buy produce from local farmers. All the fruits on sale in **Figure 1B** are readily available in European supermarkets, but to get there they have been transported thousands of **'food miles'** from the tropics. A **carbon footprint** is left behind from emissions released as oil is burnt to power the cargo ships and planes.

Figure 1 Developing world: there are no food miles here.
A: Bringing home the bananas in Brazil.
B: Market for local produce in a small town in Ecuador.

Most foodstuffs eaten in the UK come from crops grown and animals kept in temperate latitudes. However, food tastes have become wider, and more people expect fruits and vegetables to be available all year, not just in season. All of this means more winter food miles between the UK and countries with warmer climates, such as Egypt for fresh tomatoes and strawberries. They could be grown in winter in the UK in glasshouses with heating and under artificial lights, but at a much higher energy cost than under the Egyptian sun. Campaigns to encourage supermarkets, restaurants and ordinary people to buy more local produce are gathering pace. Farm shops in the countryside and farmers' markets in towns are increasing in popularity.

Where is the extra food coming from?

The answer lies in the topsoil, that thin layer which covers the surface. It is being made to work harder, to produce more. How? More intensive farming of the land is only possible with more inputs, typically applying fertilisers (to add minerals and increase yields), and spraying pesticides (to reduce competition from weeds) and insecticides (to kill off bugs that eat the crops). Production can also be extended to new land, for instance by more use of irrigation water in areas where the climate is too dry for cultivation, or by clearing rainforests as in Brazil (pages 74–5). Increased population pressure in poor developing countries is forcing many farming communities to extend production on to **marginal land**, areas of land previously considered not good enough to be worth farming. Reasons why marginal land is left until last include poor soils, steep slopes and unreliable rainfall.

How great are the risks to the environment?

Reference was made to the fragile nature of tropical soils under the rainforests on page 74. Soil erosion was covered on pages 186–7. An example of a cultivated area that can be considered marginal for crop farming is shown in **Figure 2**. What are the problems for farming in this area? What have farmers done to reduce the negative effects?

Figure 2 Colca valley in the Andes of Peru.

The big problem is **environmental degradation**, where soil fertility is reduced to the point where once productive land is turned into wasteland. It is caused by overcultivation and overgrazing as traditional good agricultural practices, such as leaving land fallow (unused) and vegetation ungrazed for some time to allow recovery, have been abandoned due to the demands of increasing populations (**Figure 3**).

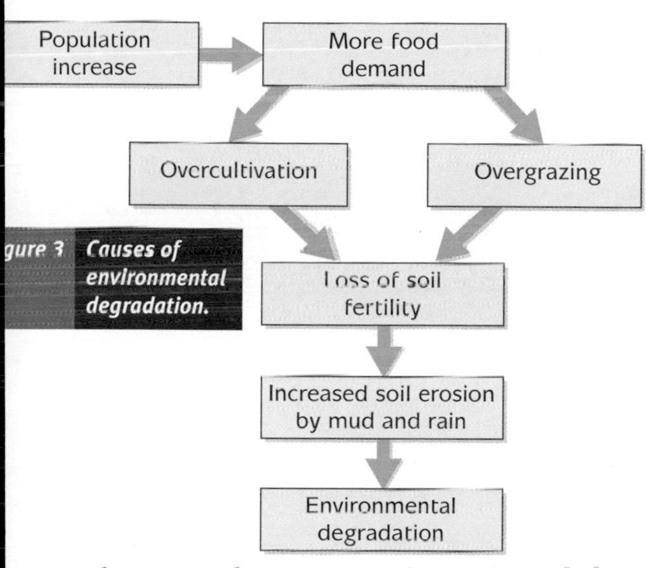

gure 3 Causes of environmental degradation.

What are the economic and social repercussions?

The people who suffer most are poor farmers who have only small plots of land. They are **subsistence farmers** working to feed themselves and their growing families, like the Masai in Kenya (page 245). All they can do is produce more from the land they have, and risk degrading the land and soil. Years of rising populations have taken away any surpluses built up in earlier good years. To buy the higher-yielding seeds, fertilisers and chemicals needed for increased output, they must borrow and go into debt. Bad seasons, due to poor rains for example, increase the debt. Trapped in a cycle of poverty, often the only way out is to sell their land to large farmers and migrate to the city.

The result is a widening wealth gap between poor and rich farmers. Once cash crop farming, associated with globalisation, replaces subsistence farming, the wealth 'gap' becomes a 'gulf'. Overseas buyers demand reliable quality and regular supply; this requires investment in fertilisers, pesticides and insecticides and water supply. All of these put it out of the reach of small farmers, unless they can organise themselves into a local cooperative and participate in fair trade (pages 204–5). What the cash crop farmers are paid for their crops, while low in UK terms, is considerably higher than local market prices. Kenya is a large out-of-season supplier of vegetables to UK supermarkets; it provides vegetables such as green beans, peas and sweetcorn, as well as tropical fruits and fresh flowers. Thousands of Kenyan farmers now depend upon this international trade continuing; so too does the government of Kenya, since these exports are worth US$500 million a year, on a par with the export value of tea and coffee. Agricultural exports account for half of Kenya's visible exports.

In turn this raises other issues:

- Should land, often the best farmland, be used to grow cash crops for export to well-fed people in rich countries, instead of growing food crops for the country's own people? Can this be justified in a time of rising populations, hunger and food shortages in developing countries?
- What about the 5000 food miles between Kenya and the UK, the cost and climate change?

Supermarket boss says 'food miles' miss point

The chief executive of a big supermarket chain will today say that shoppers' concerns over so-called 'food miles' are 'misplaced'. He will also argue that the movement to increase the amount of British food on supermarket shelves is 'flawed' because it over-estimates the UK's ability to be self-sufficient in food.

He will say that supermarkets must trade with Africa and other parts of the developing world to help lift the countries out of poverty. 'We know that because of the size of our business we can make a real difference to people in places like Africa. This is why I am concerned about some suggestions that we should be focusing more on buying British, or at least European'.

He will also argue that concerns over food miles are misjudged. Products produced locally do not always have a lower carbon footprint. He will express his concern that climate change and other green issues are 'overshadowing' poverty reduction, and that they will block the free trade benefits from globalisation to developing countries.

Figure 4 ***Newspaper report.***

FURTHER RESEARCH

Find out more in other parts of the book:

- Reasons for population growth, pages 138–9.
- Rainforest clearances for farming, pages 72 and 74–5.
- Soil erosion, pages 186–7.
- Subsistence economies, pages 72 and 186.
- Kenya's trade, **Figure 3** on page 243.
- Fair trade, pages 204–5.

ACTIVITIES

1 (a) What is meant by food miles?
(b) Explain the environmental effects of high food miles.
(c) State and explain the two arguments used in **Figure 4** to defend high food miles.
(d) Why are food miles lower in developing than in developed countries?

2 (a) Explain why soil is the world's most valuable natural resource.
(b) Draw a labelled sketch of **Figure 2** to show the environmental risks of farming in a mountain area like this.

Tackling issues associated with globalisation

Why is international cooperation needed to tackle pollution and water supply issues? What contributions can be made at the local level?

Many of the issues are truly **international**. The obvious example is atmospheric pollution from burning fossil fuels, global warming and climate change. Pollution controls need to be internationally agreed to be effective. This brings in the **political** factor. Getting countries to agree is never easy. The effectiveness of the Kyoto protocol for reducing greenhouse gas emissions was diminished by the failure of the US Senate and Bush presidency to take global warming seriously, especially as the USA is the world's largest polluter. Many hope that the new Obama presidency will show more environmental awareness. The EU has done much more, setting targets for reducing carbon emissions, increasing energy contributions from renewables and organising more recycling of waste. **Carbon trading** is another initiative to reduce the costs of globalisation. Countries or companies that have exceeded their carbon dioxide emissions targets can buy **carbon credits** from those who have stayed below their targets by consuming less fossil fuel. Look at pages 72–3 to see how carbon credits might be used to save tropical rainforests.

People in the UK, living in a generally wet climate, are less aware of the major global issue of water scarcity, and the potential for 'water wars' between countries sharing river basins. **Figure 1** shows that 88 per cent of water use in Africa is for agriculture (mainly irrigation); in Asia it is only 3 percentage points lower. Without irrigation water, not all of the world's people could be fed.

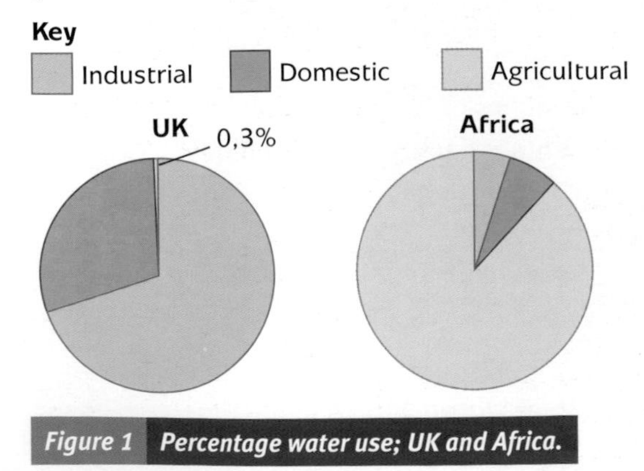

Figure 1 Percentage water use; UK and Africa.

Water supply in the Nile Basin

The Nile is the world's longest river (6695 kilometres). From its origins in Ethiopia (Blue Nile) and around Lake Victoria in Central Africa (White Nile), the river and its tributaries flow through ten countries across half the length of the continent (**Figure 2**). For centuries, Egypt viewed the Nile as its own river, its own private possession, to do with as it liked. Egypt is a populous country of 74 million and is entirely dependent on Nile water to feed its people. This is why Egypt built the Aswan High Dam, opened in 1971. It gives the whole country assured water and electricity supplies; the area under cultivation has doubled. Egyptian farmers are convinced that, if the Ethiopians ever try to stop the Nile, in their words 'Egyptians will attack and kill all Ethiopians'.

Cairo
EGYPT
Aswan High Dam
Lake Nasser
River Nile
Key
Existing dam
Dam project proposed or under construction
Khartoum
SUDAN
Blue Nile
White Nile
ETHIOP
DEMOCRATIC REPUBLIC OF CONGO
UGANDA

Egypt 74 million
GDP per head US$ 4210
Aswan High Dam supplies water and electricity to all Egypt; water resources fully used

Sudan 36 million
GDP per head US$ 19
Suffers from drought
Many old dams now silting up

Ethiopia 77 million
GDP per head US$ 760
Repeated droughts
Plans to use the great water potential of Blue Nile

Uganda 29 million
GDP per head US$ 14
Plans to make better u of its water resources

Figure 2 Nile River Basi

While Egypt irrigates millions of hectares of desert neighbouring countries of the Upper Nile, it remains among the world's poorest and driest. For a long time Ethiopia, Sudan and Uganda demanded a fairer share of Nile water. They are far behind Egypt in all aspects of development. Egypt realised that it needed to cooperate instead of just complaining, and making threats about what it would do if less Nile water reached Egypt. In this new spirit, the Nile Basin Initiative (NBI) was set up in 1999 with the aim of coordinating new developments along the Nile, boosting economies and raising living standards throughout the Basin. New dams are being built and more are planned, for both power and irrigation. Tensions remain; but it makes sense to talk because all of them are affected by variations in water levels, vicious seasonal floods on the Blue Nile and sediment flows.

Tackling international issues at the local level

Some argue that more can be done and more quickly by governments and individuals to tackle issues of global concern. Making improvements in energy efficiency is one example, as preventing pollution in the first place is better than cleaning it up later. The UK government has toughened up the **energy efficiency** standards for new buildings so that less heat is lost through walls, roofs and windows. You may have noticed the energy-efficient labels on electrical goods (such as fridges) in the shops, which is a government initiative. Combined heat and power schemes are encouraged. Next to some power stations are glasshouses heated by water that has passed through the cooling towers. Supermarkets use heat given off by their freezers to heat other parts of the store.

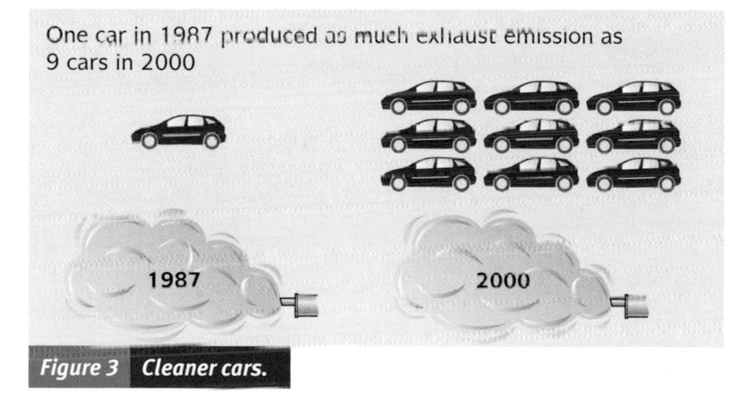

One car in 1987 produced as much exhaust emission as 9 cars in 2000
1987
2000

Figure 3 **Cleaner cars.**

There is no evidence that people in developed countries will stop using their cars; it is perhaps the last thing that most would give up. However, cars can at least be made more energy-efficient and cleaner (**Figure 3**). Today's petrol car in Europe is 90 per cent cleaner than its ten-year-old counterpart. The car companies have been highly successful at reducing emissions of pollutants, such as oxides of nitrogen, carbon monoxide and hydrocarbons, mainly through the use of catalytic converters. Today they are under pressure to achieve greater energy efficiency by reducing fuel consumption using improved engine technology.

Conservation of existing natural resources is also helped by **recycling**, the recovery and conversion of waste products into new materials. One product commonly recycled is the glass bottle (**Figure 4**). Next to supermarket bottle banks, there are often bins for waste paper (pulped down and made into new paper goods), clothes and textiles (converted into upholstery and blankets), and aluminium cans (melted down and manufactured into new containers). For its own sake, recycling is only worthwhile if it saves natural resources without consuming a large amount of energy for reprocessing. Recycling aluminium cans is very energy-efficient because reprocessing only consumes 5 per cent of the energy needed to produce aluminium from its raw material (bauxite). Under pressure from the government and EU, most local councils in the UK have introduced household collections of recyclables to reduce disposal of waste in **landfill**, which can result in land contamination by toxic metals (such as mercury and lead), water pollution and health hazards.

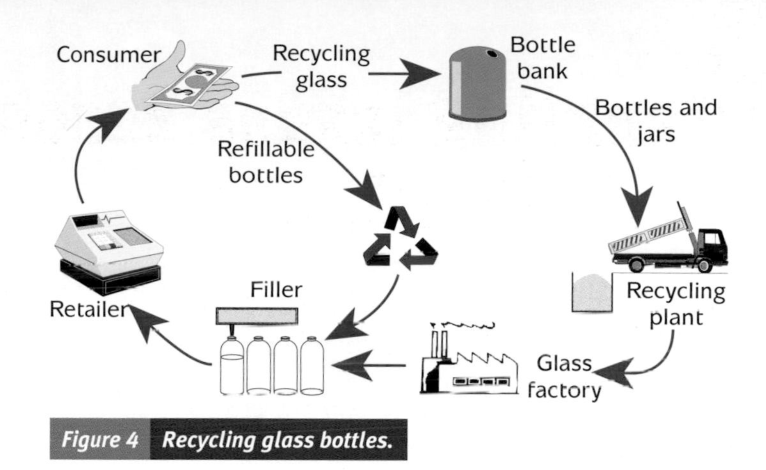

Figure 4 **Recycling glass bottles.**

However, even UK council recycling is not immune from the effects of globalisation. Much of the recycled waste is treated and compressed in the UK before being shipped out to China, to be made into new products. Since the economic gloom late in 2008, raw material prices have fallen and UK recycling companies are cutting back or going out of business.

ACTIVITIES

1 (a) Describe the differences between Egypt and its Nile Basin neighbours for (i) location (ii) wealth (iii) water use.
(b) (i) State two reasons why Nile Basin countries need to cooperate.
(ii) Why is this cooperation not easy to achieve?

2 Study **Figure 4**. Which is better – selling milk in glass bottles or plastic cartons? Explain your answer.

3 Make a large version of the table below and fill it in.

	Definition and examples	Advantages	Problems
Recycling			
Energy efficiency			
Reducing pollution			

GradeStudio

Practice GCSE question

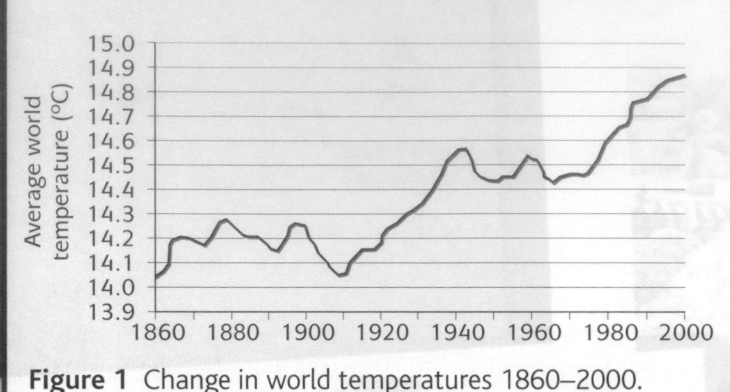

Figure 1 Change in world temperatures 1860–2000.

See a Foundation Tier Practice GCSE Question on the weblink www.contentextra.com/aqagcsegeog.

Figure 3 Drax coal-fired power station.

Figure 4 Wind turbine in UK.

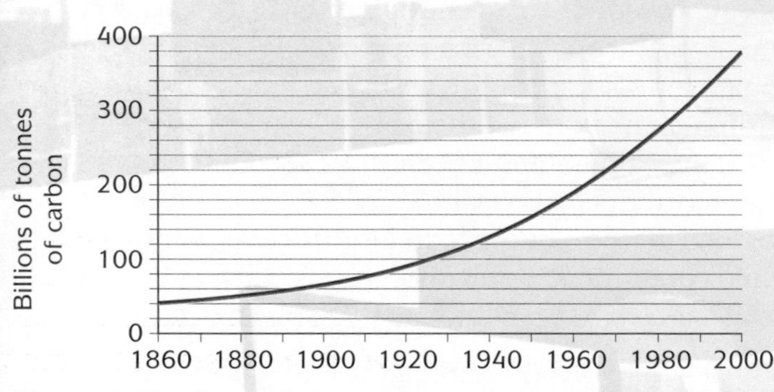

Figure 2 Total cumulative carbon emissions 1860–2000.

	An average coal-fired power station	**The average wind turbine**
Potential output per day	36 000 megawatts	30 megawatts
Energy for	840 000 homes	700 homes
Actual average output per day	36 000 megawatts	8 megawatts
Energy for	840 000 homes	190 homes

Figure 5 Fossil fuel and renewable energy sources compared.

5 (a) Study **Figures 1** and **2**.

(i) Describe the similarities and differences between the two graphs. **(4 marks)**

(ii) Explain how the two graphs might be related. **(3 marks)**

(b) (i) **Figure 3** shows the UK's largest coal-fired power station. Name two gases that are being emitted from it. **(2 marks)**

(ii) **Figure 4** shows a new wind turbine. State two advantages of this location for generating renewable energy. **(2 marks)**

(c) **Figure 5** compares output from coal-fired power stations and wind turbines in the UK.

(i) Using information from **Figure 5**, describe two advantages of fossil fuels over renewables for electricity generation in the UK. **(4 marks)**

(ii) State arguments in favour of an increased contribution from renewable energy sources in the UK and elsewhere. **(4 marks)**

(d) With reference to examples, explain why the importance of manufacturing industry is changing, declining in some countries and increasing in others. **(6 marks)**

Total: 25 marks

Exam tip

The question for this topic will be Question 5 in Paper 2.

Improve your GCSE answers

How to interpret line graphs

Line graphs are widely used for showing changes with time, either during the months of a year, or over a number of years.

What to look for in answering GCSE questions

GCSE question – Describe what **Figure 1** shows. (4 marks)

Figure 1 World population growth.

1 Look for the general **trend**
Possibilities:
- in the same direction all the time, increasing or decreasing
- a change in direction – state what the significant time is when it changes
- many changes; the line fluctuates up and down

2 Look for **speed of change** from the steepness (gradient) of the line
Possibilities:
- flat line or gently sloping – little change with time
- steep line – fast change in a short time
- line with a marked change in gradient

3 Quote and use **values** to elaborate
Possibilities:
- start and end, and size of the difference between them
- highest and lowest (peak and trough), and the time/date for each one
- range – the difference between highest and lowest

GCSE student answer. *Examiner's comment is inserted in red.*

Population is always rising up. (trend = ✓). The fastest increase is after 1950. (speed of change = ✓). Before that the population grew slowly to 1800, but did speed up a bit between about 1800 and 1950. (further detail about speed of change = ✓) In 1950 world population was less than a million, but by 2000 it was over 6 million. (values = ✗ no mark – trying to do the right thing, but no mark because the scale is misread – world population is over 6000 million or 6 billion.) The population rising was caused by an increase in birth rates. (reasons = ✗ no mark – the question does not ask for explanation, only description).

Mark total for this answer 3 out of 4.

Exam tip from this answer
- Read the scale accurately.
- Obey the command word(s) in the question.

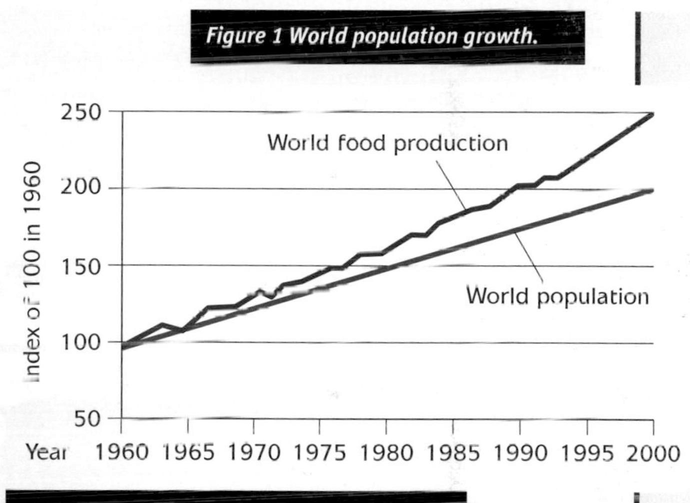

Figure 2 Trends in world food production.

GCSE question – Describe what **Figure 2** shows. (4 marks)

This graph is different because there are two lines. Use the same approach to answering as in **Figure 1** and describe separately (but perhaps more briefly) what each line shows. Then:

4 Look at the **relationship** between the two lines
Possibilities:
- one is greater than the other all the time
- the lines cross once and then get wider apart
- the lines keep crossing; sometimes one is larger, sometimes the other

5 Look for the size of the **difference** between them
- times/dates when the difference was largest and smallest
- any pattern of large and small differences

Exam tip
- Do not be put off by graphs based on an index (instead of actual values), like **Figure 2**.
- Quote and use the values in the normal way.

ExamCafé

REVISION

Key terms from the specification

Exam Café

Carbon credits – each one gives the buyer the right to emit 1 tonne of carbon into the atmosphere
Carbon footprint – emissions of carbon dioxide left behind by burning fossil fuels
Carbon trading – companies that have exceeded their carbon emissions allowance buy carbon credits from those that have not
Cash crop farming – crops grown for sale instead of farmer's own use (the opposite of subsistence farming)
De-industrialisation – declining importance of manufacturing industry
Environmental degradation – productive land turned into wasteland by damage to the soil
Food miles – distance that food travels between supplier and supermarket shelf
Global interdependence – shared need between two or more countries, located anywhere in the world, for one another's goods or services
Globalisation – increasing importance of international operations for people and companies
Marginal land – areas of land previously not considered good enough to be worth using
Renewable energy – natural source of power that will never run out
Subsistence economy – one that is based on what can be grown and provided for itself
Sustainable development – growth of activities working with the environment for a long future
Transnational corporations (TNCs) – large businesses with interests in many countries

Checklist

	Yes	If no – refer to
Do you understand what globalisation means?		page 214
Can you describe the main characteristics of TNCs and give an example?		pages 215 and 217
Can you give reasons why some companies are moving factories and call centres from the UK to Asia?		pages 218-20
Do you know why China has become the world's new economic giant?		page 221
Can you give two reasons for increasing global demand for energy?		page 222
Are you able to explain the environmental disadvantages of fossil fuels?		page 223
List some advantages and disadvantages of using renewable energy sources.		pages 224-5
How is the environment being damaged by increasing global demand for food?		page 226–7

Case study summaries

One TNC	China (economic giant)	One type of renewable energy
Location of HQ	World importance	How it works
Other locations	Types of industry	Locations for its generation
Types of businesses	Reasons for growth	Advantages and disadvantages

Chapter 13
Tourism

Tourism in an extreme environment – adventure travel to places such as Antarctica is increasing.

QUESTIONS

- **Why is tourism one of the great global growth industries?**
- **Why do tourist areas need management for continued success?**
- **Why are tourists seeking out ever more remote places to visit?**
- **Is ecotourism the way to a more sustainable future for tourism?**

Global growth of tourism

Why is tourism one of the world's great growth industries? Why is it so important for the economies of many countries?

According to the World Tourism Organisation (WTO), **tourism** involves activities that require travel from home and staying away from home for at least one night. This not only includes people going on holiday, but also people taking business trips and/or visiting friends and relatives. Most people, however, would also include day visits from home to the coast or a National Park under the heading of tourism.

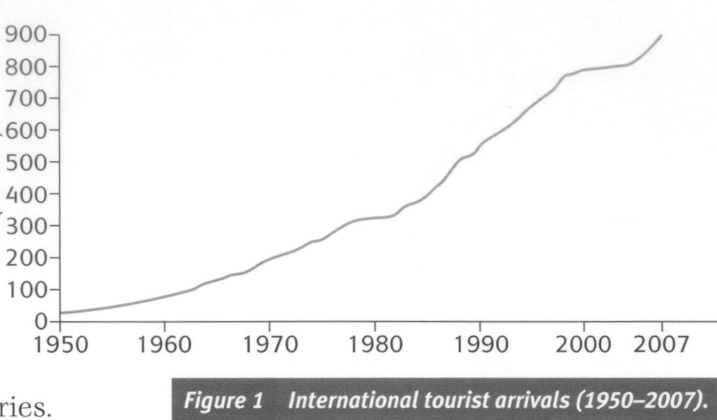

Figure 1 International tourist arrivals (1950–2007).

Tourism is one of the world's great growth industries. Its continued growth since 1950 has been truly remarkable (**Figure 1**). Global tourist numbers are now more than 900 million a year, and still rising. Adverse events, such as terrorism and economic downturns, have only tended to check rates of growth (down to 2 or 3 per cent, instead of 6 per cent, per year) rather than reduce total numbers.

Reasons for the global increase in tourism

Travelling from home costs money. **Economic** reasons for growth are very significant. Tourism growth is a clear sign of growing incomes and increased wealth. The majority of people living in the rich parts of the world have money to spare after paying for life's essentials (food, shelter and clothing). They might choose to spend some or all of it on holidays, weekends away or days out. Going away is just one of the signs of the high quality of life enjoyed by inhabitants of economically developed countries. As other countries develop economically, notably those in East and South-East Asia (pages 219–20), their populations are copying what Europeans have been doing for many years.

Figure 2 Beaches everywhere in Europe are busy on sunny summer weekends. This one is in the Netherlands.

People in rich countries have more **leisure time**. Within the European Union (EU), paid holidays are a right – an annual minimum of four weeks (including Bank Holidays) for every worker. Most people work five days a week for up to 40 hours, leaving two days a week for time off. Going away on the annual holiday is socially acceptable; indeed, it is considered a necessity by many families and included in the annual family budget. In contrast, before the 1960s half of the UK's working population was entitled to only one week's holiday or less per year.

Improvements in **infrastructure** (transport, hotels and other accommodation, and tourist services) have largely kept pace with demand. Increasing car ownership improved family mobility. For travel to other countries, the jet aircraft was a significant technological advance. Most places in the world can now be reached within 24 hours by a wide-bodied jet. The increased speed of getting there has been accompanied by dramatic reductions in the cost of air travel (relative to other costs) on charter flights and low-cost airlines. Booking a holiday has never been easier; either do-it-yourself on the Internet, or a package deal (with transport and accommodation included) from a tour operator.

Economic importance of tourism

Globally tourism is big business. Both rich and poor countries are keen to promote it because it can be a big boost to employment, as well as a major source of foreign exchange. The tourist tree in **Figure 3** shows some of the spin-offs from tourism, which include:

- increases in the number and variety of service sector jobs
- improvements in infrastructure and public services
- support for local industries (construction, food processing and handicrafts)
- increases in local and government tax revenues.

Tourism has a **multiplier effect**, encouraging the growth of services and other businesses.

Tourism is now one of the world's largest economic sectors. Export income generated by international tourism (worth about 7 per cent of the world exports of goods and services) ranks fourth after fuels, chemicals and automobiles; tourism receipts are running at more than US$3 billion per day. What is more, this income is shared by some 80 countries around the world. For many developing countries, tourism is one of the main income sources; for some it is the number one export earner, creating not only much-needed employment, but also opportunities for economic development and modernisation.

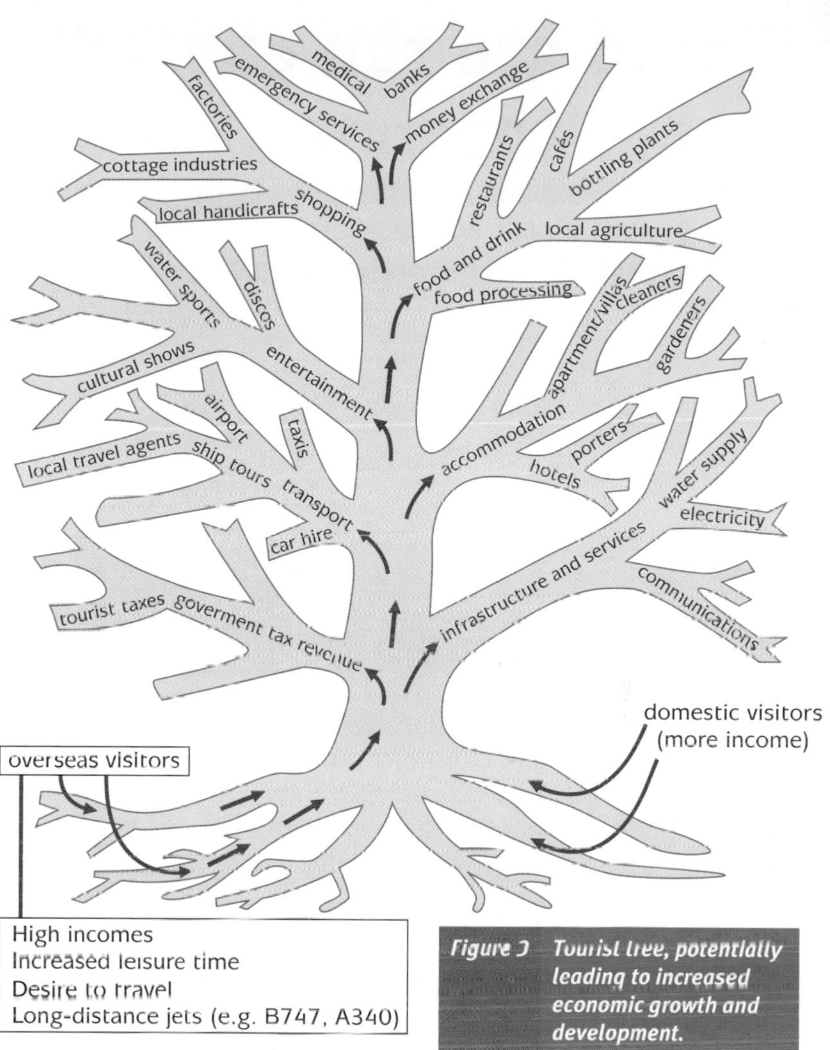

Figure 3 **Tourist tree, potentially leading to increased economic growth and development.**

Until the arrival of long-distance international tourism, many small island countries in the tropics could not see a way to increase their income. Most island states in the Caribbean and Indian Ocean welcome the growth of tourism as the one chance they have to lessen their dependence on exporting a single commodity, such as sugar cane or bananas, which is subject to wide fluctuations in world market price.

Figure 4 **A palm-fringed beach in Mauritius – warm sea water (25°C+) and average monthly temperatures always above 25°C are natural resources that do not exist in Europe.**

ACTIVITIES

1. (a) Describe what **Figure 1** shows about international tourist numbers.
   (b) State three factors to explain what **Figure 1** shows.
   (c) In your view, which one of the factors is the most important? Explain your choice.
2. What are the similarities and differences as tourist resorts between the places in **Figures 2** and **4**?
3. Describe how **Figure 3** shows that tourism in a poor developing country can (i) increase the country's income (ii) lead to better public services and higher quality of life for its people.

FURTHER RESEARCH

Find out more about the WTO at the weblink www.contentextra.com/aqagcsegeog.

Where are the environments that favour tourism?

What types of places do people choose for their holidays? Which are the most popular long-distance destinations for British people?

Rank	Country	Visitors		Examples of places visited
		millions	*% of total*	
1	Spain	13.9	20	Costa del Sol, Majorca, Ibiza and Canary Islands
2	France	11.2	16	Paris, Nice and Côte d'Azur, Brittany
3	Irish Republic	4.2	6	Dublin, Cork and visiting relatives
4	USA	3.9	5	Florida, New York and National Parks in the Rockies
5	Italy	3.4	5	Rome, Venice, Alps and Italian Lakes (e.g. Garda)

Figure 1 *Top five destinations from the UK for visits abroad in 2007.*

Looking at the list of most popular overseas destinations for UK residents offers clues about the types of environments that attract tourists, as well as suggesting some different reasons why people go abroad (**Figure 1**). Three types of environments can be identified, two physical and one human.

Although 80 per cent of visits are to places in Europe, 'long-haul' tourism from the UK has seen great growth. This is generally taken to include places outside Europe

Coastal areas

Basic attractions – the three Ss: Sun, Sand and Sea.

UK coastal resorts have sand and sea, but not guaranteed sun.

The mass summer exodus from the UK to beaches around the Mediterranean took off in the 1970s with charter flights and package holidays, especially to Spain.

Winter beach holidays are to the Caribbean, Indian Ocean and Asia.

Key geographical factor – Climate

Figure 2 *Scarborough. What suggests the typical British summer weather?*

Mountains

Basic attractions – the two Ss: Snow and Scenery.

The Scottish Cairngorms have ski facilities, but not guaranteed winter snow.

The Alps have more snow and it can be guaranteed at high levels.

The Rockies of USA and Canada have more snow and it is guaranteed.

Scenery in the Alps and Rockies is more rugged and more varied than in the UK.

Key geographical factors – Climate and Relief

Figure 3 *Hiking in the Swiss Alps (see page 109 for more details about the glacier, and pages 110–11 for information about Alpine tourism).*

Cities

Basic attractions – many and varied, but mostly human.

Historical buildings – religious, defensive public buildings, palaces.

More recent creations – bridges, opera houses, sports stadiums, museums.

Major events – sporting, cultural, historica entertainment, international events.

Uniqueness – Venice with its canals, Manhattan skyscrapers in New York, Cuzco with Inca remains, Mecca with the Holy Kaab

Key geographical factor – Human

Figure 4 *Grand Canal in Venice, the city's main highway.*

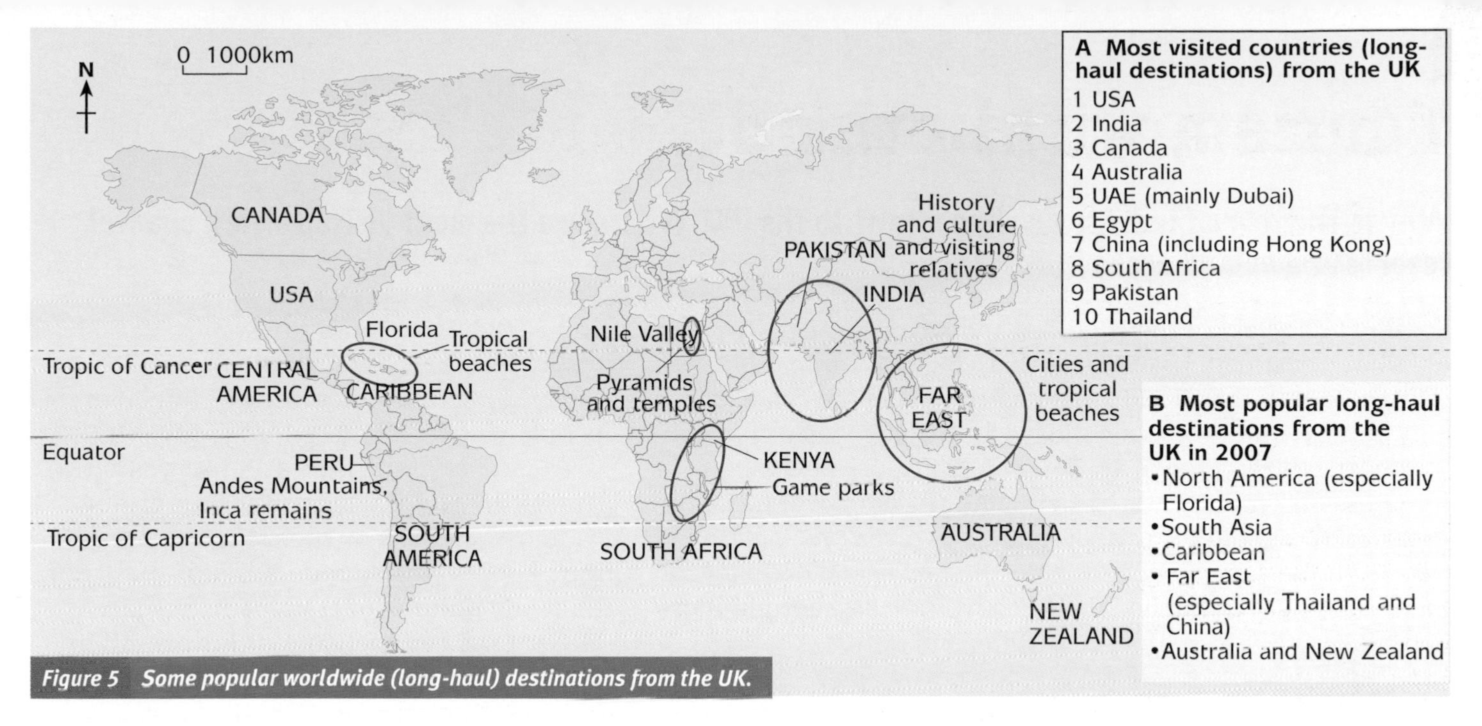

***Figure 5** Some popular worldwide (long-haul) destinations from the UK.*

and beyond Morocco and the Canary Isles, with flight times above four hours (more than 2000 air miles) from London. The most popular destinations in 2007 are named in **Figure 5B**. The USA is now established as one of the permanent top five destinations from the UK. All the other countries listed in **Figure 5A** received between 400 000 and a million visitors from the UK in 2007. Notice the large number of Commonwealth countries, showing the importance of links established in former colonial times, including familiarity with the English language, and visiting relatives and friends. Visiting relatives is the dominant type of tourism to Pakistan.

The growth of long-haul tourism partly reflects people's greater experience of travel and desire to broaden their horizons. Increasing numbers wish to see environments and experience cultures different from their own, when they can afford it. For many years Kenya dominated the African wildlife and safari market for tourists, but now it is facing stiff competition from South Africa and Tanzania. Central and South America are climbing up the long-haul destination charts as people are becoming more aware of the region's natural attractions (mountains, waterfalls, glaciers, jungles and wildlife) and monuments from earlier civilisations (Mayas, Aztecs and Incas). Having the money and time to do it are the main reasons why people from Europe and North America head for the Caribbean in winter. The peak season is December to April, which also coincides with the dry season in the Caribbean.

GradeStudio

1 Visits abroad from the UK in 2007 to various world regions (percentages): Europe 80%; North America 6.5%; Asia (including Middle East) 6%; Africa 4%; Caribbean 1.5%; Australia and New Zealand 1%; and Central and South America 1%.

a Draw a graph to show these percentages. (4 marks)

b For places visited by people from the UK, what do they suggest about the importance of

(i) distance from the UK (4 marks)

(ii) historical links with countries. (4 marks)

2

Caribbean climate data	J	F	M	A	M	J	J	A	S	O	N	D
Av. daytime max. temp. (deg. C)	30	30	30	31	31	32	32	32	32	31	31	31
Rainfall (mm)	20	18	10	37	138	114	51	92	86	168	52	25
Daily sunshine hours	8	9	9	9	8	8	9	8	8	7	9	8

a Describe the main features of the climate of the Caribbean as a tourist destination. (4 marks)

b State the physical and human reasons why the Caribbean is a popular winter tourist destination. (4 marks)

Exam tip

Look at the Exam Preparation Box on page 44. This is a guide about what to look for when describing the climate of a place.

FURTHER RESEARCH

Find out more about tourist areas named in **Figure 5** from the websites of national tourist organisations and tour companies.

Tourism within the UK

Why is the tourist industry so important to the UK? Where are the most visited cities, coastal resorts and upland areas in the UK?

So much publicity in travel guides and advertisements is given to holiday destinations overseas that you might think that tourism within the UK no longer exists. While the balance between UK and overseas holidays for British people changed in the late 1990s (**Figure 1**), the domestic tourist industry remains substantial, with an estimated worth of over £65 billion in 2007. On top of this, over 30 million overseas visitors came to the UK in the same year with an estimated spend of up to £20 billion. The UK is number six in the world for visitor numbers behind France, Spain, the USA, China and Italy.

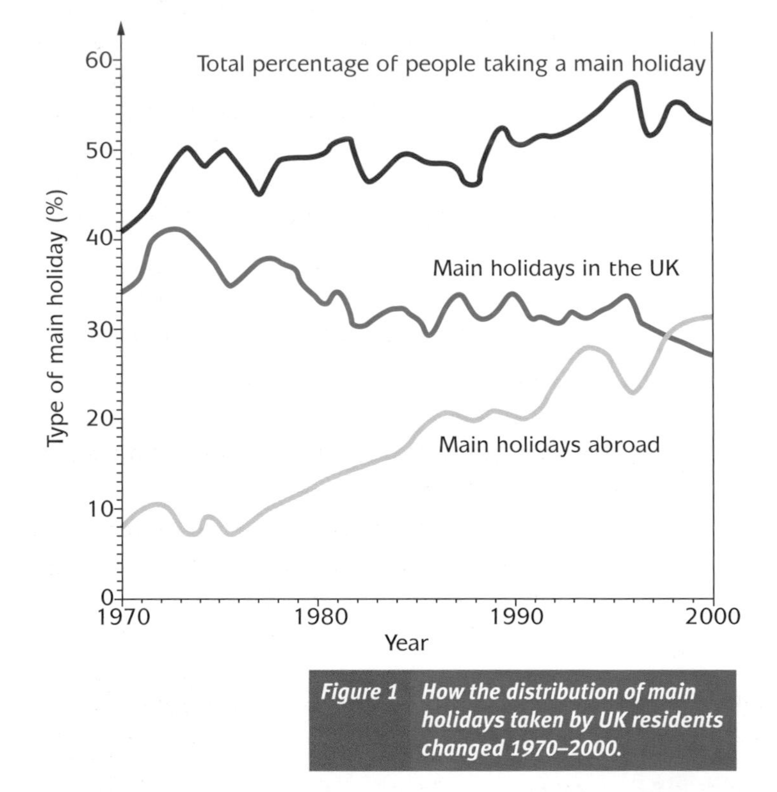

Figure 1 How the distribution of main holidays taken by UK residents changed 1970–2000.

Tourism is regarded as one of the UK's key long-term growth sectors, providing employment for over two million people, either directly or indirectly, in service sector occupations. Direct employment is 5 per cent of the total workforce. This includes workers in all types of accommodation from luxury hotels to camp sites, food and drink suppliers, transport, travel agencies and tour operators, museums and historical sites, entertainment and sport ... the list is endless. Although the greatest single concentration of tourism-related jobs is in London, they are widely dispersed in the UK, including some in remote rural and upland areas where alternative employment opportunities are limited.

Unfortunately, tourist numbers go up and down. People making their living from tourism have no control over some of the major deciding factors. Will two wet summers in 2007 and 2008 be bad news for English seaside resorts in 2009 as more British people seek the Mediterranean sun? How many people will be able to afford to go away if unemployment rises? In most years, Americans are the biggest visitor group from overseas, as in 2007 (**Figure 2A**); they are also big spenders. Their numbers fluctuate quite a lot from year to year. Most of the factors listed in **Figure 2B** are described as **external factors** – factors outside the control of the UK tourist authority. Media coverage might be the only exception.

A Country of origin (2007)

Rank	Country	Number (m)	% of total
1	USA	3.6	11
2	France	3.4	10
3	Germany	3.4	10
4	Irish Republic	2.9	9
5	Spain	2.2	7

B Factors influencing American visitor numbers to UK

Currency	weak £, strong $	✓	UK good value
	strong £, weak $	✗	UK expensive
Security	terrorist bomb attack	✗	fearful
	no recent incidents	✓	peace of mind
Economic activity	expanding economy	✓	happy to spend
	rising unemployment	✗	afraid to spend
Media coverage	good publicity about UK	✓	attracts
	bad reports/ tourist incidents	✗	repels

Figure 2 Overseas visitors to the UK.

Which are the tourist areas in the UK?

Surveys show that some 50 per cent of overseas tourists spend at least half their stay in London. Certain places in the capital are visited by virtually every overseas visitor – Trafalgar Square, Piccadilly Circus, the Tower of London, Houses of Parliament and Big Ben, London Eye, at least one of the museums ... how many more 'must

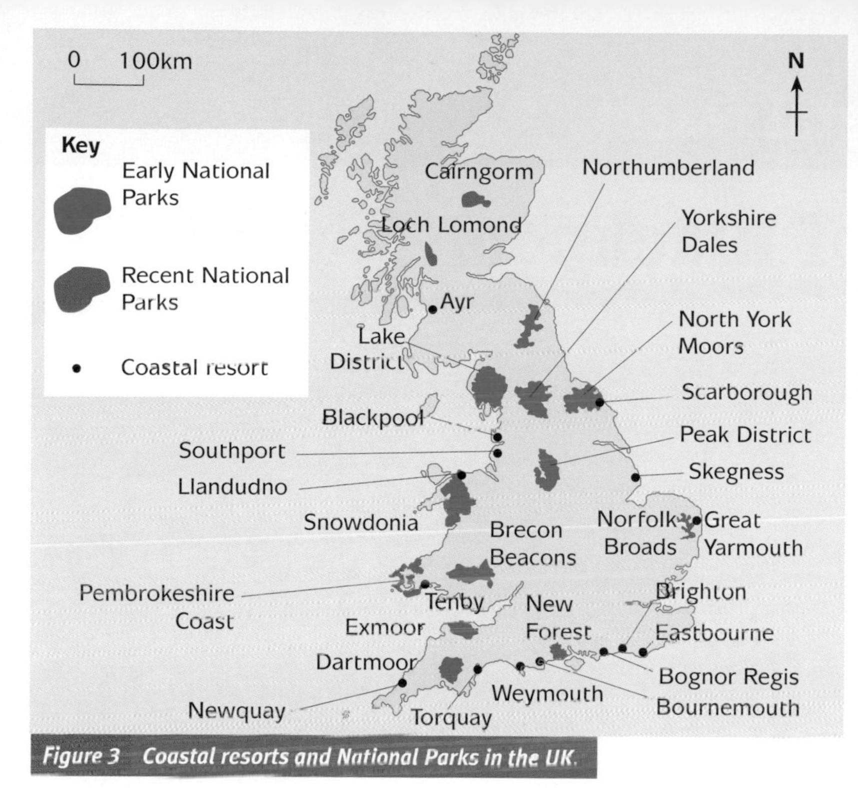

Figure 3 Coastal resorts and National Parks in the UK.

see' tourist sites in London can you name? Other inland cities and towns that are major tourist destinations (for both home and overseas visitors) are York and Edinburgh, Oxford and Cambridge, Stratford-upon-Avon and Windsor. What are their main attractions? Cities, together with coastal seaside resorts and National Parks (**Figure 3**), make up the three types of tourist environments referred to on page 236. Most of the National Parks, though not all, are in upland areas.

Although UK **coastal resorts** are in decline, about 20 per cent of UK main summer holidays are taken in such resorts. A beach is a necessity for any coastal resort; if the longshore drift is in the habit of washing it away, the local authority builds groynes to preserve it (page 125). Spectacular coastal scenery close by, such as cliffs, caves, arches and stacks, helps, as do scenic upland areas inland from the resort for day trips. Another factor of great importance is climate. The greatest concentration of large coastal resorts is along the south coast of England. Here the warmest summer weather and highest number of hours of sunshine are recorded.

Within the past 60 years, scenic environments inland have increased in popularity. Pressure of visitors and conflicts between local people and visitors in areas of great scenic beauty led to the setting up of National Parks, through The National Parks and Access to the Countryside Act of 1949. A **National Park** can be defined as 'an area of beautiful and relatively wild countryside'. Creating a National Park has two aims:

- to preserve and enhance an area's natural beauty
- to promote people's enjoyment of the countryside.

Figure 4 Improved footpath using local raw materials near Malham Cove in the Yorkshire Dales National Park.

Achieving both aims is not easy – if there are too many visitors they destroy the peaceful and beautiful countryside they are all going to see. Rules and regulations are needed and they must be enforced, which is why each park is managed by its own National Park Authority. Management tasks are a mixture of the positive and the negative:

- managing the land, undertaking conservation work, planting woodland and repairing/re-routing footpaths
- working with and advising local landowners
- controlling building and new commercial developments
- providing access and setting up facilities for visitors, such as information centres, car parks and picnic sites, while controlling where they are located.

ACTIVITIES

1 (a) Identify from **Figure 5** at least four typical features that overseas visitors expect to see.
(b) State two different reasons why a majority of overseas visitors spend some time in London.

Figure 5 Central London.

2 Study **Figure 3**.
(a) Describe the distribution of the ten early National Parks.
(b) State two ways in which the distribution of the recent National Parks is different.
(c) Explain why
(i) there are more National Parks in the north and west than in the south and east of England.
(ii) the greatest number of coastal resorts is along the south coast of England.

The rise and fall of Blackpool

Based on visitor numbers, Blackpool is still the UK's leading coastal resort. The view in **Figure 1** shows many of the features typical of a British seaside resort, like donkeys on the beach and a sea front promenade, as well as its special visitor attraction, Blackpool Tower.

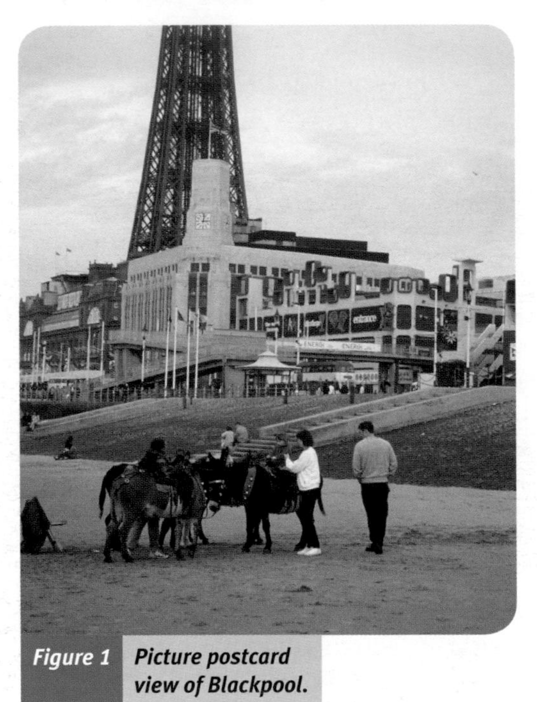

Figure 1 *Picture postcard view of Blackpool.*

All the coastal resorts named in **Figure 3** on page 239 had their origins in Victorian times, and their reasons for growth were similar (points A and B in **Figure 2**). By the middle of the nineteenth century the railway had linked Blackpool to Manchester and the other densely populated textile towns of Lancashire. Factory workers poured into Blackpool on Bank Holidays. Later, after the introduction of paid annual holidays, they spent a week there, every year. Many northern families never considered going anywhere else. Blackpool had the natural advantage of a sandy beach, which stretched for miles; by 1900, the tourist infrastructure of promenade, piers, big hotels and the Tower were all in place.

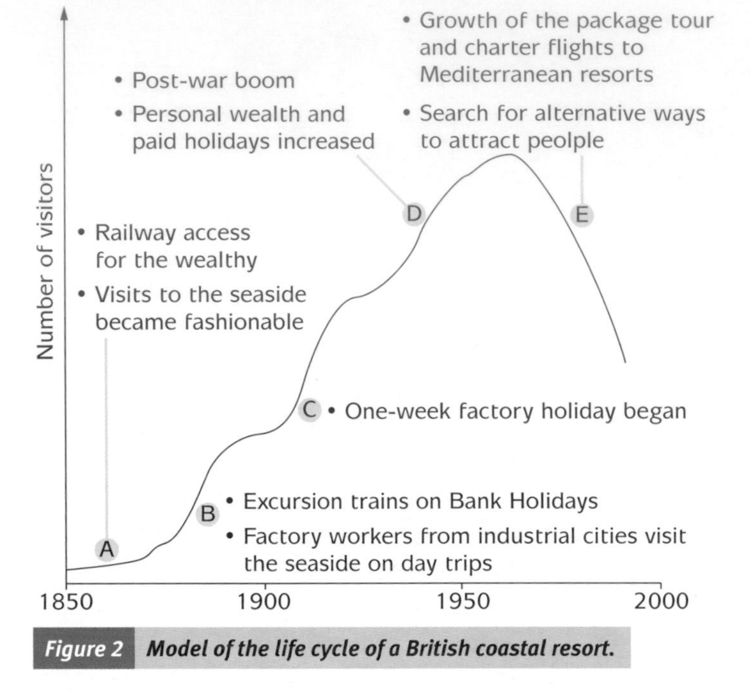
Number of visitors
• Growth of the package tour and charter flights to Mediterranean resorts
• Search for alternative ways to attract peolple
• Post-war boom
• Personal wealth and paid holidays increased
• Railway access for the wealthy
• Visits to the seaside became fashionable
A
B
C
D
E
• One-week factory holiday began
• Excursion trains on Bank Holidays
• Factory workers from industrial cities visit the seaside on day trips
1850
1900
1950
2000

Figure 2 *Model of the life cycle of a British coastal resort.*

Decline and decay in Blackpool

Growth was more or less continuous until 1960 (points C and D in **Figure 2**). However, the traditional British seaside resorts have been in decline for 40 years, ever since people discovered guaranteed summer sun and warmth in Mediterranean countries. Blackpool was badly affected. Look what happened in the decade from 1990 to 1999:

- visitor numbers per year dropped from 17 million to 11 million
- 1000 hotels ceased trading
- 300 holiday-flat premises closed
- average hotel occupancy rates fell as low as 25 per cent.

Blackpool was not exciting existing visitors enough to make them come back the following year, nor was it attracting sufficient new customers. By 2000 some bed-and-breakfast prices had fallen as low as £10 per night, which left no money for investment in improvements. A downward spiral of decline set in as some parts of town started to look very run-down. This happened despite improvements in road access after the M55 was completed as the motorway link from the M6.

Families frightened off by binge-drinking culture of 'stag nights' and 'hen parties'

Beach erosion during winter storms

Beach and sea water pollution

Unemployment out of season

Overcrowding and traffic jams on Bank Holidays

Cheap package holidays to the Mediterranean taking regular visitors away

Unreliable summer weather – wet and windy

Figure 3 Blackpool's problems.

Figure 4 *Blackpool tram and Pleasure Beach. Both still attract visitors, but the town does the traditional better than the modern*

Strategies for solving Blackpool's problems

The local authorities now recognise the need for urgent action to arrest the decline. Since 2001 serious efforts have been made to smarten up areas frequented by visitors, by pulling down old buildings and landscaping car parks. Beaches have been cleaned up and beach facilities improved, so that by 2006 three of them were flying EU blue flags. Sand extraction further south along the coast has been reduced. The 'Blackpool Illuminations', which are vital for extending the visitor season into the autumn, are being transformed by a £10 million investment after years of 'always being the same'. Other off-season events, such as conferences and festivals, are being promoted.

Attractions for both summer and winter are essential to offset the effects of the British weather. Some visitors enjoy the thrills of the Pleasure Beach including 'the Big One', once the biggest and fastest rollercoaster in the world. A new attraction, Water World, opened in 2006. There are plans to make more covered ways between the main visitor attractions and around the shops, for greater comfort in bad weather. Blackpool put in a bid to house the government's new super-casino; some dreamed of Blackpool being turned into the 'Las Vegas of the UK', attracting massive investment into the resort.

How effective have the strategies been?

The big dream was shattered when the government awarded the super-casino to Manchester (only for the plan then to be scrapped). Average occupancy rates in Blackpool holiday accommodation remain below 25 per cent. Almost everyone acknowledges that Blackpool will have to rely on day trippers and people staying two or three nights at best. Competition for day trippers is increasing, since there is now so much for people to do. 'I see places here the same as they were 20 years ago' was the comment from one recent visitor. Blackpool Pleasure Beach remains the most visited attraction in the UK, but numbers fell from 7 to 6 million between 2000 and 2005. The feeling is spreading that Blackpool has been left behind. The Labour Party deserted Blackpool for Manchester for its 2008 conference.

Size limits Blackpool's options. Smaller resorts can look for niches to move forward. Celebrity chef Rick Stein's restaurant has done wonders for the small Cornish resort of Padstow. Scarborough is diversifying its economy by aggressively promoting the growth of small business enterprises. Blackpool is too big for small solutions like these and suffers from its image. It is not a wealthy town, and has high rates of unemployment and areas of deprivation. Does the same fate await Benidorm in Spain?

ACTIVITIES

1 Draw a large version of the model in **Figure 2** with a title 'Life cycle of Blackpool'. Add labels that apply to Blackpool.

2 (a) Make a table and rearrange Blackpool's problems under the headings: Physical, Environmental, Economic and Social.
(b) Choose two problems from different headings and explain each of them more fully.

3 (a) How typical are Blackpool's problems of British coastal resorts in general?
(b) Are these problems easy or difficult to solve? Explain your answer.

4 Adopt the role of one of the people listed below. Write a short but precise letter to your local paper, outlining your views on how Blackpool should be managed in the future:
- local resident who runs an amusement arcade on the sea front
- representative of a conservation group
- member of the local council
- retired resident of Blackpool.

Mass tourism – good or bad?

What are the economic and environmental effects of tourism? Why does Kenya attract more visitors than most other African countries?

GAINS	LOSSES
A Economic • Great earner of foreign exchange. • Increases size of the domestic economy (e.g. transformation of Spain from poor Mediterranean country to wealthy Western European country) • New opportunities from the great increase in number and variety of service occupations (**Figure 3**, page 235); tourism is labour-intensive. • New infrastructure (airports, roads, water and electricity supplies) can benefit other industries. • Low-income jobs can be converted to provide a better living (e.g. fishing boat used for coastal tours, tourist fishing or snorkelling and diving trips).	***A Environmental*** • Complete destruction of environments, and resulting habitat losses, in order to build hotels, roads and airports. • Loss of rural peace and quiet, which is replaced by urban activity and noise. • Pollution problems from litter and untreated waste going into rivers and sea. • Specific local issues, e.g. divers damaging coral reefs in the tropics, pressure on frequently visited landscapes (e.g. footpath erosion), disturbance of wildlife in natural environments.
B Environmental • Greater awareness of the need for, and interest in, conservation of landscape features, vegetation and wildlife, and preservation of ancient monuments. • Income from tourism/ entrance fees may pay for management, conservation and repairs.	***B Economic*** • Some local people, notably farmers and fishermen, may lose their livelihoods. • Visitor numbers go up and down and the area's popularity may wane. • The country/tourist region might gain only a small percentage of total tourist spend. • Many jobs in tourism are seasonal, poorly paid, low status and unskilled; high-earning jobs, such as those of guides with language skills, often go to outsiders.

Figure 1 ***Consequences of mass tourism for Spain and other countries.***

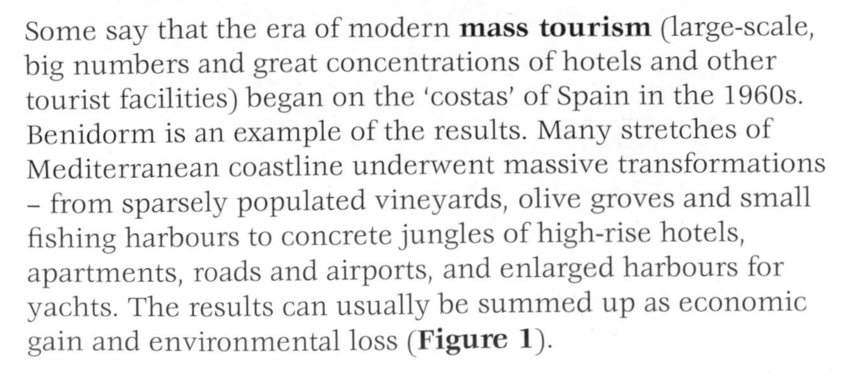

Some say that the era of modern **mass tourism** (large-scale, big numbers and great concentrations of hotels and other tourist facilities) began on the 'costas' of Spain in the 1960s. Benidorm is an example of the results. Many stretches of Mediterranean coastline underwent massive transformations – from sparsely populated vineyards, olive groves and small fishing harbours to concrete jungles of high-rise hotels, apartments, roads and airports, and enlarged harbours for yachts. The results can usually be summed up as economic gain and environmental loss (**Figure 1**).

Economic losses are greatest for local people, often the elderly, who are less able to adapt to the new economy; they suffer the greatest social loss as cultural traditions and community ties are destroyed. It is generally agreed that rich countries make more out of tourism than poor countries. A poor country can make as little as 15 per cent of the total amount paid in the UK for a holiday at a tropical beach resort, if the booking is through a UK travel agent, travel is on a UK airline and the hotel is part of a big American or European chain. Any environmental gains usually come later, after much damage has been done in the construction phase. Only then does effective management begin to protect what remains.

Some poor countries in the developing world have been able to exploit the potential economic gains from modern mass tourism more than others. One of them is Kenya (**Figure 2**). Others have never really been on the tourist map of the world, for many and varied reasons:

- inaccessible locations (e.g. landlocked countries in Africa and Asia)
- climate too extreme for comfort (too hot, cold or wet)
- lack of environments that are of interest to visitors

Figure 2 ***Kenya offers visitors the chance to see wild animals in natural environments. What tells you that this is the wet season?***

- political instability (e.g. civil wars, frequent changes in government, military rule, crime)
- government hostile to tourists (e.g. North Korea).

Why does Kenya attract large numbers of visitors?

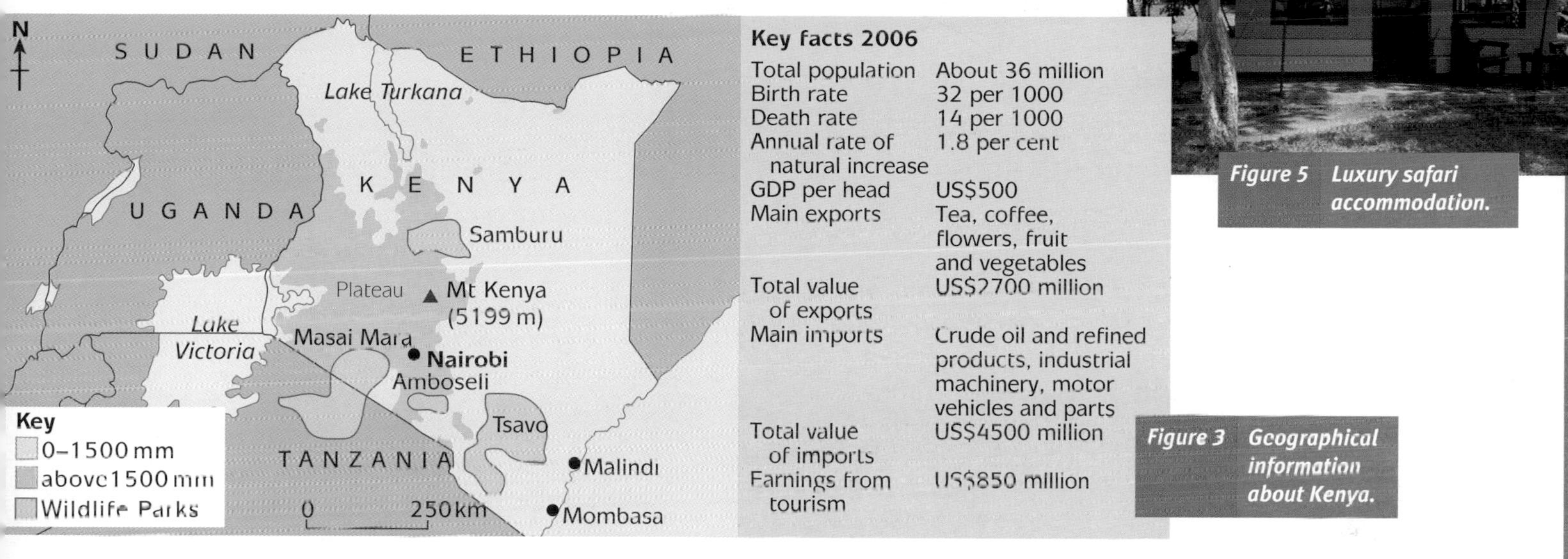

Key facts 2006

Total population	About 36 million
Birth rate	32 per 1000
Death rate	14 per 1000
Annual rate of natural increase	1.8 per cent
GDP per head	US$500
Main exports	Tea, coffee, flowers, fruit and vegetables
Total value of exports	US$2700 million
Main imports	Crude oil and refined products, industrial machinery, motor vehicles and parts
Total value of imports	US$4500 million
Earnings from tourism	US$850 million

Figure 3 Geographical information about Kenya.

Figure 5 Luxury safari accommodation.

KENYA

One of the 'Big Five' that tourists can expect to see on safari

Of all the countries in Africa, Kenya has some of the most prolific and most accessible game parks. Here you can observe some of the greatest examples of wildlife, including the 'Big Five' – elephant, lion, leopard, buffalo and rhino. Scenically stunning with vast expanses of savanna grassland and bush across the plateau of the highlands, it is blessed with beautiful mountains such as Mount Kenya, Africa's second highest peak. A tempting option for the more adventurous visitor is to take the three-day trek complete with porter, cooks and guides to the 5000-metre summit.

As to the beaches, the coastal strip from Malindi to beyond Mombasa has mile upon mile of white coral sand, lapped by the warm waters of the Indian Ocean, sheltered in parts by gently swaying palms. Why not take a trip in a glass-bottomed boat on to the reef where you can see over 240 species of fish and a wide variety of corals? And if you want something different, visit Mombasa – it is hot and dusty but there is a buzz of excitement in its colourful bazaars!

Safaris

Seeing animals in the wild, in their own natural habitat, free of any civilizing influences, is a life-enriching experience. Seeing them in Africa is truly awesome. While a bush safari is one of life's great adventures, we make sure that it is a comfortable one. You will travel in a specially adapted minibus with a guaranteed window seat, be guided by an English-speaking driver and accommodated in comfortable lodges.

Figure 4 Tourist brochure information about Kenya.

The country is fortunate in having two different environments that are both attractive to foreign visitors:

- wildlife parks on the plateau
- Indian Ocean coastline.

Although, in a typical two-week holiday to Kenya, people spend one week on safari and one week on the coast, surveys have shown that for up to 80 per cent of visitors the principal reason for choosing Kenya was the wildlife.

The key human factor was that Kenya in the 1970s was ahead of most other East African countries in protecting its wildlife within 45 National Parks and game reserves, which together cover 10 per cent of the country. It was also a leader in providing luxurious accommodation in safari lodges and clubs (**Figure 5**). Kenya was comparatively prosperous when neighbouring countries like Somalia, Ethiopia and Sudan were racked by civil wars.

ACTIVITIES

Start to put together a case study of tourism in Kenya.

1 Draw a sketch map of Kenya to show the locations of its main tourist attractions.

2 (a) State the attractions of Kenya for safari holidays.
(b) (i) Describe the climate of Mombasa on the coast.
(ii) When is the best time to visit? Explain your answer.

3 Using **Figure 3**, explain why tourist earnings are important to Kenya by working out
(i) the visible trade gap (difference between total value of exports and imports)
(ii) how much it is reduced by tourist earnings.

Kenya – positive and negative effects of tourism

Benefits – the positive effects

As a relatively poor African country, with a visible trade deficit, Kenya relies massively on inflows of foreign exchange from foreign tourists. Not only is tourism itself a labour-intensive service industry, but other economic sectors benefit from its multiplier effect. For example, tourism increases demand for goods and services in agriculture, drinks, transport, entertainment, textiles and crafts. While tourists are staying in hotels, if they spend money in locally run shops and cafés and on local services such as taxis, this money passes directly into the hands of local people who can use it to purchase food and other necessities. Money spent locally by tourists drips down through several levels of Kenyan society and extends to other members of the family before it is exhausted. Tourist revenues are vital to Kenyans in all walks of life.

INFORMATION

Tourism in 2007 – the figures

- Kenya's biggest foreign exchange earner (US$1 billion in the year)
- Tourist revenues accounted for 15 per cent of gross domestic product (GDP)
- Tourist numbers at record high: 1.8 million
- Direct employment in tourism: 250 000
- Indirect employment in tourist-related businesses: another 250 000
- Each full-time worker supports on average 7–12 other people

There is an old conservation saying 'If it pays, it stays'. The Masai Mara earns more tourist income than any of the other Kenyan game parks. Clearly, there is much more surviving wildlife in Kenya than there would have been if the game parks had not been created, especially during a time of rapidly rising populations. The reasons are explained under problem number 3 on the next page. Without the tourist dollars, the government would not have been interested, and more of the park land would have been ploughed up or used as grazing land for domestic animals.

Problems – the negative effects

In Kenya, these are a mixture of economic, environmental and social factors.

1 **Economic – visitor numbers go up and down**
Just when Kenya's tourist industry seems to be on the up and up, it crashes – as in 1997, 2002 and early 2003, and again in 2008 (**Figure 1**). Although there was some violence in Kenya during the run-up to elections in 1997, the main reasons for the drop in numbers were out of Kenya's control. Tourists from Europe and North America were frightened to travel to Africa because of the massacre of innocent tourists in Luxor, even though this was thousands of miles away, in Egypt. In 2002 a missile attack on an aircraft and a car bomb outside a Mombasa hotel that killed 13 people resulted in lost tourist income calculated at US$1 million per day in early 2003.

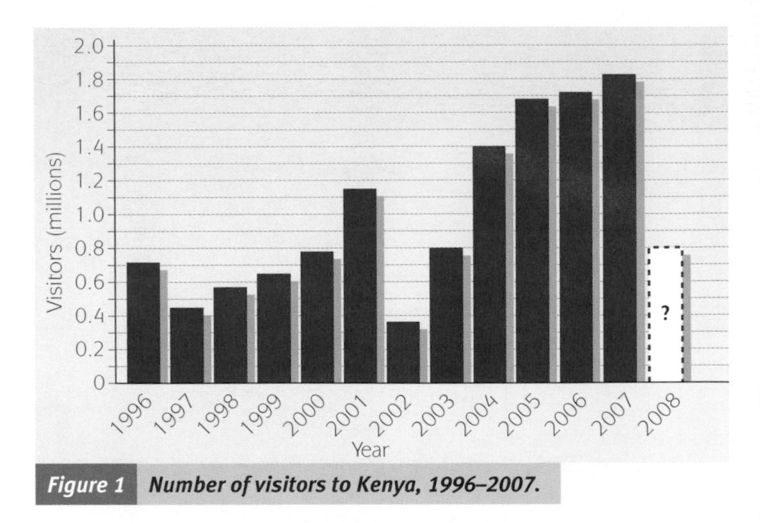

Figure 1 **Number of visitors to Kenya, 1996–2007.**

A disputed election result in December 2007 led to an eruption of tribal violence. International news images of houses being torched in the countryside and city streets full of rioting machete-wielding youths and trigger-happy police led to a tourist nightmare for Kenya, involving mass cancellations of bookings, especially after the UK Foreign Office issued its advice not to travel to Kenya. The fact that the main tourist destinations were unaffected and peaceful did not matter; the image of Kenya as a peaceful tourist destination for safari and beach holidays took a real battering. Although the violence stopped in early 2008, and the Kenya Association of Tour Operators mounted a vigorous international publicity campaign, it was a struggle to get occupancy levels in hotels up to 50 per cent in 2008; and most expect that it will take until late 2009 for 'normal service to be resumed'.

2 Environmental damage
On the reefs off Mombasa, boats drop their anchors into the coral and some tourists take pieces of coral away as souvenirs. Floating pontoons have been placed in popular diving spots for boats to hook on to; patrols have been increased and boat owners are being given better education.

Figure 2 ***Lions and tourists in the Masai Mara.***

In the game parks, drivers are keen to get as close to the animals as possible and leave the main tracks. The armies of minibuses surround and disturb animals (**Figure 2**). They churn up the ground in the wet season; it is not unknown for tourists to be trapped in bogged-down vans for up to eight hours at peak times (like Christmas) in the middle of the wet season. In the dry season, the grassland is turned into a 'dust bowl'. Particularly in the Masai Mara, there is too high a concentration of visitors in too small an area. On busy days up to 8000 visitors will be in the park, leading to queues of 70 or more safari vans at prime viewing points. Wildlife numbers in the park are declining. Numbers of wildebeest on the annual migration have dropped from 800 000 to 300 000 in 20 years. Why will other species' numbers fall as well?

The Kenya Wildlife Service (KWS) is responsible for the game parks. It is under-resourced for the important work it needs to do, and although many of the staff are well motivated, poorly paid employees with limited experience of the work are more likely to be open to bribery. Some ignore poaching. Others take no action against minibus drivers who go off the roads to get close to the wild animals to earn higher tips from their tourist passengers.

3 Social – conflicts with local people
There are also conflicts between local tribespeople, the Masai, and the Kenyan authorities. When the game parks were set up, the Masai were driven off the land to make way for wild animals. Acute shortages of grazing land, coupled with rapid population growth, have forced farmers to move closer to the edges of the parks. Elephants trample their crops. Lions eat cattle. Villagers and tribespeople are injured, or sometimes killed, by wild animals, but they are not allowed to kill them. A survey revealed that less than 2 per cent of the money spent at the world-famous Masai Mara Park benefited the local Masai people; even the high US$27 daily entry fee went direct to the government in Nairobi.

Strategies for the future

The Kenya National Tourism Master Plan emphasises the need to:

- diversify the country's tourist product range, by opening up new avenues of tourism, such as adventure activities on rivers and lakes (rafting, canoeing, sailing and cruising)
- achieve a better distribution of tourist activities throughout the country to reduce environmental pressure on tourist hot spots.

At the local level, there are environmental concerns to be addressed. Under a new programme announced in 2007, the Kenya Tourist Board aims to curb tourist numbers in over-visited parks like the Masai Mara while at the same time increasing income by more than doubling park entry fees, setting a higher minimum price level in hotels and camps, and adding a premium to be used for game park improvements.

In future the emphasis is going to be on quality not quantity; when the place is crowded, the magic of a safari is lost. There are also big hopes for **ecotourism** as a way of spreading tourist dollars among more people and increasing the involvement of tribespeople in preserving wildlife and the environment.

ACTIVITIES

Continuing the case study of tourism in Kenya.

1 State (a) the positive effects and (b) the negative effects of tourism for
(i) the economy (ii) the environment of Kenya.

2 Describe the strategies in place to reduce the negative effects:
(a) preventing damage to the environment (i) in the safari parks (ii) on the coast
(b) sharing tourist income more widely among Kenyan people.

3 Overall, is tourism good or bad for Kenya? Justify your answer.

The spread of tourism to extreme environments

Why are tourists now visiting places where once only explorers used to go? What are the dangers (for both tourists and the environments)?

Tourists are becoming more adventurous. One way in which this is shown is the types of activities that tourists now engage in – white-water rafting and kayaking, trekking and climbing in high mountains and cross-country skiing. Along with other holiday fun activities like bungee jumping, they tend not to be covered by standard holiday insurance policies (a fact not realised by many of the participants). Further evidence for the adventurousness of tourists is provided by the remote destinations now included on a tourist map of the world – Spitzbergen, Easter Island, Galapagos Islands, Bhutan, Maldives. Can you locate them? What attracts tourists to such destinations? Remote places like these, previously little visited, have now been made accessible by air travel.

Extreme environments are places where few people live, due to difficult physical conditions. Look at a world map of population distribution in an atlas and identify the areas that have very few people. Many of these areas will be extreme environments because of one or more of these natural factors – great cold (polar lands), very dry (hot deserts), great height and steepness (high mountain ranges), inland areas covered by tropical rainforests. Adventurous (and often better-off) tourists are drawn to these areas by their emptiness. Such environments appeal to these tourists because they want to see natural worlds that are totally different from the ones in which they normally live. Geographers describe them as **wildernesses**, undeveloped areas that are still primarily shaped by the forces of nature. A striking example of an extreme environment is Antarctica (**Figure 2**). Rainforest examples are included under ecotourism on page 248.

Figure 1 Macchu Picchu at the end of the Inca Trail in the Andes of Peru. So many tourists now want to do the four-day trek that the authorities have been forced to restrict numbers; similarly, numbers trekking in the Himalayas and climbing Mount Everest are controlled.

Figure 2 Hope Bay, Antarctica. What is the evidence that this is a wilderness? What is different about it that attracts tourists?

Antarctica

Antarctica offers tourists magnificent scenery, icebergs and nesting penguins by the million in summer. Visitors can have a real, close-up wildlife experience, of the type normally only seen on TV wildlife films; the seals, penguins and other birds show no fear of humans (**Figure 3**). Until recently its remoteness saved it from tourists. There are no commercial airports, only landing places, which are unusable for most of the year because of bad weather. Most visitors arrive by cruise ship, and the season is narrow – between mid-November and mid-March. Nevertheless, there has been a rapid increase in tourist numbers in the last 20 years, up to 35 million in 2008. This is partly due to the increasing size of cruise ships; however, 'penguin fever' is also blamed after successes like the *March of the Penguins* film.

Protection measures already in place

Most cruise ships carry 60–100 tourists; they sign up to and abide by the guidelines of the International Association of Antarctica Tour Operators (IAATO); however, these guidelines are not mandatory. Importantly, two of the largest ships with more than 400 passengers have not signed up. They land more than 100 visitors at a time, above the maximum allowed by the guidelines. Supervising the tourists to ensure that they 'do not go within 5 metres of penguins and other wildlife', 'do not walk on lichens' and 'do not leave litter or waste' becomes more difficult with such large numbers. Responsible tour operators hire as tourist guides and lecturers people who have worked or carried out research in the Antarctic, who instill responsibility into their passengers. This is a fragile natural environment: survival is a struggle for all life at the climatic extremes, while tourists go back to friendlier climes after a couple of weeks' holiday.

Worries about the future

With the number of visitors predicted to double in the next ten years, possible impacts include sea and coastal pollution, littering, damage to flora and fauna, and disruption of breeding patterns – since the peak tourist and peak breeding seasons coincide. The probability looms of ever larger ships, helicopters and commercial air strips. Already large cruise ships with up to 1000 passengers sail to Antarctica.

Although they do not land passengers, how much of an ecological disaster will it be if one of them hits ice and sinks? Unlike the smaller ships they are not ice-strengthened and they use heavy fuel oil, which disperses more slowly than marine fuel oil. A spill of heavy oil near the coast would see thousands of penguins coated in oil. Where would the clean-up equipment come from?

A big worry is any development of land-based tourism – the thought of people skiing and snowboarding down Antarctica's uninhabited slopes frightens environmentalists. There is a danger that future tourists will see Antarctica as a theme park instead of a very fragile nature reserve.

Figure 3 Antarctic beach, highly populated in summer.

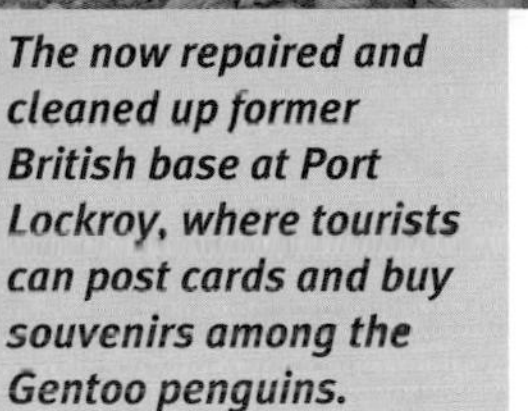

Figure 4 The now repaired and cleaned up former British base at Port Lockroy, where tourists can post cards and buy souvenirs among the Gentoo penguins.

GradeStudio

1 What are the meanings of the terms
  a adventure holidays (2 marks)
  b extreme environments. (3 marks)

2 Study **Figures 2** and **3**.
  a Describe what attracts visitors to these locations. (4 marks)
  b Explain why the risk of damage from visitors here is high. (3 marks)

3 a Describe the measures in place to reduce damage to the environment from tourists in Antarctica. (4 marks)
  b How effective are they? (3 marks)

Exam tip

Match answer length to marks available.
Which of the first three answers should be the longest?
Which one the shortest?

Ecotourism and responsible tourism

What makes ecotourism different from ordinary tourism? How widespread is it? Is it successful?

Not all tourism news spreads environmental gloom. Landscapes and wildlife habitats are being saved by tourist income, which generates the money needed for their protection and encourages governments and local people to be more conservation-minded. As the case study for Antarctica has shown, the need for stewardship and conservation in tourist areas has always been there; and as tourist numbers go up, the need becomes greater. Antarctica has been luckier than most other destinations because the stewardship began before any tourist damage had occurred. **Ecotourism** is the heading usually applied to tourism in which the natural environment is looked after, although it includes more than this (Exam Preparation Box). Another name for it is 'Green tourism'.

EXAM PREPARATION

Ecotourism

A Environmentally sound
- natural environments and wildlife safeguarded

B Socially sound
- considers the needs of, and involves, local communities

C Sustainable
- looking after today's tourist needs does not damage those of future generations

When companies use the word 'ecotourism' to describe their tours it is a powerful sales tool. However, you should always check that it is not just a sales gimmick. It may do no more than just distinguish small tours to more distant, environmentally interesting places from mass-market package tours to popular destinations. One cynic has described it as 'ordinary tourism dressed up in a politically correct manner'. It is gradually being replaced by the label **responsible tourism**, meaning tourism that protects the environment, respects local cultures, benefits local communities, conserves natural resources, and causes minimum pollution.

Essentially it is the same as ecotourism, but with objectives that are easier to check.

Rainforest ecotourism in Ecuador

Tourists are attracted to tropical rainforests for the 'jungle' experience – unique places, 'hot houses' for plants, with the greatest biodiversity of plant and animal species on Earth (see pages 70–71). Tourism is seen as one of the major ways to preserve and protect remaining rainforests. One jungle lodge, Sacha Lodge, is described on page 73 and shown there in **Figure 4**. What is the evidence that Sacha Lodge is an ecotourism lodge?

Figure 1 Location of Ecuador.

Another lodge, Kapawi Ecolodge, closer to the border with Peru, is in the heart of the tribal reserve of the Achuar people. It is a safari-style collection of wooden chalets. The Achuar granted the owners permission to build here in return for a good rent and an agreement to train them to take over the running and management by 2011. The native Indian guides are vital to the success of the tourist experience. Only they have the keen eyesight to spot and point out parrots, macaws, butterflies and other creatures of the forest, and the knowledge to identify plants and their uses. Behind the Achuar's decision to allow tourists in is the fear of the oil companies and gold prospectors, and the way such people are likely to destroy the forest, which provides the Achuar with food, tools and medicines. The present set-up gives the Achuar a fair share of the tourism benefits. Where mining companies have already arrived in eastern Ecuador, the frontier towns, carved out of the rainforest, are not pretty sights (**Figure 3**).

Figure 2 Tourist view of the rainforest.

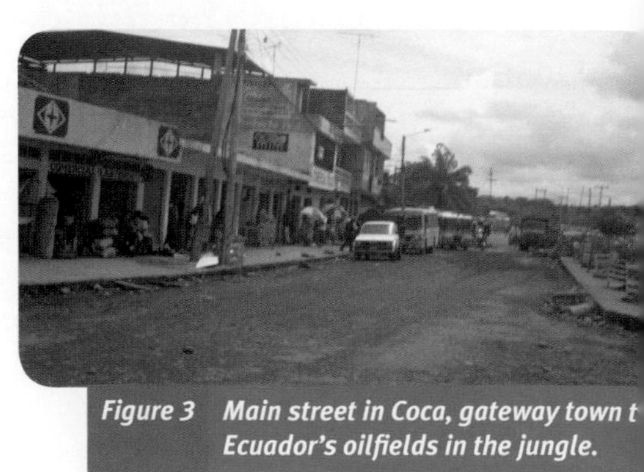

Figure 3 Main street in Coca, gateway town to Ecuador's oilfields in the jungle.

Kenya – the Masai and wildlife tourism

Figure 4 Masai tribespeople.

Three-quarters of the wildlife in Kenya is found outside the game parks, much of it on land owned by the Masai (**Figure 4**). In colonial times the Masai were driven off their traditional land to make way for wild animals in the parks. The Masai were seen as a nuisance. Now the vegetation is healthier, wildlife is more plentiful outside the parks than before, and the tourist potential of Masai land has increased.

Three tented camps, owned and run by Kenyans, have been set up in Kimana on an important migration corridor for wildlife between Amboseli and Tsavo National Parks. The Masai are paid a rent for use of their land of about £1000 per year. The Kenya Tourist Boards support the extension of small-scale camps outside game reserves. Close to the entrance of the Mara Park, 156 Masai have joined their plots together. In return for not grazing their cows and chopping down wood, they rent their land out for a number of tented camps and ecolodges in the Olare Orok complex. Each Masai landowner receives about £70 a month from the owners. Some young Masai men make money as tourist guides.

Masai communities involved in ecotourism are financially better off. The social benefits are more children being sent to school and better healthcare. However, some are said to be wasting their income on alcohol. Most adult Masai are unable to read and write, which leaves them vulnerable to cheating tour operators with contracts and leases. Outside the tourist camps, the Masai need to carry on with their traditional way of life, planting crops and keeping cattle, activities that do not fit well with encouraging wildlife. Until more of them can be convinced of the benefits of tourism, the living space for the wildlife will continue to shrink.

ACTIVITIES

1 List the distinctive characteristics of ecotourism.

2 For one example of ecotourism:
   (a) give its location and describe how it operates
   (b) state its benefits (to the environment, local economy and people)
   (c) comment on whether it will lead to sustainable development.

FURTHER RESEARCH

Find more information about Kapawi Lodge at the weblink www.contentextra.com/aqagcsegeog.

Practice GCSE question

See a Foundation Tier Practice GCSE Question on the weblink www.contentextra.com/aqagcsegeog.

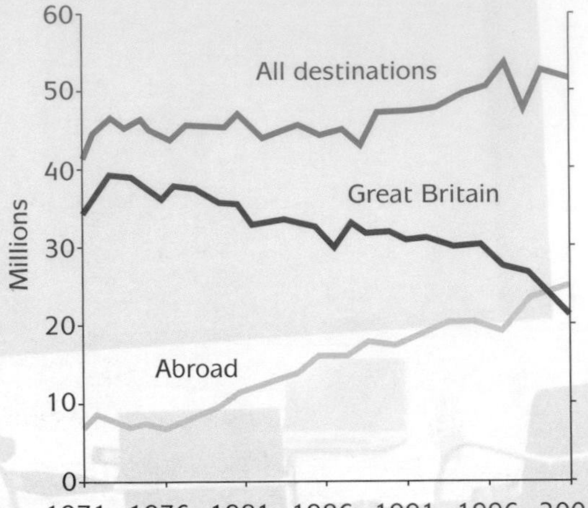

Figure 1 Holidays taken by UK residents 1971–2001.

Figure 3 Antigua in the Caribbean.

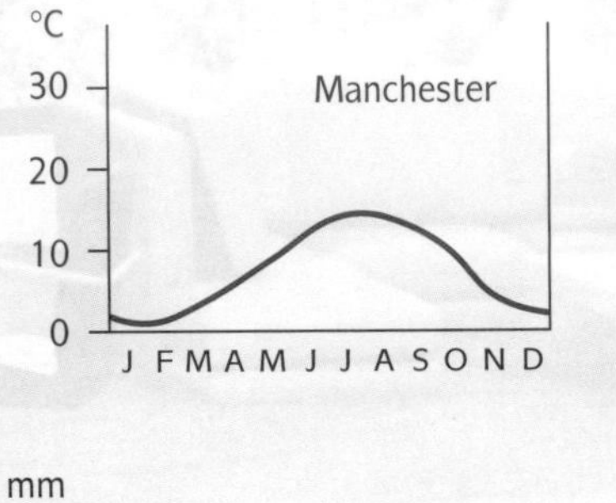

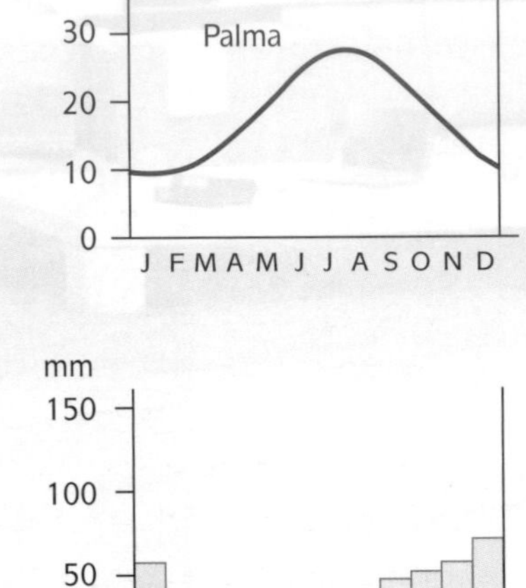

Figure 2 Climate graphs for Manchester (England) and Palma (Majorca in Spain).

6 (a) Study **Figure 1**, which shows holidays taken by UK residents 1971–2001.
  (i) Describe what it shows about the changing pattern of holidays in the UK. **(3 marks)**
  (ii) State two reasons for the trend shown by the 'All destinations' line. **(2 marks)**
(b) Study the climate graphs for Manchester and Palma in **Figure 2**.
  (i) State the differences between Manchester and Palma for distribution of rainfall during the year, and winter and summer temperatures. **(3 marks)**
  (ii) Explain how the differences in climate between the two places help to explain what **Figure 1** shows. **(4 marks)**
(c) **Figure 3** shows a holiday area in the tropics.
  (i) Describe the attractions for tourism of the area shown in **Figure 3**. **(3 marks)**
  (ii) Explain how tourism can bring economic benefits to poor developing countries. **(4 marks)**
  (iii) Choose one established tropical tourist area. Describe the negative effects of tourism on the environment and the attempts made to reduce them. **(6 marks)**

Total: 25 marks

Exam tip
The question for this topic will be Question 6 in Paper 2.

Improve your GCSE answers

Case studies – when and how to use them

Why does knowing and using case studies matter?
Because exam candidates, overall, score least well on these questions. If you use your case study knowledge well, you will already be nudging into the grade A*/A bands. If you do not use it at all, you will be struggling to reach grade C.

1 Know your tourism case studies

- There are four of them listed for this topic at the bottom of the next page. Each of them has a list of headings. These show the themes most likely to be asked in GCSE exam questions. They are based upon what the specification says.
- When revising, check that you have information for all the headings. Be prepared – think about what you will need to use it for. It might be a good idea to write a crib sheet or card for each case study, using the headings given. Look back, because you might have already done this in some of the Activity questions.

2 Know when you should use them

- There are two different types of questions:

A When case studies **must be used**, if you are to get full marks for your answer, questions will be worded like this:

Question 1 *Use a named example* to describe ... say the attractions of a UK coastal resort or National Park.

Question 2 Choose *one tourist area* in the tropics and state ... say the benefits (positive effects) of tourism for the economy.

Question 3 *Using example(s)*, explain why ... say cities (or mountains) are attractive environments for the development of tourism.

In all these questions, the instruction to use a named example is precise. Without it, the examiner will not be allowed to give your answer a top-level mark (no matter how good it is). This means that

- in a 4-mark question, the highest mark possible will be 2
- in a 6-mark question, the highest mark possible will be 4
- in an 8-mark question, the highest mark possible will be 6.

Naming and giving information about an example keeps you ahead of the pack.

B Using them to **improve** your answer, making the examiner more likely to give you full marks. Questions can be worded like this:

Question 4 Describe how the negative effects of large numbers of tourists can be reduced in ... say coastal resorts (or National Parks).

Question 5 Explain how ecotourism can benefit ... say the environment (or lives of local people).

Neither question says that an example must be used; if the answer is good enough, the examiner will give you full marks without an example. However, if you make passing references to a relevant example (and, even better, give information about it), you will be marked up by the examiner for what you write.

C Know how you should use them

Take careful note of the question, especially the command words (page 253):

Question 6 Name an example of ... say an extreme environment visited by tourists.

A name such as 'Antarctica' is enough.

Question 7 Name and locate an example of say ecotourism.

A name like 'Kapawi Ecolodge' is not enough; you need to state a location as well, 'in the rainforests in eastern Ecuador', for a full-mark answer.

Question 8 Describe the main attractions of ... say a tourist area in the tropics.

A name alone might not even be enough for a mark. Specific details of the area's attractions are what this question asks for.

Important advice

There are always some marks for making general points that fit the theme of the question. For example, if you choose coastal resort in **Question 1**, there are marks for describing the tourist attractions of coastal resorts in the UK in general, such as sandy beach and warm summer weather. Being unable to give information about a named case study loses you only some of the marks, and should not stop you giving an answer when you know that you do not have relevant case study knowledge.

ExamCafé

REVISION

Key terms from the specification

Exam Café

Adventure holidays – more active with more risk, off the beaten track, in more unusual destinations
Ecotourism – involves protecting the environment and the way of life of local people
External factor – something unrelated to tourism, which affects tourist numbers, such as the economy, currency exchange rates, political unrest, wars and terrorism
Extreme environment – difficult place for humans to live in or visit, often due to hostile climate
Fragile environment – place where wildlife and landscape are easily damaged by outsiders
Mass tourism – large numbers of visitors, often on package holidays with accommodation and travel included
Multiplier effect – spin-offs from one business growing, allowing other businesses to grow as well
National Park – area set aside to protect landscape and habitats, managed to stop visitor damage
Stewardship – entrusted to look after and manage places and areas

Checklist

	Yes	If no – refer to
Can you explain why tourism is a global growth industry?		pages 234–5
Are you able to name examples of coastal areas, mountains and cities visited by many tourists, and describe their attractions?		pages 236–7
Do you know where and why tourism is an important industry in the UK?		pages 238–9
What measures are being taken to try to ensure successful tourism in tourist areas in the UK such as coastal resorts and National Parks?		pages 239–41
Do you understand what is meant by mass tourism?		page 242
Can you state both positive and negative effects of mass tourism on the economies of countries and the environment?		pages 242–3
Can you give examples of extreme environments that are attracting more adventurous tourists?		pages 246–7
How is ecotourism different from other types of tourism?		pages 248–9

Case study summaries

UK – National Park or coastal resort	Tropical tourist area	Extreme environment	Ecotourism
Reasons for growth	Attractions	Attractions	Characteristics
Visitor strategies	Positive effects (benefits)	Impacts	Benefits (environment, local economy, people)
Their effectiveness	Negative effects (costs)	Measures in place	Sustainable development
Plans for the future	Strategies and plans	Coping in the future	

ExamCafé

About GCSE questions

Each geography GCSE examination question can be broken down into at least two parts:

1 the command words – i.e. what you are being told to do

2 the question theme – i.e. what the question is about.

Some questions specify a location or world region for the question and have a third part:

3 where – the area or areas of the world.

Example of a two-part question:

Name two volcanoes.

command words question theme

Example of a question with three parts:

State two problems in inner-city areas of cities

command words question theme

in the UK and other developed countries.

where

Command words

A • Name • Give • State • List

Name one country in which ...

Give two reasons for ...

These are simple and clear command words and need no further comment.

B These command words are asking for definitions. You are most likely to be asked to define key terms used in the specification. The meaning of these terms is explained in the Revision Section on the last page of each chapter. These and some other important geographical terms are highlighted in bold in the text. A complete glossary of over 200 geographical terms can be found on pages 259–61, arranged alphabetically for easy reference.

Top tip

Remember to look at the number of marks for the answer – absolutely vital with this type of question.

Question

What is meant by an ageing population? (3 marks)

Answer from Candidate 1 A lot of old people in the country. There are many pensioners.

same point made

Examiner comment The candidate has only attempted to make two points in a three-mark question, and can't gain more than two marks. The answer is only worth one mark because the two statements are making the same point – many old people.

Answer from Candidate 2 An ageing population is one with large numbers of old people. Most of them are over 65. They increase the percentage of old people in the country and make the population pyramid wider at the top. The government has to pay out more on pensions and spend more on care homes and healthcare.

same point made

effects of an ageing population – not needed

Examiner comment This candidate also begins the answer by making the same point twice, but then extends the answer, making two other valid points. There is just sufficient for the candidate to be given all three marks, before the answer drifts into effects.

Note that the examiner said *just sufficient* for full marks. If you know and understand the topic and want a high grade, give a little bit more information than you think is needed for full marks, in case part of your answer does not match what is in the mark scheme. For example, Candidate 2 could have extended the answer by naming an example of a country with an ageing population such as the UK.

C Describe

Describe the features of a delta.

Describe what Figure 1 shows.

(**Figure 1** may be a graph, diagram, map or photograph.)

Describe is one of the most commonly used command words in geography examinations. 'Describe' commands you to write about what is there, or its appearance, or what is shown on a graph. You are not being asked to explain. The amount of detail expected in the answer is suggested by the number of marks for the question.

Let us take one example. Describing landscapes and landforms is an important part of physical geography. When asked to *describe* a landform (volcano, corrie, etc.), you are really being asked to *say what it looks like*.

- Write about its shape, size and what it is made of.
- Be generous with your use of adjectives, such as wide/narrow, steep/gentle/flat, straight/curved, etc.

Question

Describe the physical features of the landform shown in Figure 1.

Figure 1

Approach to the answer	Possible answers
• Name the landform first	volcano (or even composite volcano)
• Describe its shape	cone-shaped
• Describe other features	lava flows on its sides
• Give a more detailed description	steep slopes on the snow-covered top of the cone slightly more gentle slopes lower down bare rock (lava) on the lower slopes

D • Explain the formation of ...

- Explain why ...
- Why have ...?
- Give reasons for ...
- Why does ..?

To answer these questions you need geographical knowledge and understanding. You are being asked to account for the appearance or occurrence of physical and human features of the Earth's surface. These command words do not usually cause problems. What causes problems is giving enough precise information.

Question

Why is population growth high in developing countries? (4 marks)

Answer Because birth rates are high and death rates are low.

Examiner comment This is the basic answer, which has only reached the first level of explanation. It is a one-mark answer. Explanations why birth rates are high and why death rates are low are also needed.

Question

Explain the formation of spits along some coasts. (5 marks)

Answer A spit forms after longshore drift moves sand along the coast. I have drawn a diagram below to show how it does this.

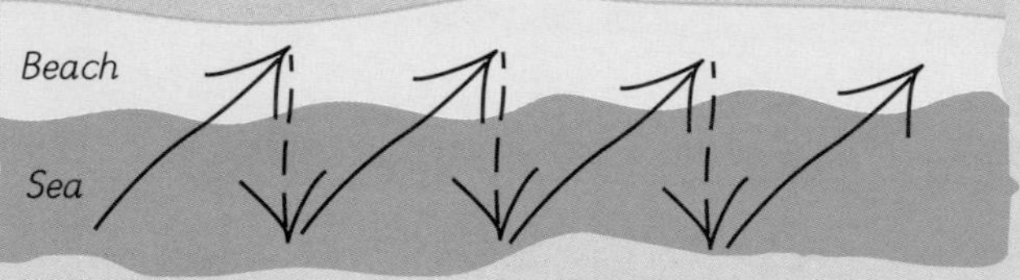

The prevailing winds take the waves on to the beach at an angle

Backwash remains at 90° to the sea resulting in transportation of material in a zig zag fashion

Longshore drift.

Examiner comment

- Everything is correct and longshore drift does play a part in the formation of a spit. The question does not ask for a diagram, but relevant points accurately made on the diagram will be credited.
- But there is only partial explanation here. How and why the spit actually forms after the longshore drift has transported sand is not explained.
- This answer is only worth two of the five marks.

Questions based on source materials

The source materials used in geography examinations are many and various. However, you will be familiar with:

- the type of source itself, because those most used in examinations will include maps, graphs, diagrams, photographs, tables of data or cartoons.

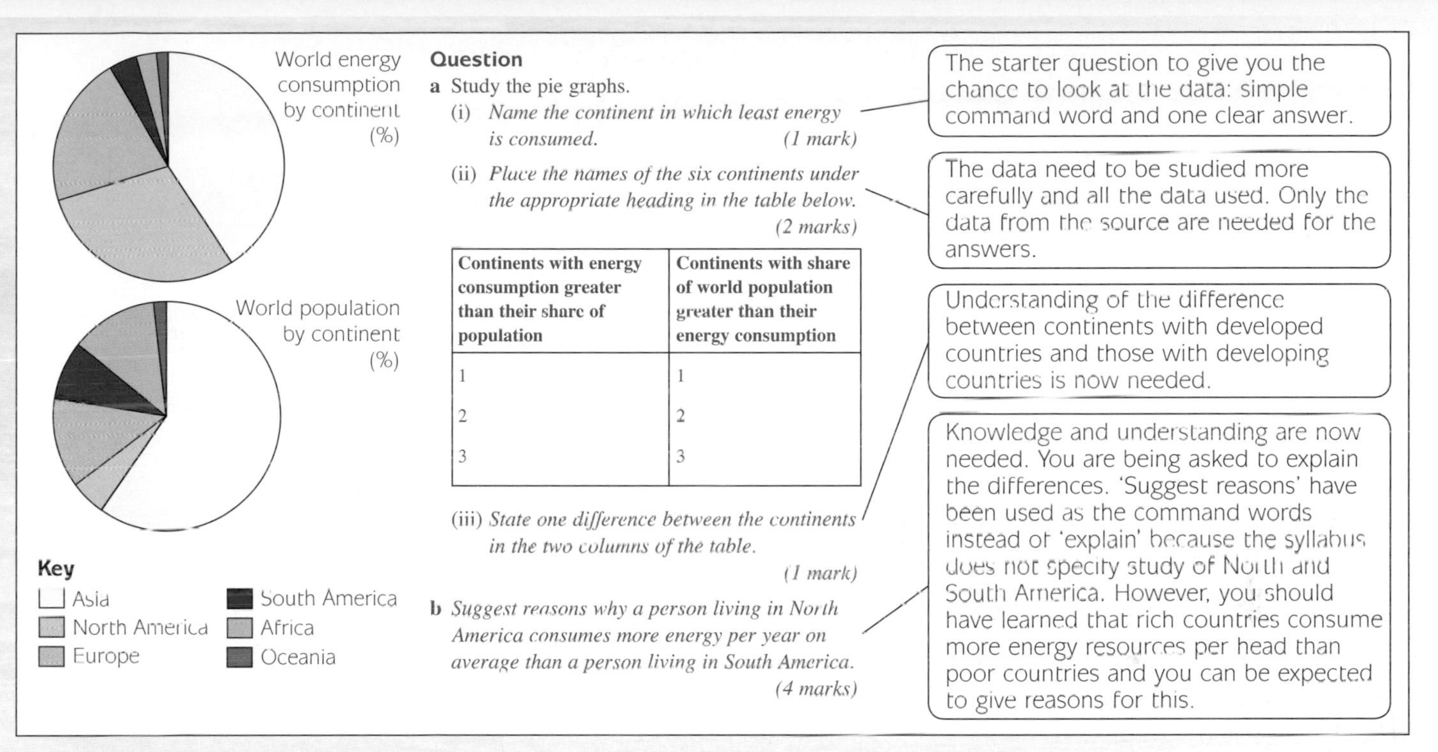

Question

a Study the pie graphs.

(i) *Name the continent in which least energy is consumed.* *(1 mark)*

(ii) *Place the names of the six continents under the appropriate heading in the table below.* *(2 marks)*

Continents with energy consumption greater than their share of population	Continents with share of world population greater than their energy consumption
1	1
2	2
3	3

(iii) *State one difference between the continents in the two columns of the table.* *(1 mark)*

b *Suggest reasons why a person living in North America consumes more energy per year on average than a person living in South America.* *(4 marks)*

The starter question to give you the chance to look at the data: simple command word and one clear answer.

The data need to be studied more carefully and all the data used. Only the data from the source are needed for the answers.

Understanding of the difference between continents with developed countries and those with developing countries is now needed.

Knowledge and understanding are now needed. You are being asked to explain the differences. 'Suggest reasons' have been used as the command words instead of 'explain' because the syllabus does not specify study of North and South America. However, you should have learned that rich countries consume more energy resources per head than poor countries and you can be expected to give reasons for this.

Figure 2

Question

Study the photograph taken in a city in a developed country

a *Name the urban zone in which the photograph was most likely to have been taken. (1 mark)*

b *Describe the evidence from the photograph that supports your choice of zone. (4 marks)*

Figure 3

- the geographical topic covered, because this is part of the syllabus.

There is only a small chance that you will have seen the data before, or the photograph, or the chosen OS map extract. This is why most questions begin with short questions looking at the source materials. **Figure 2** takes you through a question based on pie graphs.

Although most questions follow this style, in some you must use knowledge and understanding from the beginning. This makes these questions more difficult and they are likely to be used later in the examination. **Figure 3** is a question of this type, based upon photograph interpretation.

You need to make a careful study of the photograph before answering **part a** and you should understand what is meant by an urban zone. Do not rush into the answer; take a careful look at the photograph and think the answer through. When answering **part b** use evidence from the photograph to justify your choice. Do not just describe everything that you see.

Using the mark scheme to answer questions worth 6–8 marks

Question

Using the case study of an earthquake, describe its effects upon the people and economy of the area. (6 marks)

Notice how the mark scheme below refers to each of the three elements in the question:

- case study of an earthquake
- effects on the people
- effects on the economy

The command word 'Describe' is not mentioned, but all the points listed describe (rather than explain) the effects.

Mark scheme

The actual content about the earthquake will depend on the case study chosen.
Relevant points that candidates might make about chosen earthquake.

> Information needed about the earthquake chosen

Effects on people

- Primary effects – damage from earthquake to buildings, roads and bridges, deaths and injuries to people trapped in homes, offices, etc
- Secondary effects – fires causing further damage to buildings, diseases from disruption to water supplies and sewage systems, tsunamis in coastal areas, landslides in steep areas

> Effects on people; effects from earthquakes are both primary and secondary

Effects on the economy

> Effects on the economy

- Primary effects – work and output stops in factories, offices, etc; transport systems disrupted
- Secondary effects – high cost of repairs, might be many months before normal output achieved, local economy badly affected

> Before awarding the mark, examiners decide the level

Levels of response

Level 1 (Basic) **1–2 marks**
Describes only one or two general effects
Random order; not organised
No specific references to an earthquake case study

> Weak answer, lacking detail and no case study

Level 2 (Clear) **3–4 marks**
Effects are clearly stated, including both primary and secondary effects
Organised and quite effective answer
Some references to an earthquake case study

> Quite a good answer, but lacking something to stop it being Level 3, e.g. only references to effects on people, not the economy; or not enough case study detail

Level 3 (Detailed) **5–6 marks**
Effects are clearly stated for both people and economy
Precise references made to an earthquake case study

> As good an answer as can be expected; all three elements in the question covered

Student answers

Read the answers. Use the mark scheme. Decide on the level. Give each answer a mark out of six.

Candidate A

In 2001 India was hit by an earthquake. It measured 7.9 on the Richter scale. Bhachau was near the middle of the quake and almost all its buildings were destroyed. Because they were not built to withstand earthquake shocks. More than 10 000 people got crushed and killed by falling houses. People blamed the greedy builders for cutting corners and using poor building materials. In the days after the quake people tried to pick themselves up, but the economy was in a bad way. Nothing was working properly.

Examiner's verdict

Candidate A

Level 2 answer worth 3 marks. It provides part of the answer to the question.

Strengths – it refers to a case study, some information is given about it and the primary effects are stated.

Weaknesses – it uses too many words explaining why so many were killed, instead of sticking to describing the effects. There is nothing on the economic effects, except for saying that it was in a bad way.

Likely grade? D/C borderline. How could it have been improved? By giving a more direct answer to the question set.

Candidate B

Everyone is scared when the ground shakes. They try to take cover, but many get covered by rubble and debris. Sniffer dogs are used to find where people are buried alive. Bridges collapse and in America cars dropped off flyovers on to the streets below. There is rubble all over the place after an earthquake. Electricity supply gets cut off and people had nothing to eat. They had no jobs.

Candidate B

Level 1 answer worth 2 marks. This is a weak answer.

Strengths – it mentions the primary effects of an earthquake.

Weaknesses – it does not give a named case study, only a passing reference to 'America', which is not good enough. Only in the last two sentences is there a brief mention of secondary effects and the economy.

Likely grade? E. What is needed for a better answer? More knowledge.

Candidate C

My case study is the Gujarat earthquake in India in 2001. It was strong enough to destroy 90 per cent of the buildings in Bhachau near the epicentre, where at least 10 000 people were killed. Within a minute, thousands were buried under massive piles of rubble. High-rise buildings swayed and collapsed in the capital, Ahmedabad. At least 1000 people were killed there. The survivors then suffered from diseases such as cholera because water supplies were contaminated with sewage. Many months later hundreds of thousands were still living in shelters, depending on aid. The local economy was devastated. Economic losses were estimated to be at least two billion US dollars. Many local businesses were destroyed.

Candidate C

Level 3 answer worth 6 marks. This is a good answer considering the time and space available for answering – just about worth all the marks.

Strengths – it begins well by naming the case study. It mentions both primary and secondary effects. In the last two lines it refers to economic effects as well. The content throughout refers to the case study. It is well focused on describing the effects.

Weaknesses – it is stronger about the effects on people than the economy, but examiners cannot reserve full marks for 'perfect' or 'super' answers; otherwise the top marks will rarely be awarded.

Likely grade? A. Why not A*? There is no such thing as an individual A* answer. An A* grade is awarded for consistently good answers throughout the examination. Keep on writing Level 3 answers like this in other questions and you should be on target for grade A* overall.

Do you want an A* grade? Know your case studies

GCSE examination questions that discriminate most ruthlessly between grade A and A* candidates and everyone else are case study questions.

Question: For one inner city area in the UK that has been redeveloped, describe and explain how it has changed over the last 20 years. **(6 marks)**

Candidate answer A

Many inner city areas are run-down and full of problems. In the 1960s and 70s planners cleared large areas of slum housing and replaced them with tower blocks of flats. People did not like living in these and there were many problems. Only very poor families who could not afford to move out were left behind in most inner cities. No one could think of a good use for large areas of derelict land on the sides of railways, canals and old docks.

Examiner comment

This is not answering the question directly, but it is useful in setting the scene and explaining why redevelopment was needed. Only include this in your exam answer if you know you have plenty of time for answering this and all the other questions.

Candidate answer B

My chosen area is inner city Manchester. Many workers pass by it as they travel to work along Princess Parkway one of the main roads into Manchester city centre. One area is Hulme west of the city centre, where there have been great changes since 1990. Hulme people hated the crescent shaped blocks of flats. Most have now been pulled down under a government scheme called City Challenge. They have been replaced by smaller blocks of housing, with many different building styles and made of different building materials, and in different colours. There are more open spaces, the largest of these is Hulme Park. The city council listened to local residents and tried to give them the sports and play facilities they asked for.

Examiner comment

This is a direct answer to the question. The good points are:

- the area's location is made clear
- two examples of redevelopment are described
- these are also explained as well.

It might have already reached a Level 3 mark.

Candidate answer C

In the same area, but a bit closer to the Manchester city centre, is the old docklands area. This has always changed a lot in the last 20 years. Run down old dockside warehouses in Castlefields have been converted into luxury apartments. The area around them has been landscaped with parks and trees. Dirty old dock basins and canals are now filled with clean water and look good. Private developers are more likely to invest in redevelopment for wealthy people because they know there is more chance of making a good profit.

Examiner comment

There is more location information, description and explanation in this paragraph. There is now more than enough for top of Level 3 and all 6 marks. Always try to give more than enough if you can and if you have the time.

Final comment

How is a grade A* in geography awarded? It is based on the accumulation of marks from all papers. Your mark total must be high enough to place you within the top 5–7 per cent of candidates in the country who are taking that year's exam. The best two ways to keep ahead of the competition are:

- Write consistently good answers (with nothing more than the occasional weak answer along the way).
- Write good case study answers, good because precise information about named places is included.

Glossary

The terms included in the glossary are key terms from the specification and are also listed in the Exam Café at the end of each chapter.

Abrasion – waves erode coastline by throwing pebbles against cliff faces

Adventure holidays – more active with more risk, off the beaten track, in more unusual destinations

Ageing population – increasing percentage of old people (aged 65 and over)

Agri-business – type of farming that is run as a big business (no longer a way of life)

Aid – money, goods and expertise given by one country to another, either free or at low cost

Anticyclone – area of high pressure

Appropriate technology – level in terms of size and complexity that makes it suitable for local people to use

Arch – rocky opening through a headland formed by wave erosion

Arête – sharp-edged two-sided ridge on the top of a mountain

Bar – ridge of sand or shingle across the entrance to a bay or river mouth

Beach – sloping area of sand and shingle between the high and low water marks

Biodiversity – level of plant and animal variety in an ecosystem

Birth rate – number of live births per 1000 population per year

Boulder clay/till – all materials deposited by ice, usually clay containing sharp-edged boulders of many sizes

Brown earth – uniform brown-coloured soil that forms under deciduous woodland

Brownfield site – area of previously built-up land that is available to be built on again

Carbon credits – each one gives the buyer the right to emit 1 tonne of carbon into the atmosphere

Carbon footprint – emissions of carbon dioxide left behind by burning fossil fuels

Carbon trading – companies that have exceeded their carbon emissions allowance buy carbon credits from those that have not

Cash crop farming – crops grown for sale instead of farmer's own use (the opposite of subsistence farming)

Cave – hollow at the bottom of a cliff eroded by waves

Central Business District (CBD) – urban zone located in the centre, mainly shops and offices

Cliff – steep rock outcrop along a coast

Climate – average weather conditions recorded at a place over many years

Commercial farming – type of agriculture based on growing crops or rearing livestock for sale

Commuter – person who travels to work in another place every day by car or public transport

Conflict – opposing views about issues, leading to debate between people about them

Constructive wave – gently breaking wave with a strong swash and weak backwash

Continentality – influence of land surface on weather and climate

Corrie – circular hollow, high on a mountainside, surrounded by steep rocky walls except for a rock lip on the open side

Cross profiles of river valleys – V-shaped sections, changing downstream from steep to gentle

Death rate – number of deaths per 1000 population per year

De-industrialisation – declining importance of manufacturing industry

Dependency ratio – relationship between people of working and non-working ages

Depression – area of low pressure

Destructive wave – powerful wave with a weak swash and strong backwash

Development – level of economic growth and wealth of a country

Discharge – amount of water in a river at any one time

Earthquake – shaking of the ground

Ecosystem – system in which living things (plants and animals) and physical factors (climate and soils) are linked

Ecotourism – involves protecting the environment and the way of life of local people

Effects – primary (first effects) and secondary (later effects), positive (good) and negative (bad)

Environmental degradation – productive land turned into wasteland by damage to the soil

Erosion processes – wearing away the land surface by hydraulic action, abrasion, attrition and solution

External factor – something unrelated to tourism, which affects tourist numbers, such as the economy, currency exchange rates, political unrest, wars and terrorism

Extreme environment – difficult place for humans to live in or visit, often due to hostile climate

Extreme weather – weather event more severe than normally expected

Fair trade – farmers and producers in developing countries are given a fair deal by buyers in developed countries; prices paid are always higher than their costs of production

Flood plain – flat land built of silt on the sides of a river, usually in its lower course

Flooding – water covering land that is normally dry after a river bursts its banks

Fold mountains – long, high mountain range formed by upfolding of sediments

Food chain/web – nutrients and energy absorbed by plants are passed along a line of living things

Food miles – distance that food travels between supplier and supermarket shelf

Fragile environment – place where wildlife and landscape are easily damaged by outsiders

Function (of a settlement) – what it does, why it is there, e.g. capital city, port, industrial centre

Functional parts (of a settlement) – purpose of that area, e.g. residential, industrial, port area

Glacial trough – U-shaped valley, with flat floor and steep sides, formed by a valley glacier

Global climate change – variations in temperature and rainfall affecting the whole world

Global interdependence – shared need between two or more countries, located anywhere in the world, for one another's goods or services

Globalisation – increasing importance of international operations for people and companies

Gorge – steep narrow valley, with rocky sides

Gross National Income (GNI) per head – total income of the country, divided by the number of inhabitants, to give average income per person

Gross National Product (GNP) – total value of all the goods and services produced by people and companies in the country in one year

Hanging valley – tributary valley, high above the main valley floor, with a waterfall

Hard engineering strategies – strong construction methods to hold floodwater back or keep it out

Hazard – natural hazards are short-term events that threaten lives and property

Hazard (climatic) – short-term weather event that threatens lives and property

Human Development Index (HDI) – is a measure of people's quality of life using more than one measure of development, based on life expectancy, education and standard of living

Hydraulic power – erosion of rocks by the force of moving water in waves

Ice sheet – moving mass of ice that covers all the land over a wide area

Igneous rock – rock formed by volcanic activity, from magma that has cooled

Immigration – movement of people into a country from another country

Industrialisation – growth and increasing importance of manufacturing industry (making goods)

Inequalities in wealth – unfair differences/big gaps in income and development between different countries, or between regions/different groups of people within a country.

Infant mortality rate – number of child deaths under one year old per 1000 people.

Informal sector – not regular paid employment; unofficial work, often self-help small-scale services such as street sellers and shoe shiners

Infrastructure for tourism – support structures and services for visitors such as airports, hotels, electricity, tour agencies

Inner city – urban zone around the edges of the CBD, quite old

Land uses (urban) – ways in which the Earth's surface is used, e.g. houses, factories, shops, transport, parks in towns and cities

Landscape – inland scenery, with varied landforms

Latosol – deep soil, red or yellow in colour, which forms under tropical rainforest

Leaching – downward movement of minerals through soil

Levée – raised bank along the sides of a river, made of silt from river floods

Life expectancy – avarage number of years that a new-born child can expect to live

Long profile of a river – a summary of the shape and gradient of a river bed from source to mouth

Managed retreat – abandon defence of present coastline in a controlled manner

Management of problems – making changes for improvement, planning ahead to stop them occurring in the future

Management strategies – ways to control development and change, to preserve and conserve, and to plan for a sustainable future

Marginal land – areas of land previously not considered good enough to be worth using

Maritime influence – influence of the sea on weather and climate

Mass tourism – large numbers of visitors, often on package holidays with accommodation and travel included

Meander – bend in a river, usually along its middle or lower course

Measure of development – statistical way to show the size of differences in levels of economic growth and wealth between countries.

Metamorphic rock – rock that has been changed by natural agencies

Migration – movement of people either into or out of an area

Moraine – materials deposited by ice, with different names according to where they were deposited

Multicultural – when people from different ethnic, racial or religious backgrounds live together

Multiplier effect – spin-offs from one business growing, allowing other businesses to grow as well

National Park – area set aside to protect landscape and habitats, managed to stop visitor damage

Natural decrease – death rate higher than birth rate, declining population

Natural hazard – short-term event that is a danger to life and property, caused by natural events; examples are earthquakes, volcanoes and tropical storms.

Natural increase – birth rate higher than death rate, growing population

Nutrient cycling – dead remains of plants and animals are decomposed and used again

Organic farming – type of agriculture that does not use chemicals and artificial growth stimulants; farming in a natural and sustainable way

Ox-bow lake – semi-circular lake on the flood plain of a river, a cut-off meander

Plates – large rock areas that make up the Earth's crust

Population policy – national plan for population change (either to lower or increase birth rates)

Population structure – the make-up (age and sex) of a population, usually shown in a population pyramid

Precipitation – all moisture that reaches the Earth's surface from the atmosphere

Pull and push factors – circumstances that attract or drive people to migrate

Quality of life – how well a person is able to enjoy living; high quality is living comfortably (without always being wealthy) and low quality is struggling to survive.

Refugee – person forced to flee from their country or place of residence

Renewable energy – natural source of power that will never run out

Resource – something useful for human needs

Responses – actions immediately after the event or in the long-term

Ribbon lake – long and narrow lake in the floor of a glaciated valley

Rock cycle – rocks weathered, eroded and transported to sea beds, and used to form new mountains

Rural depopulation – decline in numbers living in country areas, often due to out-migration

Rural–urban fringe – area of countryside lying on the edge of the main built-up area, sometimes partly built on

Salinisation – increasing concentrations of salt in the topsoil where evaporation rates are high

Scree – pieces of rock with sharp edges, lying towards the foot of a slope

Second home – house (often in rural areas) that is not the owner's main place of residence

Sedimentary rock – rock formed by sediments laid down in the sea bed

Segregation (in urban areas) – high concentration of land uses and/or groups of people in certain areas of the city, separate from other uses/people

Soft engineering strategies – more natural ways to reduce the impact of flooding on humans, with less intervention and more preparation

Soil erosion – loss of fertile topsoil by action of wind and water

Spit – ridge of sand or shingle attached to the land, but ending in open sea

Squatter settlement – homes on land not owned by the people living there, built illegally

Stack – pillar of rock surrounded by sea, separated from the coastline

Stewardship – entrusted to look after and manage places and areas

Striations – deep grooves in surface rocks, made by the sharp edges of stones carried in the bottom of moving ice

Subsistence economy – one that is based on what can be grown and provided for itself

Subsistence farming – type of agriculture based on growing crops and rearing livestock mainly to feed the family

Suburbanised village – small settlement in the countryside that has grown with new housing and now is less like the old rural settlement it used to be

Succulent – plant that stores water in a fleshy stem to survive drought

Sustainable city – city with low use of energy and raw materials, replacement by renewables and waste recycling

Sustainable development – growth of activities working with the environment for a long future

Sustainable living – people working with the environment for a long future for their economic activities

Sustainable management – planning ahead and controlling development for a long future

Tarn – circular lake in a corrie hollow, where water is trapped by the steep sides and rock lip

Tectonic activity – movement of the large rock plates of the Earth's crust

Trade – exchange of goods and services between countries

Trade bloc – a group of countries (e.g. those in the European Union) that are linked for trade

Transnational corporations (TNCs) – large businesses with interests in many countries

Transportation processes – movement of sediment by traction, saltation, suspension and solution

Tropical revolving storm – area of very low pressure in low latitudes, with strong winds and heavy rain

Truncated spurs – higher areas on the straight rocky sides of a glaciated valley

Tsunami – giant sea wave travelling at high speed

Urban sprawl – outward spread of towns and cities into and taking over rural areas

Urbanisation – increase in the percentage of people living in urban areas

Valley glacier – a moving mass of ice confined in a valley

Volcano – cone-shaped mountain formed by surface eruptions of magma from inside the Earth

Wave-cut platform – gently sloping surface of rock, in front of cliffs, exposed at low tide

Weather – condition of the atmosphere at any one time, day-to-day variations

Weathering – breakdown of rock in the place where it outcrops (in situ)

Xerophytic – adaptations in plants that allow them to survive in a dry climate

Index

Make the Grade!

Understanding **GCSE** Geography
for AQA Specification A

Ann Bowen • John Pallister

This new edition of the best-selling AQA student book is written especially for the new 2009 AQA GCSE Geography A specification. Written by trusted, experienced authors, this book presents an unbeatable way to obtain the best possible grades.

- Exciting, engaging activities to help you develop a thorough understanding and a broad range of skills
- New Grade Studio feature to give you ongoing advice about improving your performance and help you get the grade you want!
- Motivating Exam Café provides exam preparation activities so you can go into the exam feeling fully prepared
- **ActiveBook CD-ROM** contains your whole student book in digital format

FREE! ActiveBook CD-ROM with interactive Grade Studio activities designed for students. Plus comprehensive revision and exam preparation in the popular Exam Café

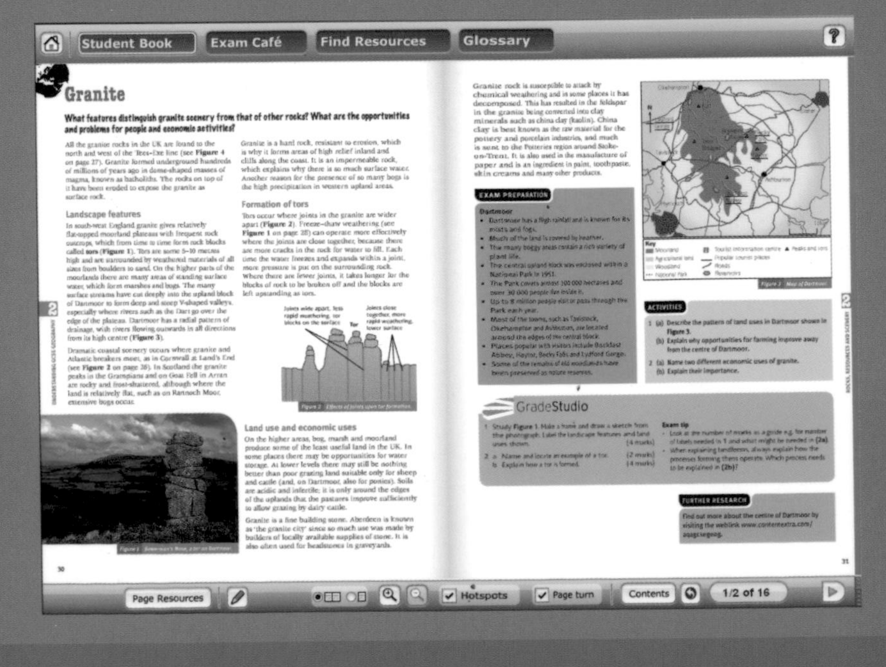

ActiveTeach provides an electronic version of the student book together with a wealth of interactive activities to provide total learning support.

ISBN 978-0-435353-30-8

Part of Pearson